APPLICATIONS OF FURROW AND MICRO IRRIGATION IN ARID AND SEMI-ARID REGIONS

Research Advances in Sustainable Micro Irrigation

VOLUME 5

APPLICATIONS OF FURROW AND MICRO IRRIGATION IN ARID AND SEMI-ARID REGIONS

Edited by

Megh R. Goyal, PhD, PE

AAP APPLE ACADEMIC PRESS

Apple Academic Press Inc.
3333 Mistwell Crescent
Oakville, ON L6L 0A2
Canada

Apple Academic Press Inc.
9 Spinnaker Way
Waretown, NJ 08758
USA

Exclusive worldwide distribution by CRC Press, a member of Taylor & Francis Group

Printed in the United States of America on acid-free paper

International Standard Book Number-13: 978-1-77188-089-3 (Hardcover)

Library and Archives Canada Cataloguing in Publication

Applications of furrow and micro irrigation in arid and semi-arid regions / edited by Megh R. Goyal, PhD, PE.

(Research advances in sustainable micro irrigation; volume 5)
Includes bibliographical references and index.
ISBN 978-1-77188-089-3 (bound)
1. Furrow irrigation. 2. Microirrigation. 3. Irrigation. 4. Arid regions agriculture. 5. Soil absorption and adsorption. I. Goyal, Megh Raj, editor II. Series: Research advances in sustainable micro irrigation; v. 5

S619.F87A67 2015 631.5'87 C2015-903528-7

Library of Congress Cataloging-in-Publication Data

Applications of furrow and micro irrigation in arid and semi-arid regions / [edited by] Megh R. Goyal, PhD, PE.

pages cm
Includes bibliographical references and index.
ISBN 978-1-77188-089-3 (alk. paper)
1. Furrow irrigation. 2. Microirrigation. 3. Irrigation. 4. Arid regions agriculture.
5. Soil absorption and adsorption. I. Goyal, Megh Raj.

S619.F87A67 2015 631.5'87--dc23 2015017825

Apple Academic Press also publishes its books in a variety of electronic formats. Some content that appears in print may not be available in electronic format. For information about Apple Academic Press products, visit our website at **www.appleacademicpress.com** and the CRC Press website at **www.crcpress.com**

CONTENTS

LIST OF CONTRIBUTORS

Hatira Abdessatar, PhD
Agronomist in Tunis, Faculty of Sciences of Tunis, Tunisia

Hédi Ben Ali, PhD
Agricultural Investment Promotion Agency, 6000 Gabès, Tunisia

Ahmed S. El-Henawy, PhD
Associate Professor of Soil Physics, Soils Science Department, Faculty of Agriculture, Kafrelsheikh University, Egypt. E-mail: aelhenawy@agr.kfs.edu; aelhenawy@yahoo.com. Tel.: +204-73232762 (Off); Mobile: +201-006342335

Hany S. Gharib, PhD
Agronomy Department, Faculty of Agriculture, Kafrelsheikh University, Egypt

Megh R. Goyal, PhD
Retired Professor in Agricultural and Biomedical Engineering, University of Puerto Rico – Mayaguez Campus; and Senior Technical Editor-in-Chief in Agriculture Sciences and Biomedical Engineering, Apple Academic Press Inc., PO Box 86, Rincon – PR – 00677 – USA. E-mail: goyalmegh@gmail.com

Moncef Hammami, PhD
Professor, High School of Agronomy of Mateur, 7030 Mateur, University of Carthage, Tunisia. Tel: 00216-71-470097; E-mail: hammami.moncef@ymail.com

Eric W. Harmsen, PhD
Professor, Department of Agricultural and Biosystems Engineering, University of Puerto Rico – Mayagüez [UPRM], Mayaguez – PR-00681, USA. E-mail: eric.harmsen@upr.edu; harmsen1000@gmail.com

Daghari Hedi, PhD
National Institute of Agronomy of Tunisia, 1082 Charles Nicole Avenue, 1082 Tunis, Tunisia

Balti Jelloul, PhD
Faculty of Sciences of Bizerte, Tunisia

M. R. Khalifa, PhD
Soils and Water Department, Faculty of Agriculture, Kafrelsheikh University, Egypt

Zayani Khemaies, PhD
Professor, High Institute of Environmental Sciences and Technology of Borj Cedria, BP 1003, Hammam Lif 2050, University of Carthage, Tunisia. E-mail: khemaies.zayani@isste.rnu.tn

Satyendra Kumar, PhD
Senior Scientist, CSSRI Zarifa Farm, Kachawa Road, Karnal – 132001, India. Tel.: 91-184-2292004; E-mail: skumar@cssri.ernet.in

Lee MacDonald, PhD
Professor, Watershed Science Program, Department of Forest, Rangeland, and Watershed Stewardship; and Department of Ecosystem Science and Sustainability, Senior Research Scientist, Natural Resource Ecology Laboratory, Warner College of Natural Resources, Colorado State University, Fort Collins, CO 80523-1472. E-mail:leemac@cnr.colostate.edu; lee.macdonald@colostate.edu

S. A. Mashali, PhD
Soils Science Department, Faculty of Agriculture, Kafr El-Sheikh University, Egypt

Mohar Singh Meena, PhD
Principal Scientist, Indian Council of Agricultural Research, Zonal Project Directorate, Zone-VI, CAZRI Campus, Jodhpur-342005 (Rajasthan) India. www.zpd6.res.in

Maalej Mohamed, PhD
National Institute of Scientific Research of Technology, B.P. 15, Hammam lif 2080, Tunisia

D. D. Nangare, PhD
Scientist (SWCE), National Institute of Abiotic Stress Management, Malegaon, Baramati, Dist Pune, Maharashtra–413115, India. Mobile: +91-9665204895; E-mail: ddnangare@rediffmail.com

P. Panigrahi, PhD
Senior Scientist, Directorate of Water Management, Chandra Sekharpur, ICAR, Bhubaneswar – 23, Odisha, India. E-mail: pravukalyan@rediffmail.com; pra73_nag@yahoo.co.in

A. Rabie, PhD
Soils Science Department, Faculty of Agriculture, Kafr El-Sheikh University, Egypt

Victor H. Ramirez, PhD
Plant Physiologist, Café Columbia. Km 4 vía Antigua, Chinchiná – Manizales, Chinchiná – Caldas – Colombia, Tel.: 57 + 8506550; E-mail: victorhugorb@gmail.com

M. M. Saffan
Soils Science Department, Faculty of Agriculture, Kafrelsheikh University, Egypt

R. K. Sharma, PhD
Water Technology Centre, Indian Agricultural Research Institute, New Delhi – 110012, India

Ajai Singh, PhD
Associate Professor and Head, Centre for Water Engineering and Management, Central University of Jharkhand, Brambe, Ranchi, India. Tel.: +91-9031450087 E-mail: ajai_jpo@yahoo.com; ajai.singh@cuj.ac.in

Gajendra Singh, PhD
Former Vice President, Asian Institute of Technology, Thailand. C-86, Millennium Apartments, Plot E-10A, Sector 61, Noida – U.P. – 201301, India, Mobile: (011)-(91) 9971087591, E-mail: professorgsingh@gmail.com

Rajbir Singh, PhD
Senior Scientist, Division of Natural Resource Management, Indian Council of Agricultural Research, Krishi Anusandhan Bhawan-II, New Delhi – 110012, India, E-mail: rajbirsingh.nrm@gmail.com; rajbir_singh1103@rediffmail.com

Eman M. K. E. Soltan, PhD
Department of Crop Physiology Research Institute, Agricultural Research Center, ARC, Egypt

A. K. Srivastava, PhD
National Research Centre for Citrus, Nagpur–10, Maharashtra, India

S. M. Youssef, PhD
Soils Science Department, Faculty of Agriculture, Kafr El-Sheikh University, Egypt

LIST OF ABBREVIATIONS

ASABE	American Society of Agricultural and Biological Engineers
DIS	drip irrigation system
DOY	day of the year
ET	evapotranspiration
FAO	Food and Agricultural Organization, Rome
FC	field capacity
FUE	fertilizers use efficiency
GPIS	gated pipe irrigation system
ICAR	Indian Council of Agriculture Research
IR	water intake rate into the soil
ISAE	Indian Society of Agricultural Engineers
LAI	leaf area index
MAD	maximum allowable depletion
MSL	mean sea level
MWD	mean weight diameter
PE	polyethylene
PET	potential evapotranspiration
PM	Penman-Monteith
PVC	poly vinyl chloride
PWP	permanent wilting point
RA	radiation
RH	relative humidity
RMSE	root mean squared error
RS	solar radiation
SAR	sodium absorption rate
SDI	subsurface drip irrigation
SRW	simulated rain water
SW	saline water
SWB	soil water balance
TE	transpiration efficiency
TEW	total evaporable water
TSS	total soluble solids
TUE	transpiration use efficiency
USDA	US Department of Agriculture
USDA-SCS	US Department of Agriculture-Soil Conservation Service

WSEE	weighed standard error of estimate
WUE	water use efficiency

LIST OF SYMBOLS

A	cross sectional flow area (L^2)
AL	average life of wells
AW	available water ($\ominus_w$%)
C	concentration of chlorine wanted, ppm
Cp	specific heat capacity of air, in J/(g·°C)
CV	coefficient of variation
D	accumulative intake rate (mm/min)
d	depth of effective root zone
D	depth of irrigation water (mm)
E	evapotranspiration rate, in g/(m^2·s)
e	vapor pressure (kPa)
e_a	actual vapor pressure (kPa)
eff	irrigation system efficiency
E_i	irrigation efficiency of drip system
E_p	pan evaporation
E_{pan}	Class A pan evaporation
EPAN	pan evaporation
ER	cumulative effective rainfall (mm)
Es	saturation vapor pressure (kPa)
e_s	saturation vapor pressure (kPa)
$e_s–e_a$	vapor pressure deficit (kPa)
ET	evapotranspiration rate (mm/year)
ET_c	crop evapotranspiration
ETc	crop-evapotranspiration (mm/day)
EU	emission uniformity
F	flow rate of the system (GPM)
F.C.	field capacity (v/v, %)
Fed	feddan
Fr	fertilization cost
G	soil heat flux at land surface, in W/m^2
gpm	gallons per minute
H	plant canopy height (m)
h	soil water pressure head (L)
I	infiltration rate at time t (mm/min)
IR	injection rate, GPH
IRR	irrigation

K	the unsaturated hydraulic conductivity (LT^{-1})
k_c	crop coefficient
K_c	crop coefficient
Kg	kilograms
K_p	pan coefficient
Ks	hydraulic conductivity
lph	liters per hour
lps	liters per second
n	number of emitters
P.W.P.	permanent wilting point (Θ_w%)
Pa	atmospheric pressure, in Pa
ppm	part per million
psi	pounds per square inch
Q	flow rate in gallons per minute
R	rainfall
r_a	aerodynamic resistance (s m^{-1})
R_e	effective rainfall depth (mm)
R_i	individual rain gauge reading (mm)
R_{MAX}	maximum relative humidity
R_{MIN}	minimum relative humidity
R_n	net radiation at the crop surface (MJ m^{-2} day^{-1})
RO	surface runoff
r_s	bulk surface resistance (s m^{-1})
Rs	incoming solar radiation on land surface
S	sink term accounting for root water uptake (T^{-1})
Se	the effective saturation
S_p	plant-to-plant spacing (m)
S_r	row-to-row spacing (m)
SU	statistical uniformity (%)
S_ψ	water stress integral (MPa day)
t	the time that water is on the surface of the soil (min)
T	time (hours)
T_{MAX}	maximum temperature
T_{MIN}	minimum temperature
TR	temperature range
V	volume of water required (liter/day/plant)
V_{id}	irrigation volume applied in each irrigation (liter $tree^{-1}$)
V_{pc}	the plant canopy volume (m^3)
W	canopy width
W_p	fractional wetted area
z	the vertical coordinate positive downwards (L)

GREEK SYMBOLS

Δ	slope of the vapor pressure curve (kPa°C^{-1})
γ	psychrometric constant (kPa°C^{-1})
θ	volumetric soil water content (L^3L^{-3})
$\boldsymbol{\theta}$(h)	the soil water retention (L^3L^{-3}),
$\boldsymbol{\theta}_s$	the saturated water content (L^3L^{-3})
θ_{vol}	volumetric moisture content (cm^3/cm^3)
λ	latent heat of vaporization (MJ kg^{-1})
λE	latent heat flux (W/mo)
ρa	mean air density at constant pressure (kg m^{-3})
$\ominus_w$	dry weight basis

PREFACE

Due to increased agricultural production, irrigated land has increased in the arid and sub-humid zones around the world. Agriculture has started to compete for water use with industries, municipalities and other sectors. This increasing demand along with increments in water and energy costs have made it necessary to develop new technologies for the adequate management of water. The intelligent use of water for crops requires understanding of evapotranspiration processes and use of efficient irrigation methods.

An article on the importance of micro irrigation in India was published online (http://www.newindianexpress.com/cities/bengaluru/Micro-irrigation-to-be-promoted/2013/08/17/article1738597.ece). Every day, a similar kind of news appears all around the world indicating that those government agencies at central/state/local levels, research and educational institutions, industry, sellers and others are aware of the urgent need to adopt micro irrigation technology that can have an irrigation efficiency up to 90% compared to 30–40% for the conventional irrigation systems. I share with the readers comments on "Scaling-up Micro-Irrigation Systems in Madagascar" (SCAMPIS) by Andriamalina R. Fenomanantsoa (project coordinator at Agriculturalists and Veterinaries without Frontiers in Madagascar) at the International Annual UN-Water Zaragoza Conference 8–10 January 2013, Water Cooperation: Making it Happen! The full version of his interview appears at: http://www.un.org/waterforlifedecade/water_cooperation_2013/madagascar.shtml. He indicates urgent implementation of micro irrigation systems in Madagascar because of water scarcity.

Micro irrigation is sustainable and is one of the best management practices. I attended the 17th Punjab Science Congress on February 14–16, 2014, at Punjab Technical University in Jalandhar. I was shocked to know that the underground water table has lowered to a critical level in Punjab. My father-in-law in Dhuri told me that his family "*bought the 0.10* acres *of land in the city for US $100.00 in 1942 AD because the water table was at 2 feet depth. In 2012, it was sold for US$200,000 because the water table had dropped to greater than 100 feet. This has been due to the luxury use of water by wheat-paddy farmers*." The water crisis is similar in other countries, including Puerto Rico, where I live. We can therefore conclude that the problem of water scarcity is rampant globally, creating the urgent need for water conservation. The use of micro irrigation systems is expected to result in water savings, and increased crop yields in terms of volume and quality. The other important benefits of using micro irrigation systems include expansion in the area under irrigation, water conservation, optimum use of fertilizers and chemicals through water, and decreased labor costs, among others. The worldwide population is increasing at a rapid rate, and it is imperative that the food supply keeps pace with this increasing population.

Micro irrigation, also known as trickle irrigation or drip irrigation or localized irrigation or high frequency or pressurized irrigation, is an irrigation method that saves water and fertilizer by allowing water to drip slowly to the roots of plants, either onto the soil surface or directly onto the root zone, through a network of valves, pipes, tubing, and emitters. It is done through narrow tubes that deliver water directly to the base of the plant. It supplies controlled delivery of water directly to individual plants and can be installed on the soil surface or subsurface. Micro irrigation systems are often used in farms and large gardens, but are equally effective in the home garden or even for houseplants or lawns.

The mission of this compendium is to serve as a reference manual for graduate and undergraduate students of agricultural, biological and civil engineering, horticulture, soil science, crop science and agronomy. I hope that it will be a valuable reference for professionals those who work with micro irrigation and water management, training institutes, technical agricultural centers, irrigation centers, agricultural extension services, and other agencies that work with micro irrigation programs.

In response from the international readers on my first textbook on *Drip/Trickle or Micro Irrigation Management* by Apple Academic Press Inc., I was motivated to bring out for the world community this ten-volume series, *Research Advances in Sustainable Micro Irrigation*. This book series will complement books on micro irrigation that are currently available on the market, and my intention is not to replace any one of these. This book series is unique because it is complete and simple, a one-stop manual, with worldwide applicability to irrigation management in agriculture. This volume is a must for those interested in irrigation planning and management, namely, researchers, scientists, educators and students.

The contribution by all cooperating authors to this book series has been most valuable in the compilation of this volume. Their names are mentioned in each chapter and in the list of contributors. This book would not have been written without the valuable cooperation of these investigators, and many of them are renowned scientists who have worked in the field of micro irrigation throughout their professional careers.

I like to thank Sandy Jones Sickels, Vice President, and Ashish Kumar, Publisher and President at Apple Academic Press, Inc, (http://appleacademicpress.com/contact.html) for making every effort to publish the book when the diminishing water resources is a major issue worldwide. Special thanks to the AAP Production staff for editing and typesetting the entire manuscript and ensuring the quality production of this book. We request that the reader offer us with your constructive suggestions that may help to improve the next edition.

I express my deep admiration to my family for understanding and collaboration during the preparation of this ten-volume book series. With the whole heart and best affection, I dedicate this book series to Jack Keller, who has been my master since 1979. He helped me to trickle on to add my drop to the ocean of service to the world of humanity. Without his advice and patience, I would not have been a *Father of Irrigation Engineering of Twentieth Century in Puerto Rico*, with zeal for service to others. My salute to him for his irrigation legacy. As an educator, there is a piece of advice to the world: "*Permit that our almighty God, our Creator and excellent Teacher, irrigate the life with His Grace of rain trickle by trickle, because our life must continue trickling on...*"

— Megh R. Goyal, PhD, PE
Senior Editor-in-Chief
June 30, 2014

FOREWORD

With only a small portion of cultivated area under irrigation and with the scope to the additional area which can be brought under irrigation, it is clear that the most critical input for agriculture today is water. It is important that all available supplies of water should be used intelligently to the best possible advantage. Recent research around the world has shown that the yields per unit quantity of water can be increased if the fields are properly leveled, the water requirements of the crops as well as the characteristics of the soil are known, and the correct methods of irrigation are followed. Significant gains can also be made if the cropping patterns are changed so as to minimize storage during the hot summer months when evaporation losses are high, if seepage losses during conveyance are reduced, and if water is applied at critical times when it is most useful for plant growth.

Irrigation is mentioned in the Holy Bible and in the old documents of Syria, Persia, India, China, Java, and Italy. The importance of irrigation in our times has been defined appropriately by N.D. Gulati: "In many countries irrigation is an old art, as much as the civilization, but for humanity it is a science, the one to survive." The need for additional food for the world's population has spurred rapid development of irrigated land throughout the world. Vitally important in arid regions, irrigation is also an important improvement in many circumstances in humid regions. Unfortunately, often less than half the water applied is used by the crop-irrigation water may be lost through runoff, which may also cause damaging soil erosion, deep percolation beyond that required for leaching to maintain a favorable salt balance. New irrigation systems, design and selection techniques are continually being developed and examined in an effort to obtain high practically attainable efficiency of water application.

The main objective of irrigation is to provide plants with sufficient water to prevent stress that may reduce the yield. The frequency and quantity of water depends upon local climatic conditions, crop and stage of growth, and soil-moisture-plant characteristics. Need for irrigation can be determined in several ways that do not require knowledge of evapotranspiration (ET) rates. One way is to observe crop indicators such as change of color or leaf angle, but this information may appear too late to avoid reduction in the crop yield or quality. Other similar methods of scheduling include determination of the plant water stress, soil moisture status, or soil water potential. Methods of estimating crop water requirements using ET and combined with soil characteristics have the advantage of not only being useful in determining when to irrigate, but also enables us to know the quantity of water needed. ET estimates have not been made for the developing countries though basic information on weather data is available. This has contributed to one of the existing problems that the vegetable crops are over irrigated and tree crops are under irrigated.

Water supply in the world is dwindling because of luxury use of sources; competition for domestic, municipal, and industrial demands; declining water quality; and losses through seepage, runoff, and evaporation. Water rather than land is one of the limiting factors in our goal for self-sufficiency in agriculture. Intelligent use of water will avoid problem of seawater seeping into aquifers. Introduction of new irrigation methods has encouraged marginal farmers to adopt these methods without taking into consideration economic benefits of conventional, overhead, and drip irrigation systems. What is important is "net in the pocket" under limited available resources. Irrigation of crops in tropics requires appropriately tailored working principles for the effective use of all resources peculiar to the local conditions. Irrigation methods include border, furrow, subsurface, sprinkler, micro/drip/trickle, and xylem irrigation.

Drip irrigation is an application of water in combination with fertilizers within the vicinity of plant root in predetermined quantities at a specified time interval. The application of water is by means of drippers, which are located at desired spacing on a lateral line. The emitted water moves due to an unsaturated soil. Thus, favorable conditions of soil moisture in the root zone are maintained. This causes an optimum development of the crop. Drip/micro or trickle irrigation is convenient for vineyards, tree orchards, and row crops. The principal limitation is the high initial cost of the system that can be very high for crops with very narrow planting distances. Forage crops may not be irrigated economically with drip irrigation. Drip irrigation is adaptable for almost all soils. In very fine textured soils, the intensity of water application can cause problems of aeration. In heavy soils, the lateral movement of the water is limited, thus more emitters per plant are needed to wet the desired area. With adequate design, use of pressure compensating drippers and pressure regulating valves, drip irrigation can be adapted to almost any topography. In some areas, drip irrigation is used successfully on steep slopes. In subsurface drip irrigation, laterals with drippers are buried at about 45 cm depth, with an objective to avoid the costs of transportation, installation, and dismantling of the system at the end of a crop. When it is located permanently, it does not harm the crop and solve the problem of installation and annual or periodic movement of the laterals. A carefully installed system can last for about 10 years.

The publication of this book series and volume 5 is an indication that things are beginning to change, that we are beginning to realize the importance of water conservation to minimize the hunger. It is hoped that the publisher will produce similar materials in other languages.

In providing this book series, Dr. Megh Raj Goyal and Apple Academic Press are rendering an important service to the farmers. Dr. Goyal, *Father of Irrigation Engineering in Puerto Rico*, has done an unselfish job in the presentation of this compendium that is simple and thorough. I know Dr. Goyal since 1973 when we were working together at Haryana Agricultural University on an ICAR research project in "Cotton Mechanization in India."

Gajendra Singh, PhD, Former Vice Chancellor, Doon University, Dehradun, India.

Adjunct Professor, Indian Agricultural Research Institute, New Delhi.

Ex-President (2010–2012), Indian Society of Agricultural Engineers.

Former Deputy Director General (Engineering), Indian Council of Agricultural Research (ICAR), New Delhi.

Former Vice-President/Dean/Professor and Chairman, Asian Institute of Technology, Thailand.

Dr. Gajendra Singh, PhD
New Delhi
December 31, 2014

BOOK SERIES: RESEARCH ADVANCES IN SUSTAINABLE MICRO IRRIGATION

Volume 1: Sustainable Micro Irrigation: Principles and Practices
Senior Editor-in-Chief: Megh R. Goyal, PhD, PE

Volume 2: Sustainable Practices in Surface and Subsurface Micro Irrigation
Senior Editor-in-Chief: Megh R. Goyal, PhD, PE

Volume 3: Sustainable Micro Irrigation Management for Trees and Vines
Senior Editor-in-Chief: Megh R. Goyal, PhD, PE

Volume 4: Management, Performance, and Applications of Micro Irrigation
Senior Editor-in-Chief: Megh R. Goyal, PhD, PE

Volume 5: Applications of Furrow and Micro Irrigation in Arid and Semi-Arid Regions
Senior Editor-in-Chief: Megh R. Goyal, PhD, PE

Volume 6: Best Management Practices for Drip Irrigated Crops
Editors: Kamal Gurmit Singh, PhD, Megh R. Goyal, PhD, PE, and Ramesh P. Rudra, PhD, PE

Volume 7: Closed Circuit Micro Irrigation Design: Theory and Applications
Senior Editor-in-Chief: Megh R. Goyal, PhD; Editor: Hani A. A. Mansour, PhD

Volume 8: Wastewater Management for Irrigation: Principles and Practices
Editor-in-Chief: Megh R. Goyal, PhD, PE; Coeditor: Vinod K. Tripathi, PhD

Volume 9: Water and Fertigation Management in Micro Irrigation
Senior Editor-in-Chief: Megh R. Goyal, PhD, PE

Volume 10: Innovations in Micro Irrigation Technology
Senior Editor-in-Chief: Megh R. Goyal, PhD, PE; Coeditors: Vishal K. Chavan, MTech, and Vinod K. Tripathi, PhD

ABOUT THE SENIOR EDITOR-IN-CHIEF

Megh R. Goyal, PhD, PE, is a retired Professor of Agricultural and Biomedical Engineering from the General Engineering Department in the College of Engineering at the University of Puerto Rico – Mayaguez Campus; and Senior Acquisitions Editor and Senior Technical Editor-in-Chief in Agricultural and Biomedical Engineering for Apple Academic Press, Inc.

He received his BSc degree in Engineering in 1971 from Punjab Agricultural University, Ludhiana, India; MSc degree in 1977 and PhD degree in 1979 from the Ohio State University, Columbus; and his Master of Divinity degree in 2001 from Puerto Rico Evangelical Seminary, Hato Rey, Puerto Rico, USA.

Since 1971, he worked as Soil Conservation Inspector (1971); Research Assistant at Haryana Agricultural University (1972–75) and the Ohio State University (1975–1979); Research Agricultural Engineer/Professor at Department of Agricultural Engineering of UPRM (1979–1997); and Professor of Agricultural and Biomedical Engineering at the General Engineering Department of UPRM (1997–2012). He spent a one-year sabbatical leave in 2002–2003 at the Biomedical Engineering Department at Florida International University, Miami, USA.

He was the first agricultural engineer to receive a professional license in Agricultural Engineering in 1986 from the College of Engineers and Surveyors of Puerto Rico. On September 16, 2005, he was proclaimed as "Father of Irrigation Engineering in Puerto Rico for the Twentieth Century" by the American Society of Agricultural and Biological Engineers (ASABE), Puerto Rico Section, for his pioneering work on micro irrigation, evapotranspiration, agroclimatology, and soil and water engineering. During his professional career of 45 years, he has received awards such as: Scientist of the Year, Blue Ribbon Extension Award, Research Paper Award, Nolan Mitchell Young Extension Worker Award, Agricultural Engineer of the Year, Citations by Mayors of Juana Diaz and Ponce, Membership Grand Prize for ASAE Campaign, Felix Castro Rodriguez Academic Excellence, Rashtrya Ratan Award and Bharat Excellence Award and Gold Medal, Domingo Marrero Navarro Prize, Adopted Son of Moca, Irrigation Protagonist of UPRM, and Man of Drip Irrigation by Mayor of Municipalities of Mayaguez/Caguas/Ponce and Senate/Secretary of Agriculture of ELA, Puerto Rico.

He has authored more than 200 journal articles and textbooks on "Elements of Agroclimatology (Spanish) by UNISARC, Colombia," and two bibliographies on "Drip Irrigation." Apple Academic Press, Inc. (AAP) has published many of his books, namely, "Biofluid Dynamics of Human Body," "Biomechanics of Artificial Organs and Prostheses," "Scientific and Technical Terms in Bioengineering and Biotechnology," "Management of Drip/Trickle or Micro Irrigation," "Evapotranspiration: Principles and Applications For Water Management," and "Sustainable Micro Irrigation Design Systems for Agricultural Crops: Practices and Theory." During 2014–2015, AAP is publishing his ten-volume set in "*Research Advances in Sustainable Micro Irrigation.*" Readers may contact him at "goyalmegh@gmail.com."

WARNING/DISCLAIMER

The goal of this compendium "***Volume 5: Applications of Furrow and Micro Irrigation in Arid and Semi-Arid Regions***" is to guide the world community on how to manage efficiently for economical crop production. The reader must be aware that the dedication, commitment, honesty, and sincerity are most important factors in a dynamic manner for a complete success. It is not a one-time reading of this compendium. Read and follow every time, it is needed. To err is human. However, we must do our best. Always, there is a space for learning new experiences.

The editor, the contributing authors, the publisher and the printer have made every effort to make this book as complete and as accurate as possible. However, there still may be grammatical errors or mistakes in the content or typography. Therefore, the contents in this book should be considered as a general guide and not a complete solution to address any specific situation in irrigation. For example, one size of irrigation pump does not fit all sizes of agricultural land and to all crops.

The editor, the contributing authors, the publisher and the printer shall have neither liability nor responsibility to any person, any organization or entity with respect to any loss or damage caused, or alleged to have caused, directly or indirectly, by information or advice contained in this book. Therefore, the purchaser/reader must assume full responsibility for the use of the book or the information therein.

The mentioning of commercial brands and trade names are only for technical purposes. It does not mean that a particular product is endorsed over to another product or equipment not mentioned. Author, cooperating authors, educational institutions and the publisher Apple Academic Press, Inc., do not have any preference for a particular product.

All web links that are mentioned in this book were active on December 31, 2014. The editors, the contributing authors, the publisher and the printing company shall have neither liability nor responsibility, if any of the web links is inactive at the time of reading of this book.

CHAPTER 1

METEOROLOGICAL INSTRUMENTS FOR WATER MANAGEMENT

LEE MACDONALD

CONTENTS

1.1 RAIN GAUGES

1.1.1 PRINCIPLES

Rain gauges should have a sharp edge near the vertical sides (inside) to minimize rain splash (drops are either in or out). It should have a funnel with a small hole to minimize evaporation losses. Sides and funnel should be smooth, hydrophobic material to minimize loss due to wetting of the surfaces conducting water. Standard height in the U.S. is about 75 cm (30 inches). In winter, the funnel is typically removed so snow doesn't accumulate on the funnel. Some gauges are heated in winter to melt and measure snow, but this causes some evaporation losses. Biggest issue is wind, as the turbulence causes undercatch. Most common shields to minimize this problem are Alter shield (strips of metal hanging around the perimeter) and Nipher shield (like the opening of a tuba). It latter will cause over catch in rain and hail. Ideal location is in a forest opening to minimize wind effects. Gauge should be located at a distance that is twice the height of the nearest object, which is an angle of 30 degrees from the top of the object (i.e., 60 m from a 30-m tall tree). If necessary, an angle of 45 degrees is acceptable. Wind effects can be minimized by installing the opening of the gauge at or near ground level, but this makes it harder to service and more susceptible to materials falling into the gauge and possible clogging it, for example, leaves.

1.1.2 TYPES

Standard rain gage gives total rain. To get the desired accuracy, the water is typically funneled into a smaller tube inside the gauge that is only 10% of the area of the rain gauge. This 10-fold increase allows precipitation to be measured to the nearest 0.1 mm or 0.01 inches.

1.1.3 RECORDING RAIN GAUGES

These provide data on rainfall intensity and these include weighing buckets, tipping buckets, a siphon gauge recording on a chart, or a storage gauge with a pressure transducer.

1.2 TEMPERATURE

1.2.1 PRINCIPLES

Measurements are made in a location that is representative. In most cases, this means in a shaded area is not subject to excessive heating or reflected radiation, such as a grassy area. The thermometer needs to be in a shelter that is painted white and well ventilated so it does not heat up because of solar radiation. Traditional wooden shelters have a double roof to minimize heating and they always open to the north in the northern hemisphere to avoid the sun shining on the thermometers at any time when observations are being made. Electronic thermometers are typically put into a stacked-plate type shelter. All shelters can result in temperatures that are too high under conditions of high radiation loads and no wind. Biggest problem is after a fresh snow because of the very high albedo.

1.2.2 MECHANICAL MEASUREMENTS

All mechanical measurements are based on principle that higher temperature leads to expansion.

1.2.2.1 THERMOMETERS (TRADITIONALLY LIQUID-IN-GLASS)

Thermometers are usually made-up of mercury or alcohol, bulb or reservoir and then a narrow tube, so small change in volume leads to a large change in distance. Therefore, read directly off the scale for regular thermometer.

1.2.2.2 MAXIMUM THERMOMETER

Maximum thermometer uses mercury, as more viscous. It has constriction so that as mercury goes up it can't come down. It stays at highest point. *Reset by shaking, but be sure to shake it so that the mercury is forced down into the bulb, not away from it, as once mercury separates it usually is not possible to get it to rejoin!* Very reliable and is generally the official means to measure maximum temperatures in the U.S., Vietnam, etc.

1.2.2.3 MINIMUM THERMOMETER

Minimum thermometer works on different principle. It has index, so fluid can go up and around it, but as temperature decreases the index is pulled down. Given this, measure the UPPER end of the index, not the lower end. Again this is the standard method worldwide.

1.2.2.4 TIMING OF READINGS

Minimum and maximum thermometers give you the minimum and maximum values since the time the thermometer was last reset. You need to be careful when you reset them. If you reset the maximum when it is still warm and the next day is colder, then it will only give you the temperature you reset it at. If you reset the minimum thermometer when it is cold and the next day is warmer, then again it will stay at the temperature when it was reset. Therefore, you want to reset the maximum thermometer at a time when you know it will get warmer (e.g., shortly after sunrise) and the minimum thermometer at a time when you know it will get colder (e.g., in the afternoon).

1.2.3 CONTINUOUS TEMPERATURE RECORDINGS

To get continuous temperature measurements, either a mechanical thermometer with a chart or an electronic thermometer is needed.

1.2.3.1 MECHANICAL RECORDING THERMOMETERS ("THERMOGRAPH")

Mechanical recording thermometers work on principle of expansion, as they typically use two pieces of metal with different expansion coefficients; change in temperature causes differential bending and this change in angle is converted by a system of rods and levers into a rise and fall on a chart. A mechanical or battery-powered clock is used to drive the chart and the rate at which the chart moves determines the resolution of the temperature measurements in time. It can range from 6 h to 30 days depending on which combinations of gears are used for the clock drive and at the base of the chart. It costs around $500.

1.2.3.2 ELECTRONIC THERMOMETERS

Electronic thermometers are very reliable and cheap, so they are the dominant means for recording temperatures. These only need a minimal power supply, a clock, a temperature sensor (voltage) and a data logger. Can purchase single-channel temperature loggers for $20 or less; now use these on refrigerated trucks to check temperatures during the time of shipping fruit or vegetables across the country or oceans; Need computer to download the data.

1.3 RELATIVE HUMIDITY

1.3.1 WET/DRY BULB METHODS

Compare the difference in temperatures between a thermometer in dry air and a thermometer with a wet bulb (usually a regular thermometer with a small cotton sock on the bulb). The cotton sock is wetted, preferably using deionized water to avoid mineral build up as the water evaporates. Evaporation from the wet bulb decreases the temperature relative to the dry bulb and the difference in temperature between the wet and dry bulb thermometers are inversely proportional to the relative humidity (i.e., lower relative humidity leads to a greater temperature difference, while higher relative humidity leads to a smaller temperature difference). Dry vs. wet bulb differences decrease with temperature (i.e., at 50% relative humidity the difference between a dry and wet bulb is greater at 40°C than at 10°C). The difference is also less with increasing elevation because the lower air pressure means the atmosphere can hold less water and the evaporation rate is less. These nonlinear controls mean that one has to use a chart to determine the relative humidity for a given difference in temperatures at a given elevation. A Colorado chart for 1500 m elevation will not work in Vietnam! Two main methods are: (1) a sling psychrometer, where the thermometers are rapidly swung in a circle (or a fan is used to blow air by the wet bulb); and (2) stationary method, where the wet bulb has a continuous water supply. Stationary method may not work as well if no wind and over time evaporation will lead to mineral deposition on the sock, so it will need to be replaced more often, providing instantaneous rather than continuous values.

1.3.2 MECHANICAL METHODS ("HYGROGRAPH")

Use the principle of expansion and contraction, usually with hairs held at high tension. As the humidity changes the hairs, expand and contract and again a system of levers transforms into the vertical movement of a pen on a chart. Cost is about $500–$1000. Many instruments combine temperature and relative humidity, so upper part of the chart is temperature and lower part is relative humidity ("hygrothermograph").

1.3.3 ELECTRONIC INSTRUMENTS

Electronic instruments are now easily available in the market and it's cheaper than a mechanical recorders. Electronic hygrographs inherently must have a thermistor to measure temperature since relative humidity is temperature-dependent.

1.4 WIND

1.4.1 VERTICAL CUP ANEMOMETER

A standard method uses a mechanical or electronic sensor to record each turn. A standard anemometer records turns per unit time, which yields wind speed. A totalizing anemometer records total number of turns, which is wind run. Wind speed can be converted into wind run by multiplying the average speed for a given period of time, or integrating shorter-term measurements for the time period of interest. It costs about $400–$1000.

1.4.2 SONIC ANEMOMETER

It measures the effect of wind on the speed of sound. This generates three-dimensional wind speeds, which are useful for measuring the upward flow of CO_2 and water vapor ("eddy covariance"), while cup anemometers only measure lateral winds (two dimensions, but only one integrated value). The cost of Sonic anemometer is around thousands of dollars.

1.4.3 LOCATION

Standard measurement height for wind speed is 2 m, but measurements are often made at 30 cm next to an evaporation pan, at 10 m, or above the vegetation canopy. For accurate wind measurements the anemometer needs to be 10 times the height of the nearest vertical element. So in a forest that is 30 m high, the clearing for an should have a radius of 10 m × 30 m, or 300 m, which converts to a diameter of 600 m! In U.S., a totalizing anemometer is almost always placed at a height of about 30 cm adjacent to an evaporation pan, as the evaporation rate is highly dependent on wind run (or wind speed). Wind direction is recorded on a 360 degree circle using a wind vane. There are mechanical wind vanes that can record on a chart, but most people now use a wind vane combined with a data logger. The reality is that most people use wind speed and ignore wind direction.

1.5 EVAPORATION

1.5.1 EVAPORATION PAN

This is the standard method of measuring the evaporation from a free water surface. This is a metal pan that holds water and one carefully measures the change in water level, usually on a daily basis and sometimes twice a day. The U.S. Weather Bureau class A pans are 120 cm in diameter, 25 cm deep and filled to 20 cm. Most evaporation pans are above the ground and therefore tend to overestimate evaporation because the meal sides absorb solar radiation. There also is a transfer of sensible heat from the air to the pan because evaporation cools the water in the pan and thus creates a temperature gradient from the air to the pan. To avoid this problem evaporation pans should be floated in a large lake, but this is usually impractical. Sunken pans are more accurate, but are rarely used because they accumulate debris and it is hard to detect a leak should it occur. Since the pans overestimate evaporation, the pan evaporation value needs to be multiplied by an empirical pan coefficient of about 0.7–0.8 to get potential evapotranspiration. The problem is that this coefficient varies seasonally and with location, so there is no single value! One also has to keep the water level relatively constant, as the evaporation rate will increase as the water becomes shallower due to the proportionally greater radiation load and greater advective heat input per unit of water depth. An adjacent rain gage is needed to subtract any rainfall from the change in water level. Errors can occur due to birds drinking or bathing in the pan and they also need to be kept clear of algae and debris. Under night-time freezing conditions the data are not reliable, as some energy needed to melt the ice before evaporation can occur. If the pan freezes solid, the welds may break and the pan may leak. The U.S. Class A pan costs about $3000, which is expensive! Daily evaporation rates are relatively low (usually 0–10 mm), so the depth measurements need to be very accurate. In U.S., people raise and lower a point gauge, which has a Vernier scale to the nearest 0.001 inch or 0.001 cm. In Vietnam, water is added from a graduated cylinder until the water level reaches a similar point as the U.S. point gauge and the volume added is divided by the relatively large pan area to get an accurate evaporation rate. Pressure transducers aren't used because they generally do not have a sufficient degree of precision and accuracy.

1.5.2 ATMOMETERS OR EVAPORIMETERS

It uses a volumetric change in a narrow glass tube with a saturated cloth or paper at the bottom. The change in water level is recorded daily. Different values will be recorded from an evaporimeter in a shelter versus one in the open because of the much lower wind speeds in the shelter. Big advantage is that these are much cheaper, as they can range from about $20 to perhaps $300. Not as widely used as they should be!

1.6 RADIATION

1.6.1 TYPES OF RADIATION

There are two types of radiations, short-wave (solar) and long-wave (infrared or thermal) radiations. Measurements are generally made by measuring the temperature increase due to radiation hitting a strip of metal that is painted black to absorb radiation. A dome can be used to restrict the type of radiation that is allowed to hit the metal strip. Measurements of solar radiation can be divided into: (1) direct, which would be measured using a long tube pointed directly at the sun ("pyrheliometer"); (2) diffuse, which comes in at all angles; and (3) reflected from the ground. Albedo is the ratio of reflected short-wave from the ground divided by the incoming short-wave. Long-wave by definition is diffuse and it is measured as incoming (from the sky) or outgoing (from the ground).

1.6.2 PYRANOMETER

It measures direct and diffuse short wave from the sky. Glass dome transmits 0.34–2.8 μm, while quartz domes transmit 0.25–4.0 μm.

1.6.3 RADIOMETER

Radiometer is a device used to measure the short wave and long wave radiations.

1.6.4 NET RADIOMETER

It measures the difference between incoming and outgoing. Values are positive during the day and negative at night.

1.6.5 PYRGEOMETER

A pyrgeometer is a device that measures the atmospheric infra-red radiation spectrum (long wave) that extends approximately from 4.5 μm to 100 μm. Pyrgeometers are frequently used in meteorology and climatology studies. The atmospheric long-wave downward radiation is of interest for research into long-term climate changes [1, 2]. The signals are generally detected using a data logging system, capable of taking high-resolution samples in the millivolt range. Pyrgeometer equation by Albrecht and Cox is given below:

$$E_{in} = \{\sigma\, \theta^4 + [(U_{emf})/S]\} \qquad (1)$$

where: σ = Stefan-Boltzmann constant, $W/(m^2 \cdot K^4)$; θ = absolute temperature of pyrgeometer detector, kelvins; U_{emf} = thermopile output voltage, V; E_{in} = long-wave radiation received from the atmosphere, $W/m^2 = E_{net} + E_{out}$; E_{net} = net radiation at sensor surface, W/m^2; E_{out} = net radiation at sensor surface, W/m^2; and S = sensitivity/calibration factor of instrument, $V/W/m^2$. The value for S is determined during calibration of the instrument. The calibration is performed at the production factory with a reference instrument traceable to a regional calibration center. As a result, the detected voltage and instrument temperature yield the total global long wave downward radiation. Pyrgeometer consists of the following major components:

- a thermopile sensor which is sensitive to radiation in a broad range from 200 nm to 100 μm.
- a silicon dome or window with a solar blind filter coating. It has a transmittance between 4.5 μm and 50 μm that eliminates solar short-wave radiation.

- a temperature sensor to measure the body temperature of the instrument.
- a sun shield to minimize heating of the instrument due to solar radiation.

1.6.6 DURATION OF SUNSHINE

Another, cruder method of measuring solar radiation is with a spherical ball. This concentrates the solar radiation onto a strip of special paper and the presence of direct solar radiation is indicated by burning a narrow strip because of the concentrated solar radiation. The thickness of the burned strip indicates the magnitude of solar radiation, but it is difficult to convert this into an amount. Hence, this is most useful for indicating the duration of sunshine (direct solar radiation) rather than the amount.

1.7 SUMMARY

This chapter includes climatic instruments to measure rainfall, temperature, relative humidity, wind, evaporation from a free water surface, sun radiation and long wave radiation.

KEYWORDS

- **evaporation**
- **long wave radiation**
- **rainfall**
- **relative humidity**
- **Stefan-Boltzmann constant**
- **sun radiation**
- **temperature**
- **wind**

REFERENCES

1. Goyal, M. R. (2013). *Management of Drip/Trickle or Micro Irrigation.* Apple Academic Press, Inc.: New York
2. Goyal, M. R., Harmsen, E. W. (2014). *Evapotranspiration: Principles and Applications for Water Management.* Apple Academic Press, Inc.: New York.

CHAPTER 2

WATER VAPOR FLUX MODELS FOR AGRICULTURE

VICTOR H. RAMIREZ and ERIC W. HARMSEN

CONTENTS

Printed with permission and modified from V. H. Ramirez and E. W. Harmsen (2011), *Water Vapor Flux in Agroecosystems: Methods and Models Review*. Chapter 1, pp. 3–48 In: *Evapotranspiration* by Leszek Labedzki (Editor), ISBN 978-953-307-251-7, InTech. Available from: http://www.intechopen.com/books/evapotranspiration/turfgrass-growth-quality-and-reflective-heat-load-in-response-to-deficit-irrigation-practices" and V. H. Ramirez and E. W. Harmsen (2014), Water Vapor Flux Models in Agriculture. Chapter 2, pp. 19–65. In: *Evapotranspiration: Principles and Applications for Water Management* by Megh R. Goyal and Eric W. Harmsen (Editors). New Jersey, USA: Apple Academic Press Inc.

2.1 INTRODUCTION

The water vapor flux in agroecosystems is the second largest component in the hydrological cycle. Water vapor flux or evapotranspiration (ET) from the vegetation to the atmosphere is a widely studied variable throughout the world. ET is important for determining the water requirements for the crops, climatic characterization and for water management. The estimation of ET from vegetated areas is a basic tool to compute water balances and to estimate water availability and requirements. In the last 60 years, several methods and models to measure the water flux in agroecosystems have been developed. The aim of this chapter is to provide a literature review on the subject and provide an overview of methods and model developed which are widely used to estimate and/or measure ET in agroecosystems.

Evapotranspiration constitutes an important component of the water fluxes of our hydrosphere and atmosphere [21] and is a widely studied variable throughout the world, due to it applicability in various disciplines, such as hydrology, climatology and agricultural science. Pereira et al. [80] has reported that the estimation of ET from vegetated areas is a basic tool for computing water balances and to estimate water availability and requirements for plants. Measurement of ET is needed for many applications in agriculture, hydrology and meteorology [102]. ET is a major component of the hydrologic water budget, but one of the least understood [120]. ET permits the return of water to the atmosphere and induces the formation of clouds, as part of a never-ending cycle. ET also permits the movement of water and nutrients within the plant, water moving from the soil into the root hairs and then to the plant leaves.

ET is a complicated process because it is the product the different processes, such as evaporation of water from the soil and water intercepted by the canopy and transpiration from plant leaves. Physiological, soil and climatic variables are involved in these processes. Symons in 1867 described evaporation as "*...the most desperate art of the desperate science of meteorology*" [69]. The first vapor flux measurements were initiated by Thornthwaite and Holzman in 1930s, but that works was interrupted by World War II [69]. In the late 1940s Penman [78] published an article called "*Natural Evaporation From Open Water, Bare Soil and Grass*" in which he combined a thermodynamic equation for the surface heat balance and an aerodynamic equation for vapor transfer. The "Penman equation" is one of the most widely used equations in the world. The equation was later modified by Monteith [67, 68] and is widely known as the "The Penman-Monteith Model." It is also necessary to introduce a review of the work of Bowen, who in 1926 [11] published the relationship between the sensible and latent heat fluxes, which is known as the "Bowen Ratio." Measurement of the water vapor flux became a common practice by means of the "Bowen Ratio Energy Balance Method" [106]. Allen et al. [5] classified the factors that affect the ET into three groups:

a. *Weather parameters*, such as radiation, air temperature, humidity and wind speed: The evapotranspirational component of the atmosphere is expressed by the reference crop evapotranspiration (ET_o) as the Penman-Monteith (FAO-56), or using direct measurements of pan evaporation data [22], or using other empirical equations;

b. *Crop factors* such as the crop type, variety and developmental stage should be considered when assessing the ET from crops grown in large, well-management fields. Differences in resistance to transpiration, crop height, crop roughness, reflection, ground cover and crop rooting characteristics result in different ET levels in different types of crops under identical environmental conditions. Crop ET under standard conditions (ET_c) refers to excellent management and environmental conditions and achieves full production under given climatic conditions (Eq. (2)); and

c. *Management and environmental conditions* (ET_{cadj}). Factors such soil salinity, poor land fertility, limited applications of fertilizers, the presence of hard or impermeable soil horizons, the absence the control of disease and pest and poor soil management may limit the crop development *etc.* and reduce the ET, (ETc_{adj} Eq. (3)).

One of the most common and fairly reliable techniques for estimating ET_0 is using evaporation pan data when adjustments are made for the pan environment [31] using the pan evaporation and the pan coefficient (K_p).

$$ET_o = K_p \times E_p \tag{1}$$

where, E_p is the pan evaporation (mm day^{-1}) and K_p is the pan coefficient, which depends on location. It is important to know or calculate pan coefficient before calculating the ET_o. Allen et al. [5] gave a methodology to calculate it. K_p is essentially a correction factor that depends on the prevailing upwind fetch distance, average daily wind speed and relative humidity conditions associated with the siting of the evaporation pan [22].

2.2 CROP WATER FLUX USING SINGLE CROP COEFFICIENTS: THE FAO APPROACH

The FAO approach is well known as the "two steps method," which is very useful for single crops and when "references" conditions are available (i.e., no crop water stress). In this case, crop evapotranspiration (ET_c) can be estimated using Eq. (2) [5, 22]:

$$ET_c = K_c \times ET_o \tag{2}$$

where, K_c is the coefficient expressing the ratio of between the crop and reference ET for a grass surface. The crop coefficient can be expressed as a single coefficient, or it can be split into two factors, one describing the affect of evaporation and the other the affect of transpiration. As soil evaporation may fluctuate daily, as a result of rainfall and/or irrigation, the single crop coefficient expresses [5] only the time-average (multiday) effects of crop ET and has been considered within four distinct stages of growth (*see* FAO #56 by Allen et al. [5]). When stress conditions exist, the effects can be accounted for by a crop water stress coefficient (K_s) as follows:

$$ET_{cadj} = K_s \times K_c \times ET_o \quad (3)$$

2.2.1 CROP COEFFICIENTS

Although a number of ET_c estimation techniques are available, the crop coefficient (K_c) approach has emerged as the most widely used method for irrigation scheduling [45]. As ET is not only a function of the climatic factors, the crop coefficients can include conditions related to the crop development (K_c) and nonstandard conditions (K_s). The K_c is the application to two concepts [5]: a. Crop transpiration represented by the basal crop coefficient (K_{cb}); and *b*. The soil evaporation, K_e, is calculated using Eq. (4):

$$K_c = K_{cb} + K_e \quad (4)$$

where, K_c is an empirical ratio between ET_c and ET_o over grass or alfalfa, based on historically measurements. The curve for K_c is constructed for an entire crop growing season and which attempts to relate the daily water use rate of the specific crop to that of the reference crop [45]. The United Nations Food and Agriculture Paper FAO #56 by Allen et al. [5] provided detailed instructions for calculating these coefficients. For limited soil water conditions, the fractional reduction of K_c by K_s depends on the crop, soil water content and magnitude of the atmospheric evaporative demand [22].

The value for K_c equals K_{cb} for conditions: the soil surface layer is dry (i.e., when $K_e = 0$) and the soil water within the root zone is adequate to sustain the full transpiration (nonstressed conditions, i.e., $K_s = 1$). When the available soil water of the root zone becomes low enough to limit potential ET_c, the value of the K_s coefficient is less than 1 [5, 44, 45].

The soil evaporation coefficient accounts for the evaporation component of ET_c when the soil surface is wet, following irrigation or rainfall [5, 45]. When the available soil water of the root zone become low enough, crop water stress can occur and reduce ET_c. In the FAO-56 procedures, the effects of water stress are accounted for by multiplying K_{cb} (or K_c) by the water stress coefficient (K_s):

$$K_c.\ K_s = (K_{cb}.\ K_s + K_e) = ET_c/ET_o \quad (5)$$

where, $K_s < 1$ when the available soil water is insufficient for the full ET_o and $K_s = 1$ when there is no soil water limitation on ET_c. Thus, to determine K_s, the available soil water within the crop zone for each day needs to be measured or calculated using a soil water balance approach [45].

The estimation of K_e using the FAO-56 method, requires the use of the soil field capacity (FC), the permanent wilting point (PWP), total evaporable water (TEW), the fraction of the soil surface wetted (f_w) during each irrigation or rain and the daily fraction of the soil surface shaded by vegetation (f_c), or conversely the unshaded fraction ($1–f_c$). Hunsaker et al., [45] reported an exponential relation between $1–f_c$ and height to the Alfalfa crop.

The measurement of K_e and K_{cb} can be made by performing a *daily water balance* and use of the following equations from FAO Paper 56 [5].

$$ET_c = (K_{cb} + K_e)\,ET_o \tag{6}$$

$$K_{cb} = (ET_c/ET_o) - K_e \tag{7}$$

The soil evaporation (E) can be calculated using the Eqs. (8) and (9):

$$E = K_e \times ET_o \tag{8}$$

and

$$K_e = E/ET_o \tag{9}$$

The soil evaporation (E) can be measured using the water balance Eq. (10):

$$E = D_{e,i-1} - (P_i - RO_i) - \frac{I_i}{f_w} + \frac{f_i}{f_{ew}} + Tew_{,i} + DP_{e,i} \tag{10}$$

where, $D_{e,\,i-1}$ is the cumulative depth of evaporation following complete wetting from the exposed and wetted fraction of the topsoil at the end of day i–1 (mm), P_i is the precipitation on day i (mm), RO_i is precipitation runoff from the surface on day i (mm), I_i is the irrigation depth on day i that infiltrates into the soil (mm), E_i is evaporation on day i (i.e., $E_i = K_e/ET_o$) (mm), $T_{ew,i}$ is depth of transpiration from the exposed and wetted fraction of the soil surface layer on day i (mm), f_w is fraction of soil surface wetted by irrigation (0.01 to 1) and f_{ew} is the exposed and wetted soil fraction (0.001 to 1).

The ratio of reference evaporation to reference transpiration depends on the development stage of the leaf canopy expressed as "δ" the dimensionless fraction of incident beam radiation that penetrates the canopy [15] mentioned by Zhang et al. [124].

$$\delta = \exp(-K.LAI) \tag{11}$$

where, K is the dimensionless canopy extinction coefficient and therefore, evaporation and transpiration can be calculated using Eqs. (12) to (14):

$$E_o = \delta.ET_o \quad (12)$$

$$T = (1-\delta).ET_o \quad (13)$$

$$K_{cb}\,K_s = 1.5 \quad (14)$$

Hunsaker [44] found that ET_c in cotton was higher when the crop was submitted to high depth of irrigation (820–811 mm) that when have low depth of irrigation level (747–750 mm), similar to the $K_{cb}K_s$ curves, obtaining higher values than the treatment with high frequency (i.e., $K_{cb}K_s$ =1.5, 90 days after planting) than the low frequency (i.e., $K_{cb}K_s$ = 1.4, 90 days after planting).

2.2.2 LIMITATIONS IN THE USE OF K_C

Katerji and Rana [52] reviewed recent literature related to K_c and found differences of ±40% between K_c values reported in the FAO-56 paper [5] and the values experimentally obtained, especially in the mid growth stage. According to the authors, these large differences are attributable to the complexity of the coefficient K_c, which actually integrate several factors: aerodynamic factors linked to the height of the crop, biological factors linked to the growth and senescence of the surfaces leaves, physical factors linked to evaporation from the soil, physical factors linked to the response of the stomata to the vapor pressure deficit and agronomic factors linked to crop management (distance between rows, using mulch, irrigation system, etc.). For this reason K_c values needs to be evaluated for local conditions

The variation in crop development rates between location and year have been expressed as correlations between crop coefficients and indices such as the thermal base index, ground cover, days after emergence or planting and growth rate (i.e., Wright and Jensen, [122]; Hunsaker, [44]; Brown et al., [14]; Nasab et al., [72]; Hanson and May [34]; Madeiros et al., [61]; and Ramírez, [87]). The K_c is well related with the growing degree grades-GDD and with the fraction of the soil cover by vegetation (f_c) (Fig. 2.1) and depends on the genotype and plant densities [87]. The Eqs. (15) and (16) are for common bean genotype Morales with 13.6 plants.m^{-2}. The Eqs. (17) and (18) are for common bean genotype SER 16, with 6.4 plants.m^{-2}.

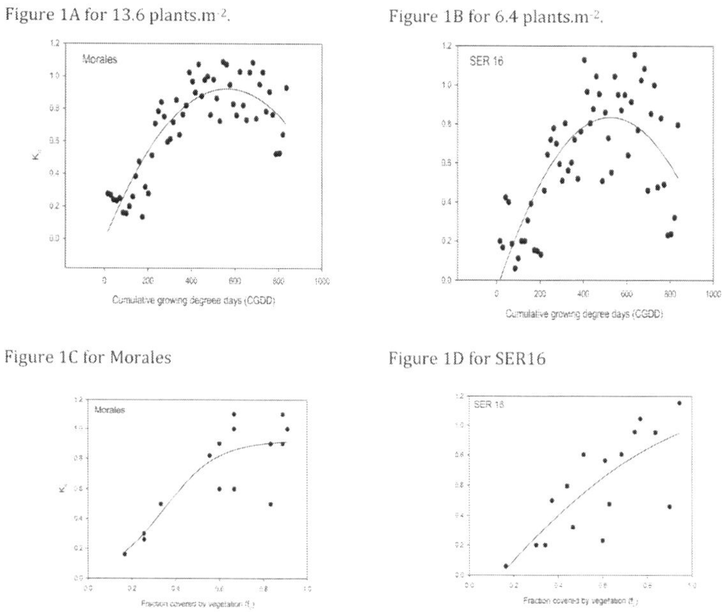

FIGURE 2.1 Crop coefficients (K_c) as related to cumulative growing degree days (CGDD) and fraction covered by vegetation (f_c) for two common bean genotypes: (**a**) Morales CGDD vs K_c, (**b**) SER 16 CGDD vs K_c, (**c**) Morales f_c vs K_c, (**d**) SER 16 f_c vs K_c. The curves were fitted from growth periods V1 to R9 (Data from: Ramirez, 2007). (These data were obtained under the project sponsored by NOAA-CREST (NA17AE1625), NASA-EPSCoR (NCC5–595), USDA-TSTAR-100 and University of Puerto Rico Agricultural Experiment Station, Mayaguez, USA).

$$K_c = -3x10^{-6} CGDD^2 + 0.0033 CGDD - 0.0533; R^2 = 0.76; p < 0.0001 \quad (15)$$

$$K_c = -1.4019 f_c^2 + 2.5652 f_c - 0.2449; R^2 = 0.70; p < 0.0003 \quad (16)$$

$$K_c = -3x10^{-6} CGDD^2 + 0.0034 CGDD - 0.0515; R^2 = 0.60; p < 0.0001 \quad (17)$$

$$K_c = -0.6726 f_c^2 + 1.90086 f_c - 0.2560; R^2 = 0.60; p < 0.0032 \quad (18)$$

2.2.3 SOIL WATER STRESS COEFFICIENT (K_S)

The soil water stress coefficient, K_s, is mainly estimated by its relationship to the average soil moisture content or matric potential in a soil layer and it can usually be estimated by an empirical formula based in soil water content or relative soil water available content [124].

The K_s is an important coefficient because it indicates the sensitivity of the crop to water deficit conditions, for example, corn grain yield is especially sensitive to moisture stress during tasselling and continuing through grain fill. Roygard et al., [94] observed that depletion of soil water to the wilting point for 1 or 2 days during tasselling or pollenization reduced yield by 22%. About 50% of stress yields reduced in 6 to 8 days. Allen et al. [5], presented the following methodology for estimating K_s:

$$K_s = \frac{TAW - Dr}{TAW - RAW} = \frac{TAW - Dr}{(1-p)TAW} \tag{19}$$

where, *TAW* is total available water and refers to the capacity of the soil to retain water available for plants (mm); Dr is root zone depletion (mm); RAW is the readily available soil water in the root zone (mm); *p* is the fraction of TAW that the crop can extract from the root zone without suffering water stress.

$$TAW = 1000(\theta_{FC} - \theta_{WP})Z_t \tag{20}$$

where, θ_{FC} is the water content at field capacity ($m^3.m^{-3}$), θ_{WP} is the water content at wilting point ($m^3.m^{-3}$) and Z_t is the rooting depth (m).

$$RAW = pTAW \tag{21}$$

Allen et al. [5] give values for different crops (FAO #56. p. 163) [5]. Roygard et al. [94] and Zhang et al. [124], reported that K_s is a logarithmic function of soil water availability (Aw) and can be estimated as follows:

$$K_s = \ln(Aw+1)/\ln(101) \tag{22}$$

where, *Aw* is calculate according to the Eq. (23):

$$Aw = 100\left(\frac{\theta_a - \theta_{wp}}{\theta_{FC} - \theta_{wp}}\right) \tag{23}$$

where, θ_a is average soil water content in the layers of the root zone depth. An example of the relationships between K_s and available soil water changes, estimated as a root zone depletion, is presented by Ramirez [87] The root zone depletion (D_r), can be calculated using the water balance Eq. (24):

$$D_{r,i} = D_{r,i-1} - (P - RO)_i - I_i + ET_{c,i} + DP_i \tag{24}$$

where, $D_{r,i}$ is the root zone depletion at the end of day i; $D_{r,i-1}$ is water content in the root zone at the end of the previous day, *i*–1; $(P\text{-}RO)_i$ is the difference between precipitation and surface runoff on day *i*; I_i is the irrigation depth on day *i*; $ET_{c,i}$ is the crop ET on day *i* and DP_i is the water loss from the root zone by deep percolation on day *i*; all the units are in mm.

The root zone depletion associated with a $K_s = 1.0$ (i.e., no water stress), was up to 10 mm for a root depth between 0 to 20 cm and up to 15 mm for a root depth of 0 to 40 cm in common beans. Fifty percent of the transpiration reduction was reached for $D_r = 22$ mm and 25 mm for the common bean genotype Morales and genotype SER 16, respectively. Transpiration ceased completely ($K_s = 0$) when Dr = 37 mm and 46 mm, respectively, for Morales and SER 16 [87].

2.3 DIRECT WATER VAPOR FLUX MEASUREMENT: LYSIMETERS

The word '*lysimeter*' is derived from the Greek root 'lysis,' which means dissolution or movement and 'metron,' which means to measure [41]. Lysimeters are tanks filled with soil in which crops are grown under natural conditions to measure the amount of water lost by evaporation and transpiration [5]. A lysimeter is the method of determining ET directly. The lysimeter are tanks buried in the ground to measure the percolation of water through the soil. Lysimeter are the most dependable means of directly measuring the ET rate, but their installation must meet four requirements for the data to be representative of field conditions [19]:

Requirement 1: The lysimeter itself should be fairly large and deep to reduce the boundary effect and to ovoid restricting root development. For short crops, the lysimeter should be at least one cubic meter in volume. For tall crops, the size of the lysimeter should be much larger.

Requirement 2: The physical conditions within the lysimeter must be comparable to those outside. The soil should not be loosened to such a degree that the root ramification and water movement within the lysimeter are greatly facilitated. If the lysimeter is unclosed on the bottom, precaution must be taken to avoid the persistence of a water table and presence of an abnormal thermal regime. To ensure proper drainage, the bottom of an isolated soil column will often require the artificial application of a moisture suction, equivalent to that present at the same depth in the natural soil [20].

Requirement 3: The lysimeter will not be representative of the surrounding area if the crop in the lysimeter is taller, shorter, denser, or thinner, or if the lysimeter is on the periphery of no-cropped area. The effective area of the lysimeter is defined as the ratio of the lysimeter ET per unit area of the surrounding field. The values of this ratio, other than unity, are caused by the in homogeneity of the surface. The maintenance of uniform crop height and density is not an easy task in a tall crop, spaced in rows. If the surface is indeed inhomogeneous, there is no adequate way to estimate the effective area from tank area overlap corrections or plant counts.

Requirement 4: Each lysimeter should have a "guard-ring" area around it maintained under the same crop and moisture conditions in order to minimize the clothesline effect. In arid climates, Thornthwaite in 1954, suggested that a "guard-ring" area of ten acres may or may not be large enough. Where several lysimeters are installed in the same field, the "guard-ring" radius may have to be about 10 times the lysimeter separation [19].

Lysimeters surrounded by sidewalks or gravel will not provide reliable data, nor will lysimeters planted to a tall crops if it is surrounded by short grass, or planted to grass and surrounded by a tall crop. Differences in growth and maturity between the lysimeters plants and surrounding plants can result in significant differences in measured ET in and outside the lysimeter [4]. The lysimeters are classified basically in two types: Weighing and Non-weighing.

2.3.1 NON-WEIGHING LYSIMETERS (DRAINAGE LYSIMETERS)

These operate on the principle that ET is equal to the amount of rainfall and irrigation water added to the system, minus percolation, runoff and soil moisture changes. Since the percolation is a slow process, the drainage lysimeters is accurate only for long periods for which the water content at the beginning exactly equals that at end. The length of such a period varies with the rainfall regime, frequency and amount of irrigation water application, depth of the lysimeters, water movement and the like. Therefore, records of drainage lysimeters should be presented only in terms of a long-period more than one day [19] and they are not useful for estimating hourly ET.

Allen et al. [4] discusses two types the non-weighing lysimeters: (a) *non-weighing constant water-table type*, which provides reliable data in areas where a high water table normally exists and where the water table level is maintained essentially at the same level inside as outside the lysimeters; (b) *non-weighing percolation type*, in which changes in water stored in the soil are determined by sampling or neutron methods or other soil humidity sensors like TDR and the rainfall and percolation are measured.

2.3.1.1 GENERAL PRINCIPLES OF A DRAINAGE LYSIMETER

Provisions are made at the bottom of the lysimeter container to collect and measure volumetrically the deep percolation. Precipitation is measured by rain gauge(s). *Evapotranspiration* is considered as the difference among *water applied, water drainage* and *soil water change* [108, 123].

When filling-in a lysimeter, the soil dug out from the pit of a lysimeter is replaced in the container, special precautions are needed to return the soil to its original status by restoring the correct soil profile and compacting the soil layers to the original density. It is desirable to have a similar soil state inside the lysimeter relative to the outside. However, if the roots are well developed and nutrients are available, as long as the water supply to the roots is unrestricted, dissimilar soil will not give significant variation in water use and yield, provided other conditions are similar [123].

Although disturbed soil in filled-in lysimeters does not pose serious problems in ET measurement, the soil can affect plant growth. Breaking up the soil, will change soil structure, aeration and soil moisture retention characteristics. The lysimeters should provide a normal rooting profile. It should be large enough to lender the effect of the rim insignificant. It can give relatively large errors in the ET measurement if the container is small. However, the greater the lysimeters area, the more costly and complicated the installation and operation becomes [123].

Installation and walls: The wall can be different materials: reinforced concrete, polyester reinforced with steel, fiberglass or plastic. The installation proceeds in the following steps: Excavation (e.g. 1 m×1 m×1.2 m) in the experimental site. Each layer of soil (e.g., 0–30 cm, 30–60 cm and 60–100 cm) is separated. Once the excavation it completed, the lysimeter is placed in the excavated hole with four wooden boards outside. Before repacking the soil layers, make a V-shaped slope at the bottom and place a 25 mm inside diameter perforated PVC pipe (horizontal). There should be a screen material placed around the perferated pipe to ovoid the soil particles from entering the pipe. Connect an access tube (25 mm PVC), approximately 1 m long (vertical). Cover the horizontal pipe with fine gravel approximately 3–5 cm thick. Fill the container with the excavated soil where each layer is repacked inside the lysimeter to match the original vertical soil state [123].

2.3.2 *WEIGHING LYSIMETER*

A weighing lysimeter is capable of measuring ET for periods as short as 10 minutes. Thus, it can provide much more additional information than a drainage lysimeter. Problems such as diurnal pattern of ET, the phenomenon of midday wilt, the short-term variation of energy partitioning and the relationship between transpiration and soil moisture tension, can be investigated only by studying the records obtained from a weighing lysimeter [5, 19, 60, 76, 92, 101, 105, 117].

Weighing lysimeters make direct measurements of water loss from a growing crop and the soil surface around a crop and thus, provide basic data to validate other water vapor flux prediction methods [23, 59, 85, 116]. The basic concept of this type of lysimeter is that it measures the difference between two mass values, the mass change is then converted into ET (mm) [47, 62].

During periods without rainfall, irrigation and drainage, the ET rate is computed as indicated by Howell [41], as:

$$ET = \left[A_l\left[\left(M_i - M_{i-1}\right)/\,A_l\right]/\,A_f\right]/\,T_i \tag{25}$$

where, ET is in units of ($mm.h^{-1}$ or $Kg.m^2$) for time interval i; M is the lysimeter soil mass, (Kg); A_l is lysimeter inner tank surface area (m^2); A_f is lysimeter foliage area (mid wall-air gap area) (m^2); *T* is the time period (h). The ratio A_f/A_l is the correction factor for the lysimeter effective area. This correction factor assumes the outside and inside vegetation foliage overlap evenly on all of the sides or edges. If there is no overlap, as occurs in short grass, the $A_f/A_l = 1.0$ [41].

Weighing lysimeters provide the most accurate data for short time periods and can be determined accurately over periods as short as one hour with a mechanical scale, load cell system or floating lysimeters [4]. Some weighing lysimeters use a weighing mechanism consisting of scales operating on a lever and pendulum principle [62]. However, some difficulties are very common like: electronic data logger replacement, data logger repair, load cell replacement, multiflex or installation etc. [62].

The measurement control in these lysimeters are important because of the following issues: (a) recalibration requirements; (b) measurement drift (e.g., slope drift, variance drift); (c) instrument problems (e.g., localized nonlinearity of load cell, load cell damage, data logger damage); (d) human error (e.g., incorrectly recording data during calibration); and (e) confidence in measurement results [62].

A load cell is a transducer that coverts a load acting on it into an analog electrical signal. The electrical signal is proportional to the load and the relationship is determined through calibration, employing linear regressions models (mV/V/mm water) and it is used to determine mass changes of a lysimeter over the period interest (e.g. day, hour, etc.).

The lysimeter characteristics can be different, for example, Malone et al. [62] built a lysimeter of the following form: 8.1 m^2 in surface area and 2.4 m depth, the lysimeter is constructed without disturbing the soil profile and the underlying fracture bedrock. The soil monolith is supported by a scale frame that includes a 200:1 lever system and a counterweight for the deadweight of the soil monolith. The gap between the soil in the lysimeters and the adjacent soil is between 5.1 cm and 7.0 cm except at the bottom slope where the runoff trough is located, this same author has given instructions for achieving a good calibration for this type of lysimeter.

Tyagi et al. [114] in wheat and sorghum used two rectangular tanks, an inner and outer tank, constructed from 5-mm welded steel plates. The dimensions of the inner tank were $1.985 \times 1.985 \times 1.985$ m^3 and those of the outer tank were $2.015 \times 2.015 \times 2.015$ m^3. The lysimeters were situated in the center of a 20-ha field. The size ratio of the outer tank to the inner tank is 1.03, so the error due to wall thickness is minimal. The effective area for crop ET was 4 m^2. The height of the lysimeter rim was maintained near ground level to minimize the boundary layer effect in and around the lysimeter. The lysimeter tank was suspended on the outer tank by four load cells. The load cells were made out of the steel shear beam type with 40,000-kg design load capacity. The total suspended mass of the lysimeter including tank, soil and water was about 14,000 kg. This provided a safety factor of 2.85. The high safety factor was provided to allow replacement of a load cell without the danger of overloading and also to account for shock loading. A drainage assembly connected with a vertical stand and gravel bedding to facilitate pumping of drainage water was provided. The standpipe also can be used to raise the water table in the lysimeter.

To calculate the ET using lysimeter, we need to employ the soil water balance (SWB) equation:

$$ET = R + I - P - Rff \pm \Delta SM \tag{26}$$

where, R is the rain, I is the irrigation, Rff is the runoff and $\pm\Delta SM$ is the soil moisture changes (all measurements are in mm). The size of the lysimeter is an important element to be considered in water vapor flux studies with this method. For example, Dugas et al. [23] evaluated small square lysimeters (<1.0 m^2) and reported significant differences in the ET estimations, basically associated with the differences in the leaf area index (LAI) inside the lysimeters, which differed among lysimeter, this problem can be addressed using LAI corrections

2.3.3 CALIBRATION OF THE WEIGHING LYSIMETER

Seyfried et al. [98] made a weighing lysimeter calibration by placing known weights on the lysimeters and then recording the resultant pressure changes. The weights used in that study were as follows: 19.9 kg for supportive blocks placed on the lysimeter, 43.4 kg for the tank, which contained the weights, and then 22.7 kg sacks of rock added in four-sack increments to a total of 24 sacks. The weight of each sack corresponded to about a 13 mm addition of water, so that weight increments were equivalent to ~52 mm and the total range was ~360 mm of water. Measurements were made when weights are added and removed.

The main arguments against the use of weighing lysimeters for monitoring water balance parameters and measuring solute transport parameters in the soil and unsaturated zone has been the discussion of potential sources of error, such as, the well known oasis effect, preferential flow paths at the walls of the lysimeter cylinders due to an insufficient fit of soil monoliths inside the lysimeters, or the influence of the lower boundary conditions on the outflow rates [25].

2.4 THE MICROMETEOROLOGICAL METHODS

For many agricultural applications, micrometeorological methods are preferred since they are generally nonintrusive, can be applied on a semi-continuous basis and provide information about the vertical fluxes that are occurring on scales ranging from tens of meter to several kilometers, depending the roughness of the surface, the height of the instrumentation and the stability of the atmosphere surface layer. Meyers and Baldocchi [65] have separated micrometeorological methods into four categories: (1) eddy covariance, (2) flux-gradient, (3) accumulation, and (4) mass balance. Each of these approaches are suitable for applications that depend on the scalar of interest and surfaces type and instrumentation availability. Some of these methods are described in the following sections of this chapter.

2.4.1. HUMIDITY AND TEMPERATURE GRADIENT METHOD

Movement of energy, water and other gases between field surface and atmosphere represent a fundamental process in the soil-plant-atmosphere continuum. The turbulent transport in the surface boundary layer affect the sensible (H) and latent (λE) heat fluxes, which along with the radiation balance, govern the evapotranspiration and canopy temperature [33].

Monteith and Unsworth [70] presented the functional form of the gradient flux equation and was applied by Harmsen et al. [36], Ramírez et al. [88] and Harmsen et al. [37]:

$$ET = \left(\frac{\rho_a \cdot c_p}{\gamma \cdot \rho_w}\right) \cdot \frac{(\rho_{vL} - \rho_{vH})}{(r_a + r_s)} \tag{27}$$

where, ρ_w is the density of water, ρ_v is the water vapor density of the air, r_a is the air density, γ is the psychometric constant, c_p is specific heat of air, r_a and r_s are aerodynamic and bulk surfaces resistances (all these variables are discussed in detail below). Here, *L* and *H* are in vertical positions above the canopy (L: low and H: High positions). For example, in small crops like beans or grass, possible values of *L* and *H* could be 0.3 m and 2 m above the ground, respectively.

Harmsen et al. [36] developed an automated elevator device (ET Station) for moving a temperature and relative humidity sensor (Temp/RH) between the two vertical positions (Fig. 2.2). The device consisted of a plastic (PVC) frame with a 12-volt DC motor (1/30 hp) mounted on the base of the frame. One end of a 2-m long chain was attached to a shaft on the motor and the other end to a sprocket at the top of the frame. Waterproof limit switches were located at the top and bottom of the frame to limit the range of vertical movement. For automating the elevator device, a programmable logic controller (PLC) was used which was composed of "n" inputs and "n" relay outputs. To program the device, a ladder logic was used which is a chronological arrangement of tasks to be accomplished in the automation process. The Temp/RH sensor was connected to the elevator device, which measured relative humidity and temperature in the up position for two minutes then changed to the down position where measurements were taken for 2 min. This process started each day at 06:00 hours and ended at 19:00 hours. When the elevator moves to the up position it activates the limit switch which sends an input signal to the PLC. That input tells the program to stop and remain in that position for two minutes. At the same time it activates an output, which sends a 5-volt signal to the control, port C2 in the CR10X data logger in which a small subroutine is executed. This subroutine assigns a "1" in the results matrix which indicates that the temperature and relative humidity corresponding to the up position. At the end of the 2-min period, the elevator moves to the down position and repeats the same process, but in this case sending a 5 volts signal to the data logger in the control port C4, which then assigns a "2" in the results matrix. All information was stored in the weather station data-logger CR-10X (Campbell Scientific, Inc.) for later downloading to a personal computer.

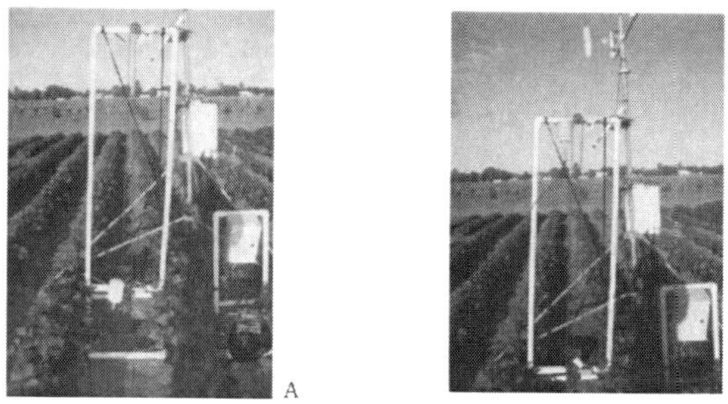

A B

FIGURE 2.2 Automated elevator device developed for moving the Temp/RH sensor between the two vertical positions: (a) Temp/RH sensor in down position and (b) Temp/RH sensor in up position. Measuring over common bean (*Phaseolus vulgaris* L.) [Picture obtained by the project sponsored by NOAA-CREST (NA17AE1625), NASA-EPSCoR (NCC5–595), USDA-TSTAR-100 and University of Puerto Rico Experiment Station].

2.4.2 THE BOWEN RATIO ENERGY BALANCE METHOD

The basis for this method is that the local energy balance is closed in such a way that the available net irradiative flux (Rn) is strictly composed of the sensible (H), latent (λE) and ground heat (G) fluxes, other stored terms such as those related to canopy heat storage and photosynthesis are negligible [65].

This method combines measurements of certain atmospheric variables (temperature and vapor concentration gradients) and available energy (net radiation and changes in stored thermal energy) to determine estimates of evapotranspiration (ET) [58]. The method incorporates energy-budget principles and turbulent-transfer theory. Bowen showed that the ratio of the sensible to latent heat flux (β) could be calculated from the ratio of the vertical gradients of temperature and vapor concentration over a surface under certain conditions.

Often the gradients are approximated from air-temperature and vapor-pressure measurements taken at two heights above de canopy. The Bowen-ratio method assumes that there is no net horizontal advection of energy. With this assumption, the coefficients (eddy diffusivities) for heat and water vapor transport, Kh and Kw, respectively, are assumed to be equal. Under advective conditions, Kh and Kw are not equal [112] and the Bowen-ratio method fails to accurately estimate ET.

Based on the assumption that Kh and Kw are equal, and by combining several terms to form the psychometric constants, the Bowen-ratio take the form of the equation 2. Although the theory for this method was develop in the 1920s by Bowen [11], its practical applications has only been possible in recent decades, due to the availability of accurate instrumentation [77]. The Bowen ratio initial concept is shown below:

$$\beta = \frac{PC_p Kh \frac{dT}{dz}}{\lambda \varepsilon Kw \frac{de}{dz}} \tag{28}$$

If it is assumed that there is no net horizontal advection of energy, Eq. (28) can be simplified as shown below:

$$\beta = \frac{PC_p \frac{dT}{dz}}{\lambda \varepsilon \frac{de}{dz}} \tag{29}$$

where, P is the atmospheric pressure (kPa), C_p is the specific heat of air (1.005 J./g◦C), ε is the ratio molecular weight of water to air = 0.622 and λ is the latent-heat flux (Jg^{-1}). Once the Bowen ratio is determined, the energy balance can be solved for the sensible-heat flux (H) and latent-heat flux (λE).

$$Rn = \lambda E + H + G \tag{30}$$

where, R_n is the net radiation, λE is the latent-heat flux, H is the sensible-heat flux and G is the soil-heat flux.

$$H = \beta \lambda E \tag{31}$$

$$\lambda E = \frac{(R_n - G)}{(1 + \beta)} \tag{32}$$

The latent heat flux can be separated into two parts: the evaporative flux E ($g\ m^{-1}\ day^{-}1$) and the latent heat of vaporization λ (Jg^{-1}), which can be expressed as a function of air temperature (T) (λ = 2,502.3–2.308 T). The latent-heat of vaporation (λ) is defined as the amount of energy required to convert 1 gram of liquid water to vapor at constant temperature T. Sensible-heat flux (H) is a turbulent, temperature-gradient driven heat flux resulting from differences in temperature between the soil and vegetative surface and the atmosphere.

The soil-heat flux (G) is defined as the amount of energy moving downward through the soil from the land surface, caused by temperature gradient. This flux is considered positive when moving down through the soil from the land surface and negative when moving upward through the soil toward the surface [111]. The soil heat flux is obtained by measuring two soil heat flux plates below the soil surface at 2 and 8 cm, soil moisture at 8 cm, and soil temperature at 6 cm between the two soil heat flux plates [15].

Because the soil-heat flux is measured below the soil surface, some of the energy crossing the soil surface could be stored in, or come from, the layer of soil between the surface and flux plate located closest to the surface, for this reason a change in storage term, S is added to the measured heat flux (equ. 33). [33]:

$$S = \left[\frac{\Delta Ts}{\Delta t}\right] d\rho_b \left(C_s + \left(WC_w\right)\right) \tag{33}$$

where, S is the heat flux going into storage (Wm^{-2}), Δt is the time interval between measurement (sec), ΔTs is the soil temperature interval between measurement, d is the depth to the soil-heat-flux plates (0.08 m), ρ_b is the bulk density of dry soil (1300 kgm^{-3}), C_s is the specific heat of dry soil (840 J./Kg◦C), W is the water content of soil (kg the water/Kg the soil) and C_w is the specific heat of water (4.190 J./Kg◦C). The soil heat flux (G) at the surface is obtained by including the effect of storage between the surface and depth, *d*, using equation 11.

$$G = \left(\frac{FX1 + FX2}{2}\right) + S \tag{34}$$

where, FX_1 is the soil-heat flux measured 1 (Wm^{-2}), FX_2 is the soil-heat flux measured 2 (Wm^{-2}). One of the requirements for using the Bowen-ratio method is that the wind must pass over a sufficient distance of similar vegetation and terrain before it reaches the sensors. This distance is referred to as the fetch, and the fetch requirement is generally considered to be 100 times the height of the sensors above the surface [16]. More detail about determination of the minimum fetch requirement is presented later in this document.

Hanks et al. (1968), described by Frank [27], reported λE/Rn of 0.16 for dry soil conditions and 0.97 for wet soil conditions; On the other hand he found λE/Rn to be lowest in grazed prairie, suggesting that defoliation changes the canopy structure and energy budget components, which may have contributed to increase water loss through evaporation compared with the nongrazed prairie treatment. Hanson and May [34], using the Bowen Ratio Energy Balance Method to measure ET in tomatoes, found that ET rates decreased substantially in respond to drying of the soil surface. Perez et al. [83] proposed a simple model for estimating the Bowen ratio (β) based on the climatic resistance factors:

$$\beta = \frac{\Delta + \gamma}{\Delta} . \frac{1 + S}{1 + C} - 1 \tag{35}$$

$$C = \frac{\gamma . r_i}{\Delta . r_a} \tag{36}$$

$$S = \frac{\gamma}{\Delta + \gamma} . \frac{r_c}{r_a} \tag{37}$$

where, r_c is the canopy resistance (s m^{-1}) based on the "big leaf concept, and r_a is the aerodynamic resistance (s m^{-1}). These resistance factors are described in detail in the next section. The factor r_i is the climatological resistance as reported by [66]:

$$r_i = \frac{\rho_a C_p VPD}{\gamma (R_n - G)} \tag{38}$$

where, ρ_a is the air density at constant pressure (Kg.m^{-3}), C_p is the specific heat of moist air at constant pressure (1004 J.Kg^{-1}°C^{-1}), VPD is the vapor pressure deficit of the air (Pa), γ is the pychrometric constant (Pa.°C^{-1}) and R_n and G are in W.m^{-2}. For homogeneous canopies, the effective crop surface and source of water vapor and heat is located at height d + z_{oh}, where d is the zero plane displacement height and z_{oh} is the roughness length governing the transfer of heat and vapor [5].

2.4.3 THE PENMAN-MONTEITH METHOD

The important contribution of Monteith and Penman's original equation was the use of resistances factors, which was based on an electrical analogy for the potential difference needed to drive unit flux systems that involve the transport of momentum, heat, and water vapor [69, 70]. The resistances have dimensions of time per unit length, as will describe later. This methodology calculates the latent heat flux using the vapor pressure deficit, the slope of the saturated vapor-pressure curve and aerodynamic resistance to heat, and canopy resistance in addition to the energy-budget components of the net radiation, soil heat flux, and sensible heat flux. Field measurements of air temperature, relative humidity, and wind speed are needed to determine these variables [11]. Eq. (21) describes the Penman-Monteith (P-M) method to estimate the λE [5, 54]:

$$\lambda E = \frac{\Delta s (R_n - G) + \rho_a C_p \frac{VPD}{r_a}}{\Delta s + \gamma \left(1 + \frac{r_s}{r_a}\right)} \tag{39}$$

where, λE, R_n, and G in fim-^{2}, VPD is vapor pressure deficit (kPa), Δs is slope of saturation vapor pressure curve (kPa ◦C^{-1}) at air temperature, ρ is density of air (Kgm^{-3}), Cp in J. Kg^{-1}◦C^{-1}, γ in kPa ◦C^{-1}, r_a is aerodynamic resistant (s m^{-1}) r_s surface resistance to vapor transport (s m^{-1}).

According to Monteith [69], the appearance of a wind-dependent function in the denominator as well as in the numerator implies that the rate of evaporation calculated from the P-M model is always less dependent on wind speed than the rate from corresponding the Penman equation when other elements of climate are unchanged. In general, estimated rates are usually insensitive to the magnitude of r_a and the error generated by neglecting the influence of the buoyancy correction is often small. In contrast, the evapotranspiration rate is usually a strong function of the surfaces resistance (r_s).

Kjelgaard and Stockle [54] say the surface resistance (r_s) parameter in the P-M model is particularly difficult to estimate due to the combined influence of plant, soil and climatic factors that affect its value. The magnitude of the stomatal resistance can be estimated in principle from the number of stomata per unit leaf area and from the diameter and length of pores, which is difficult and therefore rarely measured; therefore, the stomatal resistance is usually calculated from transpiration rates or estimated gradients of vapor concentration [69].

Knowing the value of the aerodynamic resistance (r_a) permits estimation of the transfer of heat and water vapor from the evaporating surface into the air above the canopy. The aerodynamic resistance for a single leaf to diffusion through the boundary layer surrounding the leaf, within which the transfer of heat, water vapor, *etc.*, occurs, proceeds at a rate governed by molecular diffusion. Provided the wind speed is great enough and the temperature difference between the leaf and air is small enough to ensure that transfer processes are not affecting by gradients of air density, the boundary layer resistance depends on air velocity and on the size, shape, and altitude of the leaf with respect to the air stream. In very light wind, the rates of transfer are determined mainly by gradients of temperature and therefore by density, so that the r_a depends more on the mean leaf-air temperature difference than on wind speed. According to Thom [109], the r_a for heat transfer can be determined as follows:

$$r_{ah} = \frac{\rho C\rho(T_s - T_a)}{H} \tag{40}$$

At the field level, r_a for homogeneous surfaces, such as bare soil or crop canopies, there is a large-scale analogous boundary layer resistance, which can be estimated or derived from measurements of wind speed and from a knowledge of the aerodynamic properties of the surface as is described later [69]. The r_a can be determined given values of roughness length (Z_o) and zero plane displacement height (d), that depend mainly on crop height, soil cover, leaf area and structure of the canopy [1]:

$$r_a = \frac{Ln\left[\frac{(Z_m - d)}{Z_{om}}\right] Ln\left[\frac{(Z_h - d)}{Z_{oh}}\right]}{K^2 u_z} \tag{41}$$

where, Z_m is height of wind measurements (m), Z_h is height of humidity measurements (m), d is zero displacement height (m), Z_{om} is roughness length governing momentum transfer of heat and vapor (m) is 0.123h, Z_{oh} is roughness length governing transfer of heat and vapor (m) is $0.1Z_{om}$, K is the von Karman's constant (0.41), u_z is wind speed at height *z*.

The Eq. (41) is restricted for neutral stability conditions, i.e., where temperature, atmospheric pressure, and wind speed velocity distribution follow nearly adiabatic conditions (no heat exchange). The application of the equation for short time periods (hourly or less) may require the inclusion of corrections for stability. However, when predicting ET_o in the well watered reference surface, heat exchange is small, and therefore the stability correction is normally not required [1].

Alves et al. [1] state that though this is the most used expression for r_a, in fact it is not entirely correct, since it assume a logarithmic profile from the source height ($d + Z_{oh}$) with increasing *z* in the atmosphere, using the concept to the "big leaf, Eq. (41) can be modified as follows:

$$r_a = \frac{Ln\left(\frac{z-d}{h_c-d}\right)Ln\left(\frac{z-d}{Z_{om}}\right)}{K^2 u_z} \tag{42}$$

where, h_c is the height of the crop canopy. According to Tollk et al. [110], the r_a to momentum transport in the absence of buoyancy effects (neutral stability) follows the Eq. (43):

$$r_{am} = \ln\left[\left(Zi-d\right)/Z_{om}\right]^2 / k^2 u_z \tag{43}$$

Under adiabatic conditions, the equations must be corrected using the Richardson number for stability correction, assuming similarity in transport of heat and momentum, yielding:

$$r_{ah} = r_{am}\left(1+5R_i\right) \tag{44}$$

The Ri for stability conditions is considered when ($-0.008 \leq R_i \leq 0.008$) and is calculated by:

$$R_i = \left[g\left(T_a - T_s\right)\left(Z-d\right)\right] / T_{av}.u^2{}_z \tag{45}$$

where, *g* is the acceleration of the gravity (9.8 $m.s^{-2}$), T_a is the air temperature (K), T_c is the plant canopy temperature (K), T_{av} is the average temperature taken as $((T_a+T_c)/2)$. The advantage of the R_i over other stability corrections is that it contains only experimentally determined gradients of temperature and wind speed and does not depend directly on sensible heat flux [110].

The bulk surface resistance (r_s) describes the resistance of vapor flow through transpiring crop leaves and evaporation from the soil surface. Where the vegetation does not completely cover the soil, the resistance factor should indeed include the effects of the evaporation from the soil surface. If the crop is not transpiring at a potential rate, the resistance depends also on the water status of the vegetation [5,115], and for this case they proposed the use of the following approximate:

$$r_s = \frac{r_L}{LAI_{active}} \tag{46}$$

where, LAI_{active} is 0.5 times the leaf area index (m^2 of leaf perm^2 of soil), and r_L is bulk stomatal resistance, which is the average resistance of an individual leaf, and can be measured using an instrument called a porometry, the first stomatal readings were developed by Francis Darwin who developed horn hygrometer [113].

The r_L readings are highly variable and depend on several factors, such as: crop type and development stage, the weather and soil moisture variability, the atmospheric pollutants and the plant phytohormone balance [113]. Typically to determine minimum r_L using a porometer, fully expanded, sunlit leaves near to the top of the canopy are surveyed during maximum solar irradiance (approximately solar noon under cloudless conditions) and low VPD periods [54]. This "standard value from literature or porometer measurements are hereafter identified as r_{Lmin}. In addition, r_L has been shown to increase with increasing VPD and/or reduced solar irradiance (R_s). Adjustment factors for VPD (f_{VPD}) and R_s (f_{Rs}) were empirically derived and used as multipliers of r_{Lmin}. The dependence of r_L on VPD can be represented by a linear function [46] as:

$$f_{VPD} = a + b[VPD] \tag{47}$$

where, *a* and *b* are linear regression coefficients, and f_{VPD} is equal to 1 (no adjustment) for VPD $\leq$ a threshold value, which can be taken as 1.5 kPa. The same authors presented a calibrated form of equation 47 for corn as, $f_{VPD} = 0.45+0.39(VPD)$. Kjelgaard and Stockle [54] presented a modified form of the adjustment factor:

$$f_{Rs} = \frac{R_{s\max}}{C_2 + R_s} \tag{48}$$

where, R_s and $R_{s\ max}$ are the actual and maximum daily solar irradiance (MJ m^{-2} day^{-1}) and C_2 is a fitted constant. Taking the maximum of the adjustment factors for VPD and Rs, r_{Lmin} is modified to give the r_L [54]:

$$r_L = r_{L\min} \max\left\{f_{VPD}, f_{Rs}\right\} \tag{49}$$

where, f_{VPD} and f_{Rs}, are equal to or greater than 1. Alves et al. [1] indicated that the surface resistance term (r_s) has been the most discussed in the literature. Several components to be considered here include: a) The resistances to water vapor at the evaporating surfaces: plants and their stomates (r_s^c) and soil (r_s^s); *b*) the resistance to vapor transfer inside the canopy from these evaporating surfaces up to the "big leaf (r_s^a). The resistance r_s^c, can be approximated using Eq. (50).

$$r_s^c = \frac{\left(\sum_{i=1}^{n} \frac{1}{r_{stj}}\right)^{-1}}{LAI} \tag{50}$$

where, r_{st} is the single leaf stomatal resistance (sm^{-1}), n is a leaf number. The bulk surface resistance can also be calculated using the inversion of the Penman-Monteith equation with incorporation of the Bowen ratio as follow [1, 3]:

$$r_s = r_a\left(\frac{\Delta s}{\gamma}\beta - 1\right) + \frac{\rho_a C_p VPD}{\gamma \lambda E} \tag{51}$$

Accurate prediction of r_s requires a good estimate of the Bowen ratio (β). Ramirez [87] has used the following inversion form of the Penman-Monteith equation to obtain estimates of r_s:

$$r_s = r_a.\frac{\left[\frac{\Delta(R_n - G) + \rho_a C_p\left(\frac{VPD}{r_a}\right)}{\lambda E} - \Delta - \gamma\right]}{\gamma} \tag{52}$$

Similarly these authors, analyzing the resistance concepts, concluded that the r_s of dense crops cannot be obtained by simply averaging stomatal resistance because the driving force (vapor pressure deficit) is not constant within the canopy.

Saugier [96] addressed canopy resistance (r_c), stating that it is normally a mixture of soil and plant resistances to evaporation. If the top the soil is very dry, direct soil evaporation may be neglected and r_c is approximately equal to the leaf resistance (r_L) divided by the LAI. Baldocchi et al. [10], indicated that the inverse of the 'big-leaf' model (eg., inverse of the P-M model) will be a good estimate of canopy resistance or surface resistances if certain conditions are met. These conditions include: i) a steady-state environment; ii) a dry, fully developed, horizontally homogeneous canopy situated on level terrain; iii) identical source-sink levels for water vapor, sensible heat and momentum transfer, and negligible cuticular transpiration and soil evaporation. Szeicz and Long [104] described a profile method to estimate r_s as:

$$r_s = \frac{\rho_a.C_p, VPD}{\gamma.\lambda E} \tag{53}$$

These methods can be used in the field when the rate of evapotranspiration is measured by lysimeters or calculated from the Bowen ratio energy balance method, and the temperature, humidity and wind profiles are measured within the boundary layer simultaneously. Ortega-Farias et al. [74], evaluated a methodology for calculating the canopy surface resistance ($r_{cv} \approx r_s$) in soybean and tomatoes, using only meteorological variables and soil moisture readings. The advantage of this method is that it can be used to estimate λE by the general Penman-Monteith model with meteorological reading at one level, and without r_L and LAI measurements.

$$r_s = \frac{\rho_a . c_p . VPD}{\Delta.(R_n - G)} . \frac{\theta_{FC} - \theta_{WP}}{\theta_i - \theta_{WP}} \tag{54}$$

where, θ_{FC} and θ_{WP} are the volumetric moisture content at field capacity (fraction) and wilting point (fraction), respectively, and θ_i is a volumetric soil content in the root zone (fraction) measured each day. Kamal and Hatfield [48] used the Eq. (51) to determine the surface resistance in Potato: and stated that the canopy resistance (r_c in $s.m^{-1}$; "mean stomatal resistances of crops), can be determined by dividing the r_s by the effective LAI as defined by other authors such as Hatfield and Allen [38] and for well watered crops, r_c can be can be estimated using Eq. (55).

$$r_c = \frac{0.3LAI + 1.2}{LAI} r_s \tag{55}$$

Kjelgaard and Stockle [54] discussed the estimation of canopy resistance (r_c) from single-leaf resistance (r_L, Eq. (56)), as originally proposed by Szeicz and Long [104]:

$$r_c = \frac{r_L}{LAI_{active}} \tag{56}$$

Kamal and Hatfield [48] divided the surface resistance (r_s) used in the P-M model into two components, and conceptualized an excess resistance (r_o) in series with the canopy stomatal resistance. This excess resistance was linked to the structure of the crop, particularly crop height.

$$r_s = r_c + r_o \tag{57}$$

Pereira et al. [81] stated that the surface resistance (r_s) is the sum of two components: one corresponding mainly to the stomatal resistance (r_{st}), the other to the leaf boundary layer and turbulent transfer inside the canopy (r_{ai}) (equation 58), thus, surface resistance is not a purely physiological parameter:

$$r_s = r_{st} + r_{ai} \tag{58}$$

Stomatal resistance can take values from 80 $s.m^{-1}$ to 90 $s.m^{-1}$ as a common range for agricultural crops suggested a value of 100 $s.m^{-1}$ for most arable crops [67]. Table 2.1 lists mean average values for various crops under well water conditions.

The r_L is strongly dependent on the time of day (basically due to the temporal nature of climatic conditions), for the soil moisture content and by the genotype. Figure 2.3a shows how larger differences in r_L occur, with and without drought stress, after 9: 00 am until late in the afternoon, and the most critical point is at 13: 00 hours when the highest VPD occurred. For this reason, when this variable (r_L) is not measured, appropriate parameterization is required for good water flux or ET estimation, especially under drought stress conditions. In Figure 2.3c, it is possible to see in a common bean genotype under drought stress conditions, lower r_L as compared with less drought resistance during several days with drought stress. Perrier (1975), as reported in Kjelgaard and Stockle [54], conceptualized the excess resistance (r_o) as a linear function of crop height and LAI:

$$r_o = ah_c + bLAI \tag{59a}$$

where, *a* and *b* are constants. For corn, Kjelgaard and Stockle [54] parameterized Eq. (59a) as follows:

$$r_o = 16.64h_c + 0.92LAI \tag{59b}$$

Canopy resistance can also be determined from leaf or canopy temperature since it is affected by plant characteristics, eg. Leaf area index (LAI), height, and maturity. Soil factors (available soil water-ASW, and soil solution salinity) and weather factors (R_n and wind speed) also affect the canopy resistance.

Montheith [66] showed that transpiration rate physically depends on relative changes of surface temperature and r_a, and concluded that r_a depends on the Reynolds number of the air and can be determined from wind speed, the characteristic length of the plant surface, and the kinematic viscosity of the air. An increase in r_c for Wheat was caused by a decrease in total leaf area, by an increase in the resistance of individual leaves due to senescence, or by a combination of both effects; in Sudan grass, r_c increased with plant age and a decrease in soil moisture. Van Bavel [115] studied Alfalfa throughout an irrigation cycle and found that canopy resistance increased linearly with decreasing soil water potential. Kamal and Hatfield [48] found an exponentially inverse relationship between canopy resistance and net radiation, and a linear inverse relationship between canopy resistance and available soil water.

The Drainage and Irrigation Paper-FAO 56 [5] recommended the Szeicz and Long [104] method for calculating r_s (Eq. (56)), where an average of r_L for different positions within the crop canopy, weighted by LAI or $LAI_{effective}$ is used. This method gives good results only in very rough surfaces, like forest and partial cover crops with a dry soil [67]. Alves et al. [1] concluded that r_s of dense crops cannot be obtained by simply averaging stomatal resistance (r_L) because VPD, which is the "driving force, is not constant within the canopy. Alves and Pereira [3] have stated "The PM model can be used to predict ET if accurate methodologies are available for determining the r_s that take into account the energy partitioning."

In addition to the lack of r_s values for crops, questions have been raised relative to the appropriateness of using the PM model for partial or sparse canopies because the source/sink fluxes may be distributed in a nonuniform manner throughout the field [24, 32, 55, 75]. Adequate parameterization of the surface resistance makes the P-M model a good estimator of ET [3, 74–90, 96].

Ramirez [87], reported that the daily ET estimation with the P-M model with r_s based on r_L and $LAI_{effective}$ gave a good estimation in two common bean genotypes with variable LAI, without and with moderate drought stress for both years (2006 and 2007).

Ramirez et al. [88] reported inverse relation between r_a and r_s and r_L in beans (*Phaseolus vulgaris* L), as well as those reported by Alves and Pereira [3] (Fig. 2.4), which implies that with low r_a (windy conditions), the r_L (and therefore r_s) increases. The Alves and Pereira [3] study did not measure the r_L, rather the r_s was estimated based on micrometeorological parameters.

Disparities in the measured r_s using the P-M inverse model arise from: a) imperfect sampling of leaves and the arbitrary method of averaging leaf resistance over the whole canopy, *b*) from the dependence of r_s on nonstomatal factors such as evaporation from wet soil or stems, or others and *c*) the complex aerodynamic behavior of canopies [68].

Lower LAI index (LAI <1.0) and drought stress also affect the precision in the r_s estimation [87]. Use of the $LAI_{effective}$ when LAI < 1.0 is not necessary and tends to overestimate the r_s and under-estimate the ET. Katerji and Perrier [51] found for LAI >1.0 a good agreement between measurement values of evapotranspiration over alfalfa crops using the energy balance method, and values calculated with P-M equation using variable r_s. Katerji and Perrier [50] proposed to simulate r_s using the following relation:

$$\frac{r_s}{r_a} = a\frac{r^*}{r_a} + b \tag{60}$$

where, ***a*** and ***b*** are linear coefficients that should be determined empirically, r^* ($s.m^{-1}$) is a climatic resistance [52] giving by:

$$r^* = \frac{\Delta + \gamma}{\Delta\lambda} \cdot \frac{\rho C_p VPD}{(R_n - G)} \tag{61}$$

TABLE 2.1 Average Values of the Stomatal Resistance (r_L) For Several Crops

Cover crops	rL s/m	Source	Cover crops	rL s/m	Source
Corn	200	Kirkham et al. [53]	Cassava	714, between 476 to 1428	Oguntunde [73]. This data under limited soil water conditions.
Sunflower	400	Kirkham et al. [53]	Eucalyptus	200–400	Pereira and Alves [81]
Soybean and potato	350	Kirkham et al. [53]	Maple	400–700	Pereira and Alves [81]
Sorghum	300	Kirkham et al. [53]	Crops-General	50–320	Pereira and Alves [81]
Millet	300	Kirkham et al. [53]	Grain sorghum	200	Pereira and Alves [81]
Aspen	400	Pereira and Alves [81]	Soybean	120	Pereira and Alves [81]
Maize	160	Pereira and Alves [81]	Barley	150–250	Pereira and Alves [81]
Alfalfa	80	Pereira and Alves [81]	Sugar beet	100	Pereira and Alves [81]
Clipped grass (0.15 m)	100–150	Pereira and Alves [81]	Clipped and Irrigated grass (0.10–10.12 m)	75	Pereira and Alves [81]
Common beans	170–270	Ramirez [87]	Sorghum	192	Stainer et al. [101]
Corn	264	Ramirez and Harmsen (2007).	Andes Tropical Forestry	132	Ramirez and Jaramillo [89]. (Calculated)
Coffee	149	Ramirez and Jaramillo [89]. (Calculated)	Coffee	150	Angelocci et al. [7]
Wheat	134	Howell et al. [42]	Corn	252	Howell et al. [42]
			Sorghum	280	Howell et al. [42]

Table 2.2 presents values of *a* and *b* for several crops. The Penman-Monteith model is considered as a 'single-layer' model, Shuttlewoth and Wallace [100] de-

veloped a 'double-layer' model, relying on the Penman-Monteith model concept to describe the latent heat flux from the canopy (λT) and from the soil (λE) as follows:

$$\lambda T = \frac{\Delta\left(R_n - R_{ns}\right) + \rho Cp \dfrac{VPDo}{r^{c}{}_{a}}}{\Delta + \gamma\left(1 + \dfrac{r^{c}{}_{s}}{r^{c}{}_{a}}\right)} \tag{62}$$

$$\lambda E = \frac{\Delta\left(R_{ns} - G\right) + \rho Cp \dfrac{VPDo}{r^{s}{}_{a}}}{\Delta + \gamma\left(1 + \dfrac{r^{s}{}_{s}}{r^{s}{}_{a}}\right)} \tag{63}$$

where, R_{ns} is the absorbed net radiation at the soil surface, r^{c}_{a} is the bulk boundary layer resistance of the canopy elements within the canopy, r^{c}_{s} is the bulk stomatal resistance of the canopy, r^{s}_{a} is the aerodynamic resistance between the soil and the mean canopy height, r^{s}_{s} is the surfaces resistance of the soil and VPD_{o} is the vapor pressure deficit at the height of the canopy air stream.

2.4.4 THE DOUBLE-LAYER SHULTTLEWORTH-WALLACE MODEL

The Shulttleworth-Wallace Model (S-W) assumes that there is blending of heat fluxes from the leaves and the soil in the mean canopy airflow at the height of the effective canopy source [100]. The full expression of the Shulttleworth-Wallace Model (S-W) is presented by Zhang et al. [125] as follows:

$$\lambda ET = \lambda E + \lambda T = C^{S}_{SW} PM^{S}_{SW} + C^{P}_{SW} PM^{P}_{SW} \tag{64}$$

$$PM^{S}_{SW} = \frac{\Delta A_{SW} + \left[\left(\rho CpD - \Delta r^{s}_{a}\right)\left(A_{Sw} - A^{s}_{SW}\right) / \left(r^{a}_{a} + r^{s}_{a}\right)\right]}{\Delta + \gamma\left[1 + r^{s}_{s} / \left(r^{a}_{a} + r^{s}_{a}\right)\right]} \tag{65}$$

$$PM^{p}_{SW} = \frac{\Delta A_{SW} + \left[\left(\rho CpD - \Delta r^{p}_{a} A^{s}_{SW}\right) / \left(r^{a}_{a} + r^{p}_{a}\right)\right]}{\Delta + \gamma\left[1 + r^{p}_{s} / \left(r^{a}_{a} + r^{p}_{a}\right)\right]} \tag{66}$$

$$C^{s}_{SW} = \frac{1}{1 + \left[R^{S}_{SW} R^{a}_{SW} / R^{P}_{SW}\left(R^{S}_{SW} + R^{a}_{SW}\right)\right]} \tag{67}$$

$$C_{SW}^{P} = \frac{1}{1+\left[R_{SW}^{P} R_{SW}^{a} / R_{SW}^{S}\left(R_{SW}^{P} + R_{SW}^{a}\right)\right]} \tag{68}$$

$$R_{SW}^{S} = \left(\Delta + \gamma\right) r_{a}^{s} + \gamma r_{s}^{s} \tag{69}$$

In Eq. (70), λE is the latent heat flux of evaporation from the soil surfaces (W/m²), λT the latent heat fluxes of transpiration from canopy (W/m²), r_s^p the canopy resistance (s/m), r_s^s the aerodynamic resistance of the canopy to in-canopy flow (s/m), r_s^s the soil surfaces resistance (s/m), r_a^s and r_a^s the aerodynamic resistance from the reference height to in-canopy heat exchange plane height and from there to the soil surface (s/m), respectively. A_{sw} and A_{SW}^{S} are the total available energy and the available energy to the soil (W/m²), respectively and defined in Eqs. (70)–(73):

$$R_{SW}^{P} = \left(\Delta + \gamma\right) r_{a}^{p} + \gamma r_{s}^{p} \tag{70}$$

$$Ra_{SW}^{P} = \left(\Delta + \gamma\right) r_{a}^{a} \tag{71}$$

$$A_{sw} = R_{n} - G \tag{72}$$

$$A_{SW}^{S} = R_{nsw}^{s} - G \tag{73}$$

In Eq. (73), R_{nsw}^{s} is the net radiation fluxes into the soil surface (W/m²), and can be calculated using the Beer's law:

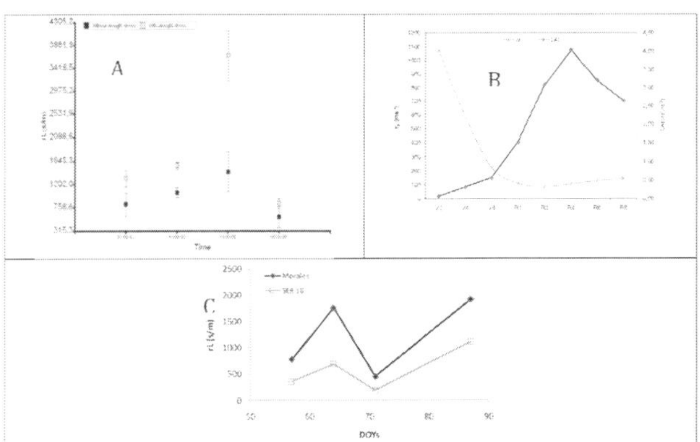

FIGURE 2.3 Relationship between (a) the changes in the stomatal resistance during the day with and without drought stress in *Phaseolus vulgaris* L. genotype 'Morales,' (b) the surfaces resistance and Leaf area index, and (c) the stomatal behavior represented in stomatal resistance (r_L) under drought stress conditions for two common bean genotypes — 'Morales' lest drought tolerant and 'SER 16' drought stress tolerant.

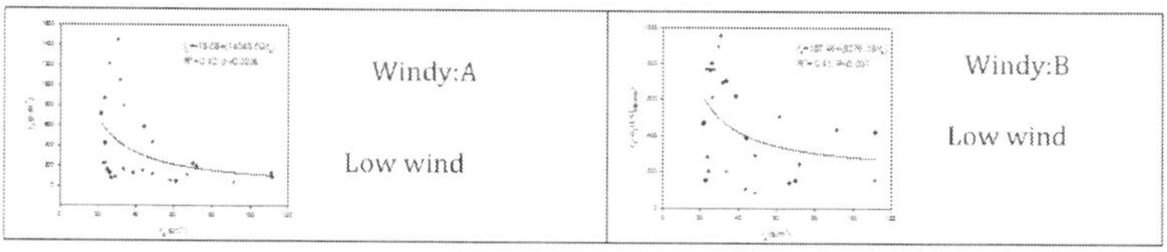

FIGURE 2.4 (a) Aerodynamic resistance (r_a) as a function of stomatal resistance (r_L); and (b) Aerodynamic resistance (r_a) as a function of measured surface resistance ($r_s = r_L/LAI_{effective}$) [87].

$$R_{nsw}^{s} = R_n.\exp^{(-c.LAI)} \tag{74}$$

In Eq. (74), c is the extinction coefficient of light attenuation (e.g., Sene, [97] indicate k=0.68 for fully grown plant, k=0 for bare soil; Zhang et al., [125] use 0.24 for vineyard crops).

The *surfaces resistance* is calculated as follows:

$$r_s^p = \frac{r_{st\,\min}}{LAI_{effective}\Pi_i F_i(X_i)} \tag{75}$$

where, $r_{st\,min}$ is the minimal stomatal resistance of individual leaves under optimal conditions. $LAI_{effective}$ is: equal to LAI for LAI ≤2.0; LAI/2 for LAI ≥4.0 and 2 for intermediate values of LAI, X_i is a specific environmental variable, and $F_i(X_i)$ is the stress function with $0.0 \leq F_i(X_i) \geq 1.0$ [46].

$$F_1(S) = \left(\frac{S}{1100}\right)\left(\frac{1100 + a_1}{S + a_1}\right) \tag{76}$$

$$F_2(T) = \frac{\left(T - T_L\right)\left(T_H - T\right)^{(TH - a_2)/(a_2 - T_L)}}{\left(a_2 - T_L\right)\left(T_H - a_2\right)^{(TH - a_2)/(a_2 - T_L)}} \tag{77}$$

$$F_2(D) = e^{-a_3 D} \tag{78}$$

$$F_4(\theta) = \left[\begin{array}{l} 1 - -if - - - - - \theta \geq \theta_F \\ \dfrac{\theta - \theta_W}{\theta_F - \theta_W} - if - -\theta_F < \theta < \theta_W \\ 0 - -if - - - - - \theta \leq \theta_W \end{array}\right[\tag{79}$$

where, S is the incoming photosynthetically active radiation flux (W/m^2), T is the air temperature (°K), θ_F is the soil moisture at field capacity (cm^3/cm^3), θ_w is the soil moisture at wilting point (cm^3/cm^3), and θ is the actual soil moisture in the root zone. (cm^3/cm^3). T_H and T_L are upper and lower temperatures limits outside of which transpiration is assumed to cease (°C) and are set at values of 40 and 0°C [39, 125]. The $a_1 = 57.67$, $a_2 = 25.78$, and $a_3 = 9.65$ that were determined by multivariate optimization [125].

TABLE 2.2 Coefficients *a* and *b* for Several Crops

Crop	*a*	*b*	Source
Grass	0.16	0.0	Katerji and Rana [52]
Tomato	0.54	2.4	Katerji and Rana [52]
Grain sorghum	0.54	0.61	Katerji and Rana [52]
Soybean	0.95	1.55	Katerji and Rana [52]
Sunflower	0.45	0.2	Katerji and Rana [52]
Sweet sorghum	0.845	1.0	Katerji and Rana [52]
Grass (Tropical climate)	0.18	0.0	Gosse (1976) in Rana et al. [91]
Grass (Mediterranean climate)	0.16	0.0	Rana et al. [91]
Alfalfa	0.24	0.43	Katerji and Perrier (1983) in Rana et al. [91]
Sorghum	0.94	1.1	Rana et al. [92]
Sunflower	0.53	1.2	Rana et al. [92]

The aerodynamic resistances [r_a^a and r_a^s] are calculated from the vertical wind profile in the field and the eddy diffusion coefficient. Above the canopy height, the eddy diffusion coefficient (K) is given by:

$$K = ku^*(z-d) \tag{80}$$

where, u* is the wind friction velocity (m/s), *k* is the van-Karman constant (0.41), *z* is the reference height (m), and *d* the zero plane displacement (m). The exponential decrease of the eddy diffusion coefficient (K) through the canopy is given as follows:

$$K = k_h.\exp\left[-n\left(1-\frac{z}{n}\right)\right] \tag{81}$$

where, k_h is the eddy diffusion coefficient at the top of the canopy (m^2/s), and *n* is the extinction coefficient of the eddy diffusion. Brutsaert (1982) cited by [125] indicated that $n = 2.5$ when $h_c < 1$ m; $n = 4.25$ when $h_c > 10$ m, linear interpolation could be used for crops with h between those values. k_h is determined as follows:

$$k_h = ku*(h_c - d) \tag{82}$$

The aerodynamic resistance r_a^s and r_a^s are obtained by integrating the eddy diffusion coefficients from the soil surface to the level of the "preferred sink of momentum in the canopy, and from there to the reference height (Shutlleworth and Gurney, 1990, cited [125]) as follows:

$$r_a^{\ a} = \frac{1}{Ku*}\ln\left(\frac{z-d}{h_c-d}\right) + \frac{h_c}{nk_h}\left[\exp\left[n\left(1-\frac{zo+d}{h_c}\right)\right]-1\right] \tag{83}$$

$$r_a^{\ s} = \frac{h_c \exp^{(n)}}{nk_h}\left[\exp\left(\frac{-nz_o{'}}{h_c}\right) - \exp\left[-n\left(\frac{z_0+d}{h_c}\right)\right]\right] \tag{84}$$

The bulk boundary layer resistance of canopy is calculated as follows:

$$r_a^{\ p} = \frac{r_b}{2LAI} \tag{85}$$

where, r_b is the mean boundary layer resistance (s/m) (e.g., Brisson et al., [13], recommend use 50 s/m).

The *soil surface resistance* r_s^s is the resistance to water vapor movement from the interior to the surface of the soil, and is strongly depending of the water content (θ_s), and is calculated using the Eq. (86) defined by Anandristakis et al. [6]:

$$r_s^{\ s} = r_s^{\ s}{}_{\min} f(\theta_s) \tag{86}$$

where, θ_s is soil volumetric water content (cm^3/cm^3), and $r_s^s{}_{\min}$ is the minimum soil surfaces resistance, that correspond with the soil field capacity (θ_{FC}) and is assumed equal to 100 s/m (e.g., [18, 125]). The $f(\theta_s)$ is expressed by Eq. (87) defined by Thompson (1981), cited by [125]

$$f(\theta_s) = 2.5\left(\frac{\theta_{FC}}{\theta_s}\right) - 1.5 \tag{87}$$

2.4.5 CLUMPING MODEL

The Clumping model is based on the Shulttleworth-Wallace model, this model separated the soil surfaces into fractional areas inside and outside the influence of the canopy, and included the fraction of canopy cover (*f*). Brenner and Incoll [12] and Zhang et al. [125] expressed the model as follows:

$$\lambda E = \lambda E^s + \lambda E^{bs} + \lambda T = f\left(C_c^s PM_c^s + C_c^p PM_c^p\right) + (1-f) C_c^{bs} PM_c^{bs} \tag{88}$$

where, λE^s is the latent heat of evaporation from soil under the plant (W/m^2); λE^{bs} is the latent heat of evaporation from bare soil (W/m^2); f is the fractional vegetative cover and the other terms are expressed as follows:

$$PM_c^p = \frac{\Delta A_c + \left[\dfrac{\left(\rho CpD - \Delta r_a^p A_c^s\right)}{r_a^a + r_a^p}\right]}{\Delta + \gamma\left[1 + \dfrac{r_s^p}{r_a^a + r_a^p}\right]} \tag{89}$$

$$PM_c^s = \frac{\Delta A_c + \left[\dfrac{\left(\rho CpD - \Delta r_a^s A_c^p\right)}{r_a^a + r_a^s}\right]}{\Delta + \gamma\left[1 + \dfrac{r_s^s}{r_a^a + r_a^s}\right]} \tag{90}$$

$$PM_c^{bs} = \frac{\Delta A_c^{bs} + \left[\dfrac{(\rho CpD)}{r_a^a + r_a^{bs}}\right]}{\Delta + \gamma\left[1 + \dfrac{r_s^{bs}}{r_a^a + r_a^{bs}}\right]} \tag{91}$$

$$C_c^s = \frac{R_c^{bs} R_c^p \left(R_c^s + R_c^a\right)}{\left[R_c^s R_c^p R_c^{bs} + (1-f) R_c^s R_c^p R_c^a + f R_c^{bs} R_c^s R_c^a + f R_c^{bs} R_c^p R_c^a\right]} \tag{92}$$

$$C_c^p = \frac{R_c^{bs} R_c^s \left(R_c^p + R_c^a\right)}{\left[R_c^s R_c^p R_c^{bs} + (1-f) R_c^s R_c^p R_c^a + f R_c^{bs} R_c^s R_c^a + f R_c^{bs} R_c^p R_c^a\right]} \tag{93}$$

$$C_c^{bs} = \frac{R_c^s R_c^p \left(R_c^{bs} + R_c^a\right)}{\left[R_c^s R_c^p R_c^{bs} + (1-f) R_c^s R_c^p R_c^a + f R_c^{bs} R_c^s R_c^a + f R_c^{bs} R_c^p R_c^a\right]} \tag{94}$$

$$R_c^s = (\Delta + \gamma) r_a^s + \gamma r_s^s \tag{95}$$

$$R_c^p = (\Delta + \gamma) r_a^p + \gamma r_s^p \quad (96)$$

$$R_c^{bs} = (\Delta + \gamma) r_a^{bs} + \gamma r_s^{bs} \quad (97)$$

$$R_c^a = (\Delta + \gamma) r_a^a \quad (98)$$

where, A_c, A_c^s, A_c^s and A_c^{bs} are energy available to evapotranspiration, to the plant, to soil under shrub and bare soil (W/m^2) respectively, r_a^{bs} the eddy diffusion resistance from in-canopy heat exchange plane height to the soil surface (s/m), r_s^{bs} the soil surfaces resistance of bare soil (s/m). The *Available energy* for this model, the net radiation (R_n) is divided into net radiation in the plant (R_n^s) and the net radiation in the soil (R_n^s). If the energy storage in the plant is assumed to be negligible, then:

$$R_{nc}^s = R_n \exp^{(-CLAI/f)} \quad (99)$$

$$R_{nc}^p = R_n - R_{nc}^s \quad (100)$$

$$A_c^s = R_{nc}^s - G^s \quad (101)$$

$$A_c^{bs} = R_n - G^{bs} \quad (102)$$

$$A_c^p = R_{nc}^p \quad (103)$$

where, R_{nc}^p and R_{nc}^s are the radiation absorbed by the plant and the radiation by the soil (W/m^2) respectively, G^s and G^{bs} are the soil heat flux under plant and bare soil (W/m^2), respectively, C is the extinction coefficient of light attenuation according for Sene [97] is equal to 0.68 for fully grown plant. The resistance for the bare soil surfaces r_s^{bs} can be calculated equally as in the S-W model, mentioned before. The aerodynamic resistance between the bare soil surface and the mean surfaces flow height (r_a^b) can be calculated assuming that the bare soil surface is totally unaffected by adjacent vegetation so that is aerodynamic resistance equal to r_a^b and defined for:

$$r_a^b = \ln \frac{\left(\frac{Z_m}{Z'_o}\right)^2}{k^2 U_m} \quad (104)$$

where, Z_m is the mean surface flow height (m), and could be assumed equal to $0.75h_c$, and u_m is the wind speed at the Z_m (m/s). According with Zhang et al. [125], the aerodynamic resistance (r_a^{bs}) varies between r_a^b and r_a^s as *f* varies from 0 to 1, and the functional relationship of this change is not known.

2.4.6 COMBINATION MODEL

Theoretical approaches to surface evaporation from the energy balance equation combined with sensible heat and latent heat exchange expressions give the following equation for actual evapotranspiration [81]:

$$ET = \frac{\Delta}{\Delta + \gamma}\left[\left(R_n - G\right) + \frac{\rho Cp}{\Delta} Hu\left(VPDa - VPDs\right)\right] \tag{105}$$

where, $(R_n - G)$ = Available energy (MJ/m^2) for the canopy consisting of net radiation, R_n and the soil heat flux, *G*; *H*(*u*) = exchange coefficient (m/s) between the surface level and a reference level above the canopy but taken inside the conservative boundary sublayer; *VPDs* and *VPDa* (kPa) = vapor pressure deficits (VPD) for the surface level and the reference level, respectively; ρ = atmospheric density (kg/m^3); *Cp* = specific heat of moist air (J./kg°C); Δ = slope of the vapor pressure curve (Pa/°C); and γ= psychrometric constant (Pa/°C). To obtain evapotranspiration with the Eq. (105), it is not an easy task to estimate *VPDs*, representing the vapor pressure deficit at the evaporative surface. If *VPDs* can be associated with a surface resistance term (r_s). Therefore, ET can be calculated directly from the flux equation:

$$ET = \frac{\rho Cp}{\gamma}\frac{VPDs}{r_s} \quad (87?) \tag{106}$$

and

$$r_a = \frac{1}{Hu} \tag{107}$$

where, r_a can be calculated using the equations discussed later in this chapter. Two main solutions can be defined for the Eq. (105) using climatic data:

1. The case of full water availability corresponding to saturation at the evaporative surface. Then *VPDs* = 0 and r_s becomes null. Eq. (105) then gives the maximum value for ET, the potential evaporation (EP), which depends only on climatic driving forces:

$$EP = \frac{\Delta(R_n - G) + \rho Cp F(u) VPDa}{\Delta\lambda} \tag{108}$$

where, F(u) = 1/r_a. The combination the equations can get:

$$ET = \frac{EP}{(1 + \frac{\gamma}{\Delta + \gamma}\frac{r_s}{r_a})} \tag{109}$$

2. The case for equilibrium between the surface and the reference levels corresponds to VP*Ds* = VP*Da*. In this case, the evapotranspiration is referred to as the equilibrium evaporation (*Ee*):

$$Ee = \frac{\rho Cp}{\gamma}\frac{VPDa}{r_e} \tag{110}$$

where, r_s was renamed r_e, termed the equilibrium surface resistance, indicating that the term, in this case, represents the surface resistance for equilibrium evaporation. The value for r_e depends predominately on climatic characteristics although these characteristics are influenced by R_n and G of the vegetative surface. For purposes here, the r_e term can be called the climatic resistance for the surface:

$$r_e = \frac{\rho Cp}{\gamma}\frac{\Delta+\gamma}{\Delta}\frac{VPDa}{R_n - G} \tag{111}$$

EP can be estimated:

$$EP = Ee\left(1 + \frac{\gamma}{\Delta+\gamma} + \frac{r_e}{r_a}\right) \tag{112}$$

and ET can be estimate using:

$$ET = \frac{EP_{(36)}}{\left(1 + \frac{\gamma}{\Delta+\gamma}\frac{r_s}{r_a}\right)} \tag{113}$$

2.4.7 PRISTLEY AND TAYLOR MODEL

Pristley and Taylor [84], proposed to neglect the aerodynamic term and replace the radiation term by a dimensionless coefficient (α):

$$ET = \alpha\frac{\Delta}{\Delta+\gamma}\left(R_n - G\right) \tag{114}$$

where, ET is water flux under references conditions (well watered grass) in mm.day^{-1}; R_n and G are net radiation and soil heat flux respectably in mm.day^{-1}; Δ and γ in kPa.°C^{-1}. The term α is given as 1.26 for grass field in humid weather conditions, and was adopted by Pristley and Taylor [84] for wet surfaces. However α ranges from 0.7 to 1.6 for various landscape situations [26]. According with Zhang et al. [124], the term α can be calculated as follows:

$$\alpha = \frac{\lambda E(\Delta+\gamma)}{\Delta(R_n - G)} = \frac{\Delta+\gamma}{\Delta(1+\beta)} \tag{115}$$

Also the term, α, sensible heat flux at the soil moisture changes [29, 30, 124], and can be estimated using a model given below:

$$\alpha = k\left[1 - \exp\left(-c\frac{\theta - d}{\theta_{FC}}\right)\right] \tag{116}$$

where, *k*, *c* and *d* are parameters of the model, θ is the actual volumetric soil moisture content ($cm^3.cm^{-3}$) and θ_{FC} is the volumetric moisture content at field capacity ($cm^3.cm\text{-}^3$).

2.4.8 EDDY COVARIANCE METHOD

The eddy covariance method is, in general, the most preferred because it provide a direct measure of the vertical turbulent flux across the mean horizontal streamlines, provided by fast sensors (~10 Hz) [65]. Realizing the limitation of the Thornthwaite-Holzman type of approach, Swinbank (1951) cited by Chang [19] was the first to attempt a direct measurement by the so-called eddy correlation technique. The method is based on the assumption that the vertical eddy flux can be determined by simultaneous measurements of the upward eddy velocity and the fluctuation in vapor pressure. Actually is a routinely technique for direct measurement of surfaces layer fluxes of momentum, heat, and traces gases (CO_2, H_2O, O_3) between the surfaces and the turbulent atmosphere [63].

This system recognizes that the transport of heat, moisture, and momentum in the boundary layer is governed almost entirely by turbulence. The eddy correlation method is theoretically simple using an approach to measure the turbulent fluxes of vapor and heat above the canopy surface. The eddy correlation fluxes are calculated and recorded in a 30 min or less temporal resolution. Assuming the net lateral advection of vapor transfer is negligible, the latent heat flux (evapotranspiration) can be calculated from the covariance between the water vapor density (ρ_v) and the vertical wind speed (*w*):

$$\lambda E = \lambda \overline{w'\rho_v'} \tag{117}$$

where, λE is the latent heat flux (W m^{-2}), is the latent heat of vaporization (J. kg^{-1}), $_v$‘ is the fluctuation in the water vapor density (kg m^{-3}), and w’ is the fluctuation in the vertical wind speed (m s^{-1}). The over bar represents the average of the period and primes indicate the deviation from the mean values during the averaging period. According to Weaver [118], the eddy correlation method depends on the relations between the direction of air movement near the land surface and properties of the atmosphere, such as temperature and humidity. The sensible heat flux can be calculated from the covariance of air temperature and the vertical wind speed.

$$H = \rho_a C_p \overline{w'T'} \tag{118}$$

where, H = the sensible heat flux (W m^{-2}), $_a$ the air density (kg m^{-3}), $C_{p=}$ the specific heat of moist air (J. kg^{-1}°C^{-1}) and T' = the fluctuation in the air temperature (°C).

The fine wire thermocouples (0.01 mm diameter) are not included in the eddy correlation system. The air temperature fluctuations, measured by the sonic anemometer, are corrected for air temperature fluctuations in estimation of sensible heat fluxes. The correction is for the effect of wind blowing normal to the sonic acoustic path. The simplified formula by Schotanus et al. [99] is as follows:

$$\overline{w'T'} = \overline{w'T_s'} - 0.51\left(\overline{T + 273.15}\right)\overline{w'q'} \tag{119}$$

where, w'T' is rotated covariance of wind speed and sonic temperature (m°C s^{-1}), *T* is air temperature (°C) and q is the specific humidity in grams of water vapor per grams of moist air.

Two Eddy covariance systems are used to measure the water vapor fluxes, the open path and close path. According to Anthoni et al. [8] the Open-path eddy covariance systems require corrections for density fluctuations in the sampled air [64, 119] and in general closed-path system require incorporation of a time lag and corrections for the loss of high frequency information, due to the air being drawn through a long sampling tube [64, 71]. The most common correction in the eddy covariance system is described by Wolf et al. [121] as: i) Coordinate rotation, ii) Air density correction, and iii) Frequency-dependent signal loss.

Estimation of turbulent fluxes is highly dependent on the accuracy of the vertical wind speed measurements. Measurement of wind speed in three orthogonal directions with sonic anemometer requires a refined orientation with respect to the natural coordinate system through mathematic coordinate rotations [103]. The vector of wind has three components (*u, v, w*) in three coordinate directions (*x, y, z*). The *z*-direction is oriented with respect to gravity, and the other two are arbitrary. Baldocchi et al. [9] provide procedures to transform the initial coordinate system to the natural coordinate system. Described in details by Sumner [103], the coordinate system is rotated by an angle η about the z-axis to align *u* into the *x*-direction on the *x-y* plane, then rotated by an angle θ about the *y*-direction to align w along the *z*-direction. The resultant forces $\overline{v}$ and $\overline{u}$ are equal to zero, and $\overline{u}$ is pointed directly to the air stream. When θ was greater than 10 degrees, the turbulent flux data should be excluded based on the assumption that spurious turbulence was the cause of the excessive amount of the coordinate rotation:

$$\cos\theta = \sqrt{\frac{\left(u^2 + v^2\right)}{\left(u^2 + v^2 + w^2\right)}} \tag{120}$$

$$\sin\theta = \frac{w}{\sqrt{\left(u^2+v^2+w^2\right)}} \tag{121}$$

$$\cos\eta = \frac{u}{\sqrt{\left(u^2+v^2\right)}} \tag{122}$$

$$\sin\eta = \frac{v}{\sqrt{\left(u^2+v^2\right)}} \tag{123}$$

The latent heat and sensible heat fluxes are computed from the coordinate rotation-transformed covariance:

$$\left(\overline{w'\rho_v'}\right)_r = \overline{w'\rho_v'}\cos\theta - \overline{u'\rho_v'}\sin\theta\cos\eta - \overline{v'\rho_v'}\sin\theta\sin\eta \tag{124}$$

$$\left(\overline{w'T_s'}\right)_r = \overline{w'T_s'}\cos\theta - \overline{u'T_s'}\sin\theta\cos\eta - \overline{v'T_s'}\sin\theta\sin\eta \tag{125}$$

After the coordinate rotation, the final latent heat flux can be estimated from Eq. (117) and the following correction of air density (C_{air}) [119] and correction of oxygen (CO_2) [107]:

$$C_{air} = \frac{\overline{\rho_v H}}{\overline{\rho C_p}\left(T+273.15\right)}\lambda \tag{126}$$

$$C_{O2} = \frac{FK_o\overline{H}}{K_w\left(T+273.15\right)}\lambda \tag{127}$$

where, F is a factor used in krypton hygrometer correction that accounts for molecular weights of air and oxygen, and atmospheric abundance of oxygen and is equal to 0.229 g°C J.$^{-1}$, K_o is the extinction coefficient of hygrometer for oxygen, estimated as 0.0045 m^3 g^{-1} cm^{-1}, K_w is the extinction coefficient of hygrometer for water and is 0.149 m^3 g^{-1} cm^{-1}, provided the manufacturer:

With the measured four flux components from the energy balance equation, the energy balance should be closed, however, this is not practically the case. A tendency to underestimate energy and mass fluxes has been a pervasive problem with the eddy covariance technique [33]. Ham and Heilman [33] reported closure of 0.79 for priarine locations and 0.96 for forest. Ramirez and Harmsen (2007-Data without publication) indicated 0.71 for grass and 0.75 for corn.

The errors in eddy covariance method are associated with: 1. Accuracy of the R_n and G measurements (errors are often 5 to 10%); 2. The length scale of the eddies responsible for transport (if is larger, the frequency response and sensor separation error may have been smaller); 3. Sensor separation and inadequate sensor response (can underestimate by 15% of λE by [33] and 10% reported by Laubach and McNauhton [57]; and 4. Ham and Heilman [33] conclude "The inherent tendency to underestimate fluxes when using eddy covariance may be linked to the errors caused by sensor separation and inadequate frequency response of the sensors. The correction proposed by Massman and Lee [64] is difficult to implement for the nonspecialist because they require calculation of cospectra using high-frequency (10 Hz) data, and also is required expertize experience to interpret the cospectra properly."

The "energy balance closure is corrected using the Bowen ratio [56] as follows:

$$H = \beta \times \lambda E \tag{128}$$

$$\lambda E = Rn - G - H \tag{129}$$

where, : β and λE are due to eddy covariance system, R_n and G are measured.

The Massman Analytical Formulae for Spectral Corrections to Measured Momentum and Scalar Fluxes for Eddy Covariance Systems: Massman [63] developed an analytical method for frequency response corrections, based on the procedure developed by Horst [43]:

For *Stable atmospheric conditions* ($0<\zeta\leq 2$):

Fast-response open path system:

$$\frac{Flux_m}{Flux} = \left[\frac{ab}{(a+1)(b+1)}\right]\left[\frac{ab}{(a+p)(b+p)}\right]\left[\frac{1}{(p+1)}\right]\left[1-\frac{p}{(a+1)(a+p)}\right] \tag{130}$$

Scalar instrument with 0.1–0.3s response:

$$\frac{Flux_m}{Flux} = \left[\frac{ab}{(a+1)(b+1)}\right]\left[\frac{ab}{(a+p)(b+p)}\right]\left[\frac{1}{(p+1)}\right]\left[1-\frac{p}{(a+1)(a+p)}\right]\left[\frac{1+0.9p}{1+p}\right] \tag{131}$$

Unstable atmospheric conditions ($\zeta\leq 0$):

$$\frac{Flux_m}{Flux} = \left[\frac{a^\alpha b^\alpha}{(a^\alpha+1)(b^\alpha+1)}\right]\left[\frac{a^\alpha b^\alpha}{(a^\alpha+p^\alpha)(b^\alpha+p^\alpha)}\right]\left[\frac{1}{(p^\alpha+1)}\right]\left[1-\frac{p^\alpha}{(a^\alpha+1)(a^\alpha+p^\alpha)}\right] \tag{132}$$

where, the subscript m refers to the measurement flux, $a = 2\pi \int x\tau_h$; $b = 2\pi \int x\tau_b$; $p = 2\pi \int x\tau_c$; and τ_h and τ_b are the equivalent time constants associated with trend removal (τ_h) and block averaging (τ_b). For relatively broad coespectra with relatively shallow peaks, such as the flat terrain neutral/stable, such as flat terrain coespectrum: α=0.925; and for sharper, more peaked coespectra, such as the stable terrain coespectra: α=0.925 [49].

These approximations are clearly easier to employ than numerical approaches and are applicable even when fluxes are so small as to preclude the use of *in situ* methods. Nevertheless, this approach is subject at the next conditions: i) horizontally homogeneous upwind fetch, ii) the validity of the coespectral similarity, iii) sufficiently long averaging periods, and preferably, iv) relatively small corrections [63].

2.4.9 THE INFRARED SURFACE TEMPERATURE METHOD

The infrared surface temperature has also been used for the estimation of the sensible heat flux (H) using the resistance model [2]:

$$H = \rho.Cp\frac{To - Ta}{r_a} \tag{133}$$

where, ρ is air density (Kgm^{-3}), Cp specific heat at constant pressure ($Jkg^{-1}°C^{-1}$), To is the temperature at surface level (°C), Ta is the temperature at the reference level (°C), and r_a is the aerodynamic resistance to heat flux between the surface and the reference level ($s.m^{-1}$), the latent heat flux (λE) can be computed as the residual term in the energy balance:

$$\lambda E = Rn - G - H = Rn - G - \rho.Cp\frac{To - Ta}{r_a} \tag{134}$$

Alves et al. [2] say the radioactive surface temperature has a several drawbacks. Thermal radiation received by the instrument can originate from the leaves but also from de soil, and the measurement can be highly dependent on crop cover, inclination of radiometer and sun height and azimuth, especially en partial cover crops, the first one lies in the use of an adequate value of r_a. The variable d is zero plane displacement height (m), Z_{oM} and Z_{OH} are the roughness lengths (m) for momentum and heat respectively, k is the von Karman constant, u_z is the wind speed (ms^{-1}) at the reference height z (m), and ψM and ψH are the integrated stability functions for describing the effects of the buoyancy or stability on momentum transfer and heat between the surface and the reference level.

The necessary instruments are: Wind speed and direction sensor at (0.85 and 1.46 m), psychrometer at the same height that wind sensor, a net radiometer placement a 1.5 m and infrared thermometer perpendicular to the rows the crop, and positioned at an angle of 60° below horizontal to view the top leaves of the plants at 0.40 m distance [2]. The sensible heat flux [H] is calculated with the flux applied to levels Z_1 and Z_2:

$$H = \rho Cp\frac{T1 - T2}{[ra]_1^2} \tag{135}$$

where, $[ra]_1^2$ is the aerodynamic resistance to heat flux between the two levels, and is computed using the Eq. (136):

$$[ra]_1^2 = \frac{\ln\left(\frac{Z2-d}{Z1-d}\right)}{ku*} \tag{136}$$

where, u* = the friction velocity, obtained in the process of determining aerodynamic parameter [d] and Z_{oM} from the win profile measurements.

The air temperature at the surface level (To) is calculated using Eq. (137). The stability conditions can be calculated using the Richardson number.

$$To = Ta + \frac{Hr_a}{\rho Cp} \tag{137}$$

2.4.9.1 FETCH REQUIREMENTS

The air that passing over a surface is affected by the field surfaces feature [93]; the minimal fetch requirement was estimated based on the thickness of the internal boundary layer (δ in m) and a roughness parameter (Z_o in m) for each genotype considering the minimal and maximal crop height during the grown season. The δ was calculated using the Eq. (138) proposed by Monteith and Unsworth [70]:

$$\delta = 0.15.L^{4/5}.Z_o^{\ 1/5} \tag{138}$$

where, *L* is the distance of traverse (fetch) across a uniform surface with roughness Z_o. The Z_o for crops is approximately one order of magnitude smaller than the crop height h, and can be calculated according with Rosenberg et al. [93] as follows:

$$\text{Log}_{10} Z_o = 0.997 \log_{10} h - 0.883 \tag{139}$$

As a factor of safety, a height to fetch of 1: 50 to 1: 100 is usually considered adequate for studies made over agricultural crop surfaces [5, 93] but may be too conservative and difficult to achieve in practice. Alves et al. [1] obtained full profile development using a 1: 48 fetch relation in Wheat and lettuce. Heilman et al. [40] found that for Bowen-Ratio estimates a fetch 1:20 was sufficient over grass, and Ham and Heilman [32] and Frithschen and Fritschen [28] obtained similar results.

2.4.9.2 STABILITY CORRECTION

The gradient method need a stability correction, one of the most used is the Monin-Obukhov stability factor (ζ) described by [17, 86, 93]:

$$\xi = \frac{(-k.z.g.H)}{\left(\rho_a.C_p.T_a.u*^3\right)} \tag{140}$$

where, K is von Karman's constant, z is height of wind and air temperature measurements (m), g is the gravitational constant (9.8 m.s^{-2}), H = β.λE, T_a is air temperature (°K), u* is the friction velocity given by Kjelgaard et al. [55] without the stability correction factor:

$$u^* = \frac{k.u_z}{\ln\left(\frac{z - d + Z_{om}}{Z_{om}}\right)} \tag{141}$$

Flux with a negative sign for ζ indicating unstable conditions and needs to be excluded. For flux under unstable conditions the λE is over R_n (Fig. 2.5a); For the flux with negative ζ are excluded and λE is lower than R_n (Fig. 2.5b). Payero et al. [77] indicated that fluxes with incorrect sign and $\beta \approx -1$should not be considered when estimated the energy balance components by the energy balance Bowen ratio method. The negative ζ corresponds to negative β (Fig. 2.6).

The Richardson number (Ri) is represented by the Eq. (45), also is well known as stability factor [2, 110] and represent the ratio of the buoyancy – "thermal effect" to "mechanical –wind shear" [86]. Negative values indicate instability conditions where surfaces heating enhances buoyancy effects, and positive Ri values indicate a stable conditions where temperature near the surfaces are cooler than away from the surfaces.

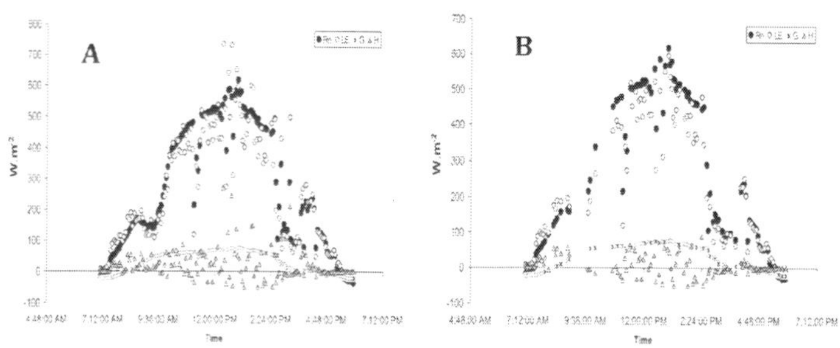

FIGURE 2.5 Energy balance components measured by Bowen ratio method in grass: (a) without stability correction and (b) with stability correction.

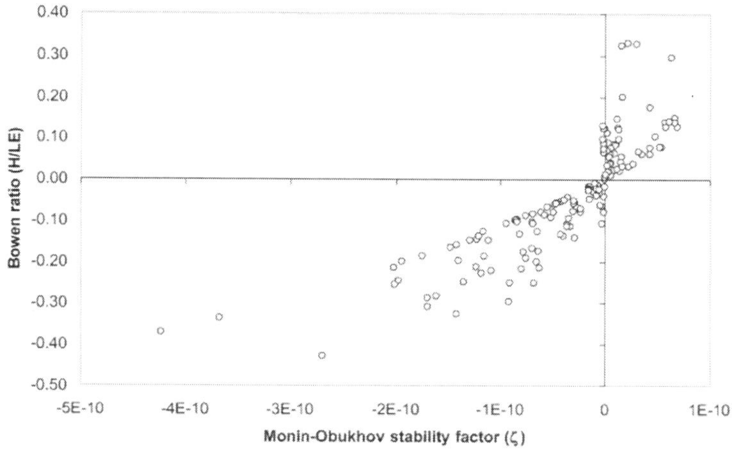

FIGURE 2.6 Relationship between Bowen ratio (β) and the Monin-Obukhov stability factor.

2.5 SUMMARY

The water vapor flux in the agroecosystems is the second largest component in the hydrologycal cycle. Water vapor flux from the vegetation to the atmosphere is a widely studied variable throughout the world, due to it applicability in various disciplines such as hydrology, climatology, and agricultural science. The evapotranspiration is important to calculate the water requirement to the crops, to made climatic characterizations and water management. The estimation of evapotranspiration from vegetated area is a basic tool to compute water balances and to estimate water availability and requirements and also to estimate agroclimatic and hydrologic indices. During the last 60 year several methods and models to measure the water flux in agroecosystems has been developed, the aim of this first part of the review is to make a review from de mass balance methods and models in the water flux estimation and the application of the two steps model and the direct transpiration measurements techniques. This chapter provides a revision of these methods and model with special application to crops and covered areas.

KEYWORDS

- **aerodynamic resistance**
- **calibration**
- **clumping model**
- **combination model**

- **crop coefficients**
- **cumulative growing degree days**
- **drainage Lysimeter**
- **eddy covariance method**
- **energy balances.**
- **evapotranspiration**
- **fetch requirements**
- **fluxes**
- **humidity and temperature method**
- **latent heat fluxes**
- **leaf Area Index [LAI]**
- **micrometeorological method**
- **Pristley–Taylor model**
- **resistances**
- **sensible heat flux**
- **Shulttleworth-Wallace model**
- **soil heat flux**
- **stability correction**
- **stomatic resistance**
- **surfaces resistance**
- **the Bowen ratio energy balance**
- **the infrared surface temperature method**
- **the Penman-Monteith reference evapotranspiration method**
- **the Penman-Monteith general evapotranspiration method**
- **water stress coefficient**
- **water vapor deficit**
- **water vapor flux**
- **weighing lysimeter**

REFERENCES

1. Alves, I., Perrier, A., Pereira, L. S., (1998). Aerodynamic and Surface Resistant of Complete Cover Crops: How Good Is the "Big Leaf"? *Trans. Am. Soc. Agric. Eng., 41(2),* 345–351.

2. Alves, I., Frontes, J. C., Pereira, L. S. (2000). Evapotranspiration Estimation from Infrared Surface Temperature. I: The Performance of the Flux Equation. *Trans. Am. Soc. Agric. Eng., 43(3),* 591–598.
3. Alves. L., Pereira, L. S., (2000). Modeling Surface Resistance from Climatic Variables? *Agric. Water Manag., 42,* 371–385.
4. Allen, G. R., Cuenca, H. R., Jensen, E. M., Blatchley K. R., Erpenbeck M. J. (1990). Evapotranspiration and Irrigation Water Requirements. *American Society of Civil Engineers*. New York (USA). 332p.
5. Allen, G. R., Pereira, S. L., Raes, D., Smith, M. (1998). Crop Evapotranspiration: Guidelines for Computing Crop Water Requirements. Food and Agricultural Organization of the United Nations (FAO) Report No 56. Rome. 300p.
6. Anandristakis, M., Liakatas, A., Kerkides, P., Rizos, S., Gavanonis, J., Poulavassilis. A., (2000). Crop Water Requirements Model Tested For Crops Grown in Greece. Agricultural. *Water Management, 42,* 371–385.
7. Angelocci, L. R., BruninI, O., Magalhaes, A. C. (1983). Variation of Stomatal Resistance and Water Vapor Diffusion Associated with the Water Energy Status in Young Coffee Farm. 18 pages. In: Ninth Water – Plant Symposium at Federal University of Vicosa, Brazil. July 25–28.
8. Anthoni, A., Unsworth, M., Law, B., Irvine, J., Baldochi, D., Kolle, O., Knohl, A., Schulze, E. D. (2001). Comparison of Open-Path and Closed-Path Eddy Covariance System. Max-Planck Institute for Biogeochemie, Jena, D-0(7745). Germany. Pages 2.
9. Baldocchi, D., Hicks, B., Meyers, T. P. (1988). Measuring Biosphereatmosphere Exchanges of Biologically Related Gases With Micrometeorological Methods. *Ecology 69(5),* 1331–1340.
10. Baldocchi, D., Luxmoore, R. J., Hatfield, J. L. (1991). Discerning the Forest From the Trees: An Assay on Scaling Canopy Stomatal Conductance. *Agric. For. Meteorol. 54,* 197–226.
11. Bowen, I. S. (1926). The Ratio of Heat Losses by Conduction and by Evaporation from Any Water Surface. *Physical Review, 27,* 779–788.
12. Brenner, A. I., Incoll, L. D. (1997). The Effect of Clumping and Stomatal Response on Evaporation from Sparsely Vegetation Shrunblands. *Agric. For. Meteorol., 84,* 187–205.
13. Brisson, N., Itier, B., L'Hotel, J. C., Lorendeau, J. Y. (1998). Parameterization of the Shuttleworth-Wallace Model to Estimate Daily Maximum Transpiration for Use in Crop Models. *Ecological Modeling, 107,* 159–169.
14. Brown, P. A., Mancino, C. F., Joung, M. H., Thompson, T. L, Wierenga P. J., Kopec, D. M. (2001). Penman Monteith crop coefficients for use with desert turf system. *Crop Sci., 41,* 1197–1206.
15. Campbell Scientific, INC (1987–1998). Bowen Ratio Instrumentation Instruction Manuel. 23 p. IN: www.campbell.com.
16. Campbell, G. S. (1977). *An Introduction to Environmental Biophysics*. Springer-Verlag, New York, USA.
17. Campbell, G. S. (1985). *Soil Physics with Basic Transport Models for Soil-Plant Systems.* Dev. Soil Science No. 14. 150 pp. Elsevier.
18. Camilo, P. J., Gurney, R. J. (1986). A Resistance Parameter for Bare Soil Evaporation Models. *Soil Sci., 141,* 95–106.
19. Chang, J. H. (1968). *Climate and Agriculture and Ecological Survey*. University of Hawaii. Ed. Aldine Publishing Company. Chicago (USA). 304p.
20. Coleman, E. A. (1946). A Laboratory of Lysimetric Drainage Under Controlled Soil Moisture Tension. *Soil Sci., 62,* 365–382.
21. Conroy, J. W., Wu, J., Elliot, W. (2003). Modification of the evapotranspiration routines in the WEPP model: Part I. ASAE Annual International Meeting. Las Vegas Nevada, USA: 27 to 30 July. 16p. In: http: //www.pubs.asce.org/WWfidisplay.cgi?8801815
22. Doorenbos, J., Pruitt, W. O. (1977). *Guidelines for Predicting Crop Water Requirements*. Food and Agricultural Organization of the United Nations (FAO). Publication No. 24. Rome. 300p.

23. Dugas, W. A., Dan, R., Ritcchie, J. T. (1985). A Weighing Lysimeter for Evapotranspiration and Root Measurements. *Agron. J., 77,* 821–825.
24. Farahami, H. J., Bausch, W. C. (1995). Performance of Evapotranspiration Models for Maize-Bare Soil To Close Canopy. *Trans. ASAE, 38,* 1049–1059.
25. Fank, J. (2008). Monolithic Field Lysimeter—A Precize Tool To Close the Gap Between Laboratory and Field Scaled Investigations. *Geophysical Research Abstracts, 10,* 8.
26. Flint, A. L., Child, S. W. (1991). Use The Pristley-Taylor Evaporation Equation for Soil Water Limited Conditions In A Small Forest Clear-Cut. *Agric. For. Meteorol., 56,* 247–260.
27. Frank, A. B. (2003). Evapotranspiration From Northern Semiarid Grasslands. *Agron. J., 95,* 1504–1509.
28. Fritschen, J. L., Fritschen, L. C. (2005). Bowen Ratio Energy Balance Method. In: *Micrometeorology in Agric. Sys.* Agronomy Monograph No, 47. American Society of Agronomy, Crop Science Society of America, Soil Science Society of America, Madison-W1–5(3711), USA, 397–405.
29. Grago, R. D., Brutsaert, W. (1992). A Comparison of Several Evaporation Equations. *Water Resources Research, 32,* 951–954.
30. Grago, R. D. (1996). Comparison of the Evaporative Fraction and the Pristley-Taylor Method for Parameterization Day Time Evaporation. *Water Resources Research, 32,* 1403–1409.
31. Grismer, M. E., Orang, M., Snyder, R., Matyac. R. (2002). Pan Evaporation to Reference Evapotranspiration Conversion Methods. *Journal of Irrigation and Drainage Engineering, 128(3),* 180–184.
32. Ham, M. J., Heilman, L. J. (1991). Aerodynamic and Surface Resistance Affecting Energy Transport in a Sparse Crop. *Agric. For. Meteorol., 53,* 267–284.
33. Ham, M. J., Heilman, L. J. (2003). Experimental Test of Density and Energy-Balance Corrections on Carbon Dioxide Flux as Measured Using Open-Path Eddy Covariance. *Agron. J., 95,* 1393–1403.
34. Hanson, R. B., May, D. M. (2004). Crop Coefficients for Drip-Irrigated Processing Tomatoes. In: ASAE/CSAE Annual International Meeting. Fairmont Chateau Laurier, The Westing, Government Centre, Ottawa, Ontario, Canada, 12p.
35. Harmsen, E. W., Gonzaléz A., Winter. J. A., (2004). Re-evaluation of Pan Evaporation Coefficients at Seven Locations in Puerto Rico, By Agric, University P.R., *88(3–4),* 109–122.
36. Harmen, W. W., Ramirez, B. Gonzalez, V. H. E. J., Dukes, D. M., Jia, X. (2006). Estimation of short-term actual crop evapotranspiration. Proceedings of the 42nd Annual Meeting of the Caribbean Food Crops Society, July 9–14, San Juan, Puerto Rico. Pages 12.
37. Harmsen, E. W., Ramirez, B. V. H., Dukes, Jia, M. D., X., Perez, A. L., Vasquez, R. (2009). A ground-base method for calibrating remotely sensed temperature for use in estimating evapotranspiration. *WSEAS Transactions and Environment and Development, 1(5),* 13–23.
38. Hatfield, J. L., Allen, G. R. (1996). Evapotranspiration Estimates Under Deficient Water Supplies. *Journal of Irrigation and Drainage Engineering, 122,* 301–308.
39. Harris, P. P., Huntingford, C., Cox, P. M., Gash, J. H. C., Malhi, Y. (2004). Effect of Soil Moisture On Canopy Conductance of Amazonian Rainforest. *Agric. For. Meteorol., 122,* 215–227.
40. Heilman, J. L., Brtittin, C. L., Neale, C. M. U. (1989). Fetch Requirements From Bowen Ratio Measurements of Latent and Sensible Heat Fluxes. *Agric. For. Meteorol., 44,* 261–273.
41. Howell, T. A. (2005). Lysimetry. In: *Soil and the Environment.* Edited by Elsevier. Pages 379–386.
42. Howell, Steiner, T. A., Scheider, J. L. A. D., Evett, S. R., Tolk. A. J. (1994). Evapotranspiration of Irrigated Winter Wheat, Sorghum and Corn. ASAE Paper No. 94–2081. ASAE, St. Joseph, MI.
43. Horst, T. W. (1997). A Simple Formula For Attenuation of Eddy Covariance Flux Measurements. Boundary Layer Meteorology, *82,* 219–233.

44. Hunsaker, D. J. (1999). Basal crop coefficients and water use for early maturity cotton. Trans. American Society of Agricultural Engineers, *42(4),* 927–936.
45. Hunsaker, D. J., Pinter, P. J. Jr. Cai, H. (2002). Alfalfa Basal Crop Coefficients for FAO-56 Procedures in the Desert Regions of the South Western Trans, US, *American Society of Agricultural Engineers, 45(6),* 1799–1815.
46. Jarvis, P. G. (1976). The Interpretations of the Variations in Leaf Water Potential and Stomatal Conductance Found in Canopies in the Field. Phil Transaction Royal Society London Bulletin, *273,* 593–610.
47. Jhonson, R. S., Williams, L. E., Ayars, J. E., Trout. T. J. (2005). Weighing Lysimeter Aid Study of Water Relations In Trees and Vine Crop. *Calif. Agric., 59(2),* 133–136.
48. Kamal, H. A., Hatfield, L. J. (2004). Canopy Resistance As Affected Soil and Meteorological Factors In Potato. *Agron. J., 96,* 978–985.
49. Kaimal, J. C., Wyngaard, Izumi, J. C., Y., Cote, O. R. (1972). Deriving Power Spectra From A Three-Component Sonic Anemometer. *Journal Applied Meteorology, 7,* 827–837.
50. Katerji, N., Perrier, A. (1983). Modelling the Real Evapotranspiration In Alfalfa Crop: Role of the Crop Coefficients. *Agronomie, 3(6),* 513–521 (in: French).
51. Katerji, N., Perrier, A. (1985). Determinations of the Canopy Resistance To The Water Vapor Diffusion Of The Six Canopy Cover. *Agricultural Meteorology, 34,* 105–120 (in: French).
52. Katerji, N., Rana, G. (2006). Modelling Evapotranspiration of Six Irrigated Crops Under Mediterranean Climate Conditions. *Agricultural and Forest Meterology, 138,* 142–155.
53. Kirkham, M. B., Redelfs, M. S., Stone, L. R., Kanemasu, E. T. (1985). Comparison of Water Status and Evapotranspiration Of Six Row Crops. *Field Crops Res., 10,* 257–268.
54. Kjelgaard, J. F., Stokes, C. O. (2001). Evaluating Surface Resistance for Estimating Corn and Potato Evapotranspiration with the Penman-Monteith Model. *Trans. American Society of Agricultural Engineers, 44(4),* 797–805.
55. Kjelgaard, J. K., Stockle, C. O., Villar Mir, J. M., Evans, R. G., Campbell, G. S. (1994). Evaluation Methods to Estimate Corn Evapotranspiration from Short-Time Interval Weather Data. *Transaction of ASAE, 37(6),* 1825–(1833).
56. Kosugi, Y., Katsuyama. M. (2007). Evapotranspiration OVER JAPANESE CYPRESS FOREST II, Comparison of the Eddy Covariance And Water Budget Methods. *Journal of Hydrology, 334,* 305–311.
57. Laubach, J., McNauhton, A. (1999). A Spectrum-Independent Procedure For Correcting Eddy Flux Measurements With Separated Sensors. *Boundary Layer Meteorology, 89,* 445–467.
58. Lloyd, W. G. (1992). Bowen-Ratio measurements. In: *Evapotranspiration measurements of native vegetation, Owens Valley, California.* U.S. Geological Survey, Water Resources Investigations Report 91–4159. Pages 5–18.
59. López-Urrea, R., Martin de Santa Olalla, F., Febeiro, C., Moratalla. A. (2006). Testing Evapotranspiration Equations Using Lysimeter Observations In A Semiarid Climate. *Agric. Water Manag., 85(1–2),* 15–26.
60. Loos, C., S., Gainyler., E., Priesack. (2007). Assessment of Water Balance Simulations For Large-Scale Weighing Lysimeters. Journal of Hydrology, *335(3–4),* 259–270.
61. Madeiros, G. A., Arruda, F. B., Sakai, E., Fujiwara, M. (2001). The Influence of the Crop Canopy on Evapotranspiration and Crop Coefficients of Beans (Phaseolus vulgaris L.). *Agric. Water Manag., 49,* 211–234.
62. Malone, R. W., Stewardson, D. J., Bonta, J. V., Nelsen, T. (1999). Calibration and Quality Control of the Coshocton Weighing Lysimeters. *Transaction American Society of Agricultural Engineers, 42(3),* 701–712.
63. Massman, W. J. (2000). A Simple Method For Estimate Frequency Response Corrections For Eddy Covariance Systems. *Agric. For. Meteorol., 104,* 185–198.

64. Massman, W. J., Lee, X. (2002). Eddy Covariance Flux Corrections And Uncertainties In Long-Term Studies of Carbon And Energy Exchange. *Agric. For. Meteorol.*, *113*, 121–144.
65. Meyers, P. T., Baldocchi, D. D. (2005). Current Micrometeorological Flux Methodologies With Applications In Agriculture. Pages 381–396. In: *Micrometeorology in Agric. Sys.* Agronomy Monograph # 47. American Society of Agronomy, Crop Science Society of America, Soil Science Society of America, 677 S., Segoe Rd., Madison, W15(3711). USA.
66. Monteith, L. J. (1965). Evaporation and the Environment. *Symp. Soc. Exper. Biol.*, *19*, 205–234.
67. Monteith, L. J. (1981). Evaporation and Surface Temperature. *Quarterly Journal of the Royal Meteorology Society*, *107(451)*, 1–27.
68. Monteith, J. L. (1995). Accommodation Between Transpiring Vegetation and the Convective Boundary Layer. *Journal of Hydrology*, *166*, 251–263.
69. Monteith, L. J. (1997). Evaporation Models. Pages 197–234. In: *Agric. Sys. Modelling and Simulation*, Edited by Robert M., Peart and R., Bruce Curry. University of Florida, Gainesville, Florida.
70. Monteith, J. L., Unsworth, M. H. (1990). *Principles of Environmental Physics*. 2nd ed., Edward Arnold Publisher. 291 pages.
71. Moore, C. J. (1986). Frequency Response Corrections For Eddy Correlation Systems. *Boundary Layer Meteorology*, *37*, 17–35.
72. Nasab, B. S., Kashkuli, H. A., Khaledian, M. R. (2004). Estimation of crop coefficients for sugarcane (ratoon) in Haft Tappeh of Iran. ASAE/CSAE Annual International Meeting, Fairmont Chateau Laurier – The Westing, Government Centre Ottawa, Ontario, Canada. 1–4 August. Pages 4.
73. Oguntunde, G. P. (2005). Whole-Plant Water Use And Canopy Conductance of Cassava Under Limited Available Soil Water And Varying Evaporative Demand. *Plant Soil*, *278*, 371–383.
74. Ortega-Farias, S. O., Antonioletti, R., Brisson, N. (2004). Evaluation of the Penman-Monteith Model For Estimating Soybean Evapotranspiration. *Irrig. Sci.*, *23*, 1–9.
75. Ortega-Farias, S. O., Olioso, A., Fuentes, S., Valdes, H. (2006). Latent Heat Flux Over a Furrow-Irrigated Tomato Crop Using Penman-Monteith Equation with a Variable Surfaces Canopy Resistance. *Agric. Water Manag.*, *82*, 421–432
76. Parton, W. J., Lauenroth, W. K., Smith, F. M. (1981). Water Loss From A Short Grass Steppe. *Agricultural Meteorology*, *24*, 97–109.
77. Payero, J. O., Neale, C. M. U., Wright J. L., Allen. R. G. (2003). Guidelines for validating Bowen ratio data. *American Society of Agricultural Engineers*, *46(4)*, 1051–1060.
78. Penman; H. L. (1948). Natural Evaporation from Open Water, Bare Soil and Grass. Proceedings of the Royal Society of London. Series, A., *Mathematical and Physical Sciences*, *193(1032)*, 120–145.
79. Penman, H. L. (1956). Evaporation: An Introductory Survey. *Netherlands Journal Agricultural Science*, *1*, 9–29, 87–97, 151–153.
80. Pereira, S. L., Perrier, A., Allen, G., R., Alves, I. (1996). Evapotranspiration: Review of Concepts and Future Trends. In: *Evapotranspiration and Irrigation Scheduling*. Proceedings of the International Conference. November 3–6. San Antonio, Texas, 109–115.
81. Pereira, S. L., Perrier, A., Allen, G. R., Alves, I. (1999). Evapotranspiration: Concepts and Future Trends. Journal Irrigation and Drainage Engineering, *125(2)*, 45–51.
82. Pereira, S. L., and Alves, I. (2005). Crop Water Requirement. Pages 322–334. In: *Soil and Environment*, by Elsevier.
83. Perez, P. J., Castellvi, F., Martinez-Cob, A. (2008). A Simple Model For Estimation The Bowen Ratio From Climatic Factors For Determining Latent And Sensible Heat Flux. *Agric. For. Meteorol.*, *148*, 25–37.
84. Priestley, C. H. B., Taylor, R. J. (1972). On the Assessment of the Surfaces Heat and Evapotranspiration Using Large-Scale Parameters. *Monthly Water Review*, *100(2)*, 81–92.

85. Prueger, H. J., Hatfield, J. L., Aase, J. K., Pikul, Jr. J. L. (1997). Bowen-Ratio Comparison with Lysimeter Evapotranspiration. *Agron. J., 89(5),* 730–736.
86. Prueger, H. J., Kustas, P. W. (2005). Aerodynamic methods for estimating turbulent fluxes. 407–436. In: *Micrometeorology in Agric. Sys.,* Agronomy Monograph no, 47. American Society of Agronomy, Crop Science Society of America, Soil Science Society of America, 677 S., Segoe Rd., Madison, W1, 5(3711). USA.
87. Ramírez, B. V. H. (2007). Plant-Water Relationships for Several Common Bean Genotypes (Phaseolus vulgaris L.) with and Without Drought Stress Conditions. M.Sc. Thesis, Agronomy and Soils Department, University of Puerto Rico—Mayaguez Campus. 190 pages.
88. Ramirez, B. V. H., Harmsen, W. E., Porch, G. T. (2008). Estimation of actual Evapotranspiration Using Measured and Calculated Values of Bulk Surfaces Resistance. Proceedings of the World Environmental and Water Res. Congress. Honolulu (Hawaii). 10 pages.
89. Ramírez, B. V. H., Jaramillo, R. A. (2008). Modifications in the Superficial Hydrology Associated at the Cover Changes in the Colombian's Tropical Andes. In: *XIV Colombian Soil Society Meeting,* 13 pages (in Spanish).
90. Rana, G., Katerji, N., Mastrorilli, M., El Moujabber, M El., Brisson. N., (1997a). Validation of a Model of Actual Evapotranspiration For Water Stressed Soybeans. *Agric. For. Meteorol., 86,* 215–224.
91. Rana, G., Katerji, N., Mastrorilli, M., El Moujabber, M. (1997b). A Model for Predicting Actual Evapotranspiration Under Soil Water Stress in a Mediterranean Region. Theoretical Applied Climatology, *56,* 45–55.
92. Ritchie, J. T., Burnett, E. (1968). A Precision Weighing Lysimeter for Row Crop Water Use Studies. *Agron. J., 60(5),* 545–549.
93. Rosenberg, J. N., Blad, B. L., Verma, S. B. (1983). *Microclimate—The Biological Environment.* John Wiley and Sons. 495 pages.
94. Roygard, F. R., Alley, M. M., Khosla, R. (2002). No-Till Corn Yield and Water Balance in the Mid-Atlantic Coastal Plain. *Agron. J., 94,* 612–623.
95. Saugier, B. (1977). Micrometeorology on Crops and grasslands. Environmental effects on Crop Physiology. Proceeding of a Symposium held at Long Ashton Research Station University of Bristol. Academic Press. 39–55.
96. Saugier, B., Katerji, N. (1991). Some Plant Factors Controlling Evapotranspiration. *Agric. For. Meteorol., 54,* 263–277.
97. Sene, K. J. (1994). Parameterization for Energy Transfer from a Sparse Vine Crop. *Agric. For. Meteorol., 71,* 1–18.
98. Seyfried, M. S., Hanson, C. L., Murdock, M. D., Van Vactor. S. (2001). Long-term Lysimeter Database, Reynolds Creek Experimental Watershed, Idaho, United States. *Water Resources Research, 37(11),* 2853–2856.
99. Schotanus, P., Nieuwstadt, F. T. M., de Bruin, H. A. R. (1983). Temperature Measurement with a Sonic Anemometer and Its Application To Heat and Moisture Fluxes. Boundary-Layer Meteorology, *50,* 81–93.
100. Shuttleworth, W. J., Wallace, J. S. (1985). Evaporation From Sparse Crops—An Energy Combination Theory. *Quarterly Journal of the Royal Meteorology Society, 111,* 839–855.
101. Steiner, J. L. A., Howell, T. A., Schneider, A. D. (1991). Lysimeter Evaluation of Daily Potential Evapotranspiration Models For Grain Sorghum. *Agron. J., 83(1),* 240–247.
102. Suleiman, A., Crago, R. (2004). Hourly and Daytime Evapotranspiration from Grassland Using Radiometric Surface Temperature. *Agron. J., 96,* 384–390.
103. Sumner, D. M. (2001). Evapotranspiration From A Cypress And Pine Forest Subjected To Natural Fires in Volusia County, Florida, 1998–1999. U.S. Geological Survey Water-Resources Investigations Report 01–4245. Washington, DC.

104. Szeicz, C., and Long, L. F. (1969). Surface Resistance of Crop Canopies. *Water Recourses Research, 5(8),* 622–633.
105. Takhar, H. S., Rudge, A. J. (1970). Evaporation Studies in Standard Catchments. *Journal of Hydrology, 11(4),* 329–362.
106. Tanner, C. B. (1960). Energy Balance Approach to Evapotranspiration From Crops. *Soil Science Society of American Proceedings, 24(1),* 1–9.
107. Tanner, B. D., and Greene, J. P. (1989). Measurements of Sensible Heat Flux and Water Vapor Fluxes Using Eddy Correlation Methods. Final report to U.S. Army Dugway Proving Grounds. DAAD 09–8, D-0088.
108. Teare, L. D., H., Schimmulepfenning., Waldren, R. P., (1977). Rainout Shelter and Drainage Lysimeters to Quantitatively Measure Drought Stress. *Agron. J. 65(4),* 544–547.
109. Thom, A. S. (1975). Momentum, Mass and Heat Exchange of Plant Communities. 57–109. In: *Vegetation and the Atmosphere.* Edtied by Monteith, J. L., Academic Press: New York.
110. Tolk, Howell, A. J., Steiner, A. T., L. J., Krieg, R. D. Aerodynamic Characteristics of Corn as Determined by Energy Balance Techniques. *Agron. J.,* (1995). *87,* 464–473.
111. Tomilson, A. E. (1994). Instrumentation, Methods and Preliminary Evaluation of Evapotranspiration for a Grassland in the Arid Lands Ecology Reserve, Beton Country – Washington. May–October (1990). U. S. Geological Survey. Water-Resources Investigations Report 93–4081 and Washington State Department of Ecology. Pages 32.
112. Tomilson, A. E. (1997). Evapotranspiration for Three Sparce – Canopy Sites in the Black Rock Valley – Yakima Country – Washington. March (1992) to October (1995). U.S. Geological Survey. Water Resources Investigations Report 96–4207 and Washington State Department of Ecology. Pages 88.
113. Turner, C. N. (1991). Measurement and Influence of Environmental and Plant Factors on Stomatal Conductance in the Field. *Agric. For. Meteorol., 54,* 137–154.
114. Tyagi, N. K., Sharma, D. K., Luthra, S. K. (2000). Evapotranspiration and Crop Coefficients of Wheat and Sorghum. *Journal of Irrigation and Drainage Engineering, 126(4),* 215–222.
115. Van Bavel, C. H. M. (1967). Changes in Canopy Resistance to Water Loss From Alfalfa Induced by Soil Water Depletion. *Agric. For. Meteorol., 4,* 165–176.
116. Vaughan, P. J., Trout, T. J., Ayars, J. E. (2007). A Processing Method for Weighing Lysimeter Data and Comparison to Micrometeorological *ETo* Predictions. *Agric. Water Manag., 88(1–2),* 141–146.
117. Von Unold, G., Fank, J. (2008). Module Design of Field Lysimeter For Specific Application Needs. *Water Air Soil Pollution Focus, 8,* 233–242
118. Weaver, L. H. (1992). Eddy-Correlation Measurements: Evapotranspiration Measurements of Native Vegetation for Ownes Valley – California. U.S. Geological Survey. Water-Resources Investigations Report 91–4159. Pages 25–33.
119. Webb, E. K., Pearman, G. I., Leuning, R. (1980). Correction of Flux Measurements For Density Effects Due to Heat and Water Vapor Transfer. *Quarterly Journal of the Royal Meteorology Society, 106,* 85–100.
120. Wilson, D., Reginato, R., Hollet, J. K. (2002). Evapotranspiration Measurements of Native Vegetation for Ownes Valley – California. U.S. Geological Survey. Water Resources Investigations Report 91–4159. Pages 1–4.
121. Wolf, A., Saliendra, N., Akshalov, K., Johnson, D. A., Laca, E. (2008). Effect of Different Eddy Covariance Balance Closure and Comparisons with the Modified Bowen Ratio System. *Agric. For. Meteorol., 148,* 942–952.
122. Wright, J. L., Jensen, M. E. (1978). Development and Evaluation of Evapotranspiration Models for Irrigation Scheduling. *Trans ASAE, 21(1),* 88–96.

123. Xingfa, H., Viriyasenakul, V., Dechao, Z. (1999). Design, Construction and Installation of Filled-In Drainage Lysimeter and Its Applications. Proceeding of 99 International Conference on Agricultural Engineering. Beijing, China, December. II: 162–167.
124. Zhang, Y., Liu, C., Yu, Q., Shen, Y., Kendy, E., Kondoh, A., Tang, C., Sun, H. (2004). Energy Fluxes and Pristley-Taylor Parameter Over Winter Wheat and Maize in the North China Plain. *Hydrology Process*, *18,* 2235–2246.
125. Zhang, B., Kang, S., Li, F., Zhang. L. (2008). Comparison of Three Evapotranspiration Models to Bowen Ratio-Energy Balance Method For Vineyard in an Arid Desert Region of North-West China. *Agric. For. Meteorol.*, *148,* 1629–1640.

CHAPTER 3

QUALITY OF IRRIGATION WATER RESOURCES: EGYPT

A. S. EL-HENAWY, M. R. KHALIFA, and M. M. SAFFAN

CONTENTS

The authors thank Tanta University, Egypt, for financial support for this research project.

3.1 INTRODUCTION

Throughout the world, irrigated agriculture faces challenge of using less water, in many cases of poor quality, to provide food and fiber for growing population. Water resources in Egypt are limited. Consequently improving irrigation system, increasing water use efficiency, and reuse of drainage water for irrigation are must. Water supply from irrigation canals is not sufficient enough, especially in the North of Nile Delta, therefore, farmers use drainage water in irrigated the fields [2].

The mandatory use of supplemental irrigation water from sources other than the Nile needs knowledge for the factors that govern the water consumption. The suitability of any water for irrigation is determined by the amount and kind of salts present and content of some heavy metals and soil properties [6].

In recent years the use of wastewater in irrigation is considered a major source for heavy metals for the soil and plant, especially in arid and semiarid zones, where crop production depends mainly on irrigation. The recycled water provides the soils with heavy metals, which may exceed the permissible limits for safe consumption by animals and humans [11, 13, 17].

This chapter focuses on the following two research studies. (1) To evaluate the canal irrigation and drainage water in three Districts of Egypt (Beiala, El-Hamoul and El-Borullus) in Kafr El-Sheikh Governorate. (2) To evaluate the effects of some heavy metal and micronutrient contents in soil and selected field crops.

3.2 MATERIALS AND METHODS

During October 1999 to June 2000, a field survey field was conducted in three districts of Egypt (Beiala, El-Hamoul and El-Borullus), each with 10 irrigation canals and 2 farms with clover and wheat crops, which were located adjacent to each irrigation canal.

Water samples were collected from each irrigation canal six times per month during the irrigation duration (between off and on). These water samples were analyzed to determine EC, pH and soluble Ca^{++}, Na^{+} and Mg^{++} using the methods described by Klute [15] and then SAR (Sodium adsorption ratio) was calculated. Soluble heavy metals and micronutrients in water samples (Zn, Mn, Cd, Ni, Co and Cu) were determined using procedures using atomic absorption spectrophotometer [16].

At harvesting, representative samples of grains and straws for wheat and shoots of clover plants were oven dried (70 °C) and wet digested in $HClO_4$ + H_2SO_4 mixture according to Chapman and Pratt [7]. Concentration of Zn, Mn, Cd, Ni, Co and Cu were determined with an atomic absorption spectrophotometer [16]. Composite soil samples were taken at the depth of 0–30 cm to determine extractable Zn, Mn, Cd, Ni, Co and Cu using atomic absorption spectrophotometer [16]. The data were subjected to statistical analysis using Irristat program. Mean values were compared using Duncan's multiple range tests.

3.3 RESULTS AND DISCUSSION

3.3.1 *EVALUATION OF WATER RESOURCES*

3.3.1.1 TOTAL SOLUBLE SALTS (EC, DS/M) AND SODIUM ADSORPTION RATIO (SAR)

Tables 3.1 to 3.3 indicate that irrigation water sources and duration of irrigation rotations had a highly significant effect on the values of EC and SAR of irrigation waters in Beiala and El-Borullus districts, whereas in El-Hamoul district only canals of irrigation had affected the these parameters. The observations also indicate that values of EC and SAR of the irrigation water in most of irrigation canals in districts of Beiala and El-Hamoul ranged from 0.35 to 0.63 dS/m for EC and from 1.40 to 2.29 for SAR, respectively. These values indicate: good water quality based on classification by Ayers and Westcot [6]; and water classification class (C_2–S_1). This implies that this irrigation water is suitable for irrigation without causing any detrimental effects in these soil types and crops (clover and wheat).

On the other hand, EC values of irrigation water ranged from 0.94 to 2.08 dS/m in drain No. 5 (Biealla district), Kitchener drain, Bahr El-Mansoura, Fom El-Khal-leg, El-Hallab canals (El-Hamoul district) and all irrigation canals of El-Borullus district. These range of EC indicate high salinity according to Ayers and Westcot [6] classification and it lies in class C_3–S_1.

This implies that high salinity water with low sodicity hazard can cause increasing salinity problems. This may be due to because the irrigation water in these canals is considered a mixture of drainage, waste and freshwaters. Therefore, when this water is used for irrigation, it must be adequately controlled and managed with good tillage, addition of amendments and good cropping. Finally, EC and SAR values of irrigation canals in the three districts can be arranged in the ascending order: El-Borullus > El-Hamoul > Beiala districts. Regarding, pH values of irrigation water in all irrigation canals for the three districts was about eight implying that it was slightly alkaline. Our results agree with those reported by other investigators [11, 14].

TABLE 3.1 Mean Values of Chemical Properties and Soluble Heavy Metals of Different Irrigation Water Canals in Beiala District (Kafr El-Sheikh Governorate), During 1999/2000 Season

Name of canal C	Irrigation rotation, R	Chemical properties			Soluble heavy metals, ppm					
		pH	EC, dS/m	SAR	Mn	Zn	Cu	Cd	Ni	Co
Drain no. 5	On	8.17	1.18^{a}	5.67^{a}	0.060	0.081	0.001	0.006	N.D	0.028
	Off	8.20	1.36^{a}	4.60^{a}	0.013	0.075	0.000	0.001	N.D	0.001
Ebshan canal	On	8.20	0.38^{b}	1.51^{b}	0.029	0.051	0.001	0.001	N.D	0.001
	Off	8.16	0.47^{bc}	1.53b	0.036	0.078	0.001	0.001	N.D	0.000
El-Sharkawia canal	On	8.18	0.35^{b}	1.53^{b}	0.00	0.048	0.001	0.010	N.D	0.042
	Off	8.28	0.42^{C}	1.62^{bc}	0.035	0.073	0.000	0.002	N.D	0.001
Bahr terra	On	8.50	0.36^{b}	1.54^{b}	0.007	0.055	0.000	0.003	N.D	0.000
	Off	8.31	0.37^{C}	1.41^{C}	0.030	0.056	0.000	0.002	N.D	0.001
Fouda canal	On	8.23	0.37^{b}	1.67^{b}	0.028	0.065	0.001	0.012	N.D	0.042
	Off	8.32	0.38^{C}	1.42^{C}	0.029	0.067	0.000	0.001	N.D	0.001
Garrd El-Agamy canal	On	8.24	0.35^{b}	1.52^{b}	0.020	0.053	0.000	0.009	N.D	0.042
	Off	8.32	0.41^{C}	1.80^{bc}	0.046	0.081	0.001	0.002	N.D	0.001
Bahr El-Nour	On	8.29	0.35^{b}	1.42^{b}	0.020	0.051	0.000	0.001	N.D	0.001
	Off	8.30	0.37^{C}	1.53^{bc}	0.034	0.067	0.001	0.000	N.D	0.000
Bahr Beiala	On	8.27	0.35^{b}	1.55^{b}	0.029	0.050	0.001	0.001	N.D	0.000
	Off	8.29	0.39^{C}	1.56^{bc}	0.024	0.066	0.001	0.001	N.D	0.000
El-Shorafa canal	On	8.12	0.44^{b}	2.17^{b}	0.033	0.057	0.001	0.000	N.D	0.000
	Off	8.18	0.63^{b}	2.61^{b}	0.029	0.075	0.000	0.009	N.D	0.000
Marrzoka canal	On	8.22	0.38^{b}	1.68^{b}	0.034	0.078	0.001	0.002	N.D	0.000
	Off	8.27	0.54^{bc}	2.29^{bc}	0.020	0.075	0.001	0.000	0.00	0.001
F. Test	C	NS	*	*	NS	NS	NS	NS	-	NS
	R	NS	*	NS	NS	*	NS	NS	-	NS
	CxR	NS	NS	NS	NS	NS	NS	NS	-	NS

Note: NS = not significant; N.D. = no data; C = canal; R = irrigation rotation; C × R = interaction.

3.3.1.2 SOLUBLE HEAVY METALS AND MICRONUTRIENTS CONTENTS IN WATER SOURCES

The data are shown in Tables 3.1 to 3.3 for different sources of irrigation water according to content of heavy metals. These observations indicate that there was no significant effect of both irrigation canals and irrigation rotations on the concentrations of soluble heavy metals and micronutrients in all districts. However, Zn-concentration in the irrigation canals (Beiala and El-Hamoul districts) and Cd-concentration (El-Hamoul and El-Borullus districts) were significantly affected only with irrigation rotations.

On the other hand, concentration of Mn, Zn and Cd elements were high in irrigation canals of the three districts, especially in case of period of irrigation rotation (on) for El-Wallda, Bahr El-Mansour, Fom El-Khallieg, Kitchener drain and El-Hallab canals (El-Hamoul district). This is because the water in these irrigation canals is agriculture drainage water mixed with fresh water or drainage water mixed with wastewater from human activity. Meanwhile, concentration of these elements was high in period of irrigation rotation (off) in irrigation canals of Fara Terra-2, El-Khashaa, Balteem El-Gedida and Neyhaite Bahr terra (El-Borullus district), Kitchener drain and El-Kafr El-Shareki (El-Hamoul district).

TABLE 3.2 Mean Values of Chemical Properties and Soluble Heavy Metals in Irrigation Water from Canals in El-Hamoul District (Kafr El-Sheikh Governorate), During 1999/2000 Season

Name of canal, C	**Irrigation rotation, R**	**Chemical properties**			**Soluble heavy metals, ppm**					
		pH	**EC, dS/m**	**SAR**	**Mn**	**Zn**	**Cu**	**Cd**	**Ni**	**Co**
Ketshenar drain	On	7.89^{b}	1.30^{a}	4.17^{a}	0.039	0.082^{ab}	0.000	0.041^{a}	N.D	0.022^{ab}
	Off	8.29	1.27^{b}	4.77^{ab}	0.004	0.094^{ab}	0.001	0.00^{a}	N.D	0.000^{s}
El-Kafr El-Sharki canal	On	8.24	0.44^{b}	1.42^{b}	0.039	0.052^{ab}	0.001	0.001^{ab}	N.D	0.069^{a}
	Off	8.21	0.49^{C}	2.08^{c}	0.00	0.092^{ab}	0.015	0.00^{a}	N.D	0.000^{a}
Bahr El-Banawan canal	On	8.25	0.37^{b}	1.49^{b}	0.043	0.045^{b}	0.002	0.009^{ab}	N.D	0.028^{ab}
	Off	8.17	0.41^{c}	1.70^{c}	0.001	0.090^{ab}	0.000	0.00^{a}	N.D	0.000^{a}
Ragheeb canal	On	8.26	0.36^{b}	1.40^{b}	0.044	0.054^{ab}	0.001	0.065^{b}	N.D	0.000^{b}
	Off	8.25	0.38^{C}	1.64^{c}	0.004	0.131^{a}	0.015	0.001^{a}	N.D	0.001^{a}
El-Ganabia El-Sabaa	On	8.24	0.39^{b}	1.44^{b}	0.030	0.052^{ab}	0.001	0.012^{ab}	N.D	0.001^{b}
	Off	8.25	0.40^{c}	1.61^{c}	0.027	0.082^{b}	0.001	0.001^{a}	N.D	0.000^{a}
Zouba canal	On	8.29	0.37^{b}	1.43^{a}	0.034	0.042^{b}	0.001	0.00^{ab}	N.D	0.029^{ab}
	Off	8.28	0.39^{c}	1.71^{C}	0.020	0.063^{b}	0.001	0.002^{a}	N.D	0.001^{a}

TABLE 3.2 *(Continued)*

Name of canal, C	Irrigation rotation, R	Chemical properties			Soluble heavy metals, ppm					
		pH	EC, dS/m	SAR	Mn	Zn	Cu	Cd	Ni	Co
El-Wallda canal	On	8.24	0.37[b]	1.39[b]	0.055	0.045[b]	0.000	0.014[ab]	N.D	0.012[ab]
	Off	8.17	0.41[C]	1.62[c]	0.020	0.087[b]	0.001	0.003[a]	N.D	0.000[a]
Bahr El-Mansour	On	7.97	0.94[a]	3.83[a]	0.004	0.081[ab]	0.001	0.010[ab]	N.D	0.000[b]
	Off	8.08	1.22[a]	4.34[b]	0.044	0.074[b]	0.001	0.001[a]	N.D	0.000[a]
Fom El-Khaleg	On	8.12	1.08[a]	4.35[a]	0.023	0.079[ab]	0.000	0.015[ab]	N.D	0.028[ab]
	Off	8.10	1.74[a]	6.54[a]	0.050	0.067[b]	0.000	0.001[a]	N.D	0.001[a]
El-Hallab canal	On	8.09	1.09[a]	4.58[a]	0.034	0.091[a]	0.001	0.015[ab]	N.D	0.028[ab]
	Off	8.20	1.17[b]	4.58[b]	0.053	0.068[b]	0.000	0.001[a]	N.D	0.000[a]
F. Test	C	N.S	**	**	N.S	N.S	N.S	N.S	-	N.S
	R	N.S	N.S	N.S	N.S	**	N.S	*	-	*
	C *x* R	N.S	N.S	N.S	N.S	**	N.S	N.S	-	N.S

Note: NS = not significant; N.D. = no data; C = canal; R = irrigation rotation; C × R = interaction.

Finally, the concentration of most tested heavy metals and micronutrients in irrigation water of most canals were less than the safe limits, recommended by Cottenie et al. [9] and Alloway [4]. The concentration was 2.0 ppm of Zn, 0.2 ppm of Mn, 0.2 ppm of Cu, 0.01 ppm of Cd, 0.2 ppm of Ni and 0.05 ppm of Co. However, co-concentration in El-Kafr El-Sharki canal (El-Hamoul district) and Mn-concentration in El-Khashaa and Balteem El-Gadida canals (El-Borullus district) exceeded the safe limits. These results are in agreement with those reported by and El-Henawy [11] and El-Wakeel [12].

TABLE 3.3 Mean Values of Chemical Analysis and Soluble Heavy Metals of Different Irrigation Water Canals in El-Borullus District (Kafr El-Sheikh Governorate), 1999/2000 Season

Name of canal, C	Irrigation rotation, R	Chemical properties			Soluble heavy metals, ppm					
		pH	EC, dS/m	SAR	Mn	Zn	Cu	Cd	Ni	Co
Branch of Terra 1 canal	On	8.11	1.21[cd]	4.16[a]	0.024	0.091	0.002	0.003	N.D	0.001
	Off	8.09	1.47[C]	4.60[c]	0.025	0.080	0.001	0.001	N.D	0.001

TABLE 3.3 *(Continued)*

Name of canal, C	Irrigation rotation, R	Chemical properties			Soluble heavy metals, ppm					
		pH	EC, dS/m	SAR	Mn	Zn	Cu	Cd	Ni	Co
Terra 2 canal	On	8.12	1.14[d]	4.34[a]	0.021	0.081	0.001	0.004	N.D	0.001
	Off	8.01	1.81[ab]	6.79[abc]	0.304	0.080	0.001	0.001	N.D	0.001
El-Maga-zz canal	On	8.10	1.16[d]	4.16[a]	0.019	0.081	0.000	0.008	N.D	0.023
	Off	7.97	1.93[ab]	5.88[bc]	0.033	0.084	0.000	0.005	N.D	0.000
El-Hellmyia canal	On	8.10	129[bcd]	4.63[a]	0.020	0.086	0.000	0.014	N.D	0.001
	Off	8.09	1.88[ab]	6.89[abc]	0.110	0.069	0.000	0.001	N.D	0.000
Terra 4 canal	On	8.08	1.19[cd]	3.98[a]	0.009	0.090	0.001	0.005	N.D	0.001
	Off	8.20	1.69[bc]	6.00[bc]	0.030	0.085	0.001	0.001	N.D	0.000
El-Kha-shaa canal	On	8.15	1.61[ab]	6.48[a]	0.027	0.119	0.001	0.011	N.D	0.001
	Off	8.16	2.06[a]	6.80[abc]	0.339	0.091	0.000	0.014	N.D	0.000
El-ganabia El-gharbia canal	On	8.23	1.51[abc]	5.80[a]	0.012	0.097	0.002	0.010	N.D	0.001
	Off	8.41	1.87[ab]	7.30[abc]	0.043	0.0102	0.001	0.001	N.D	0.001
El-Nahda canal	On	8.20	1.65[a]	6.35[a]	0.026	0.095	0.002	0.015	N.D	0.001
	Off	8.11	1.96[ab]	8.55[ab]	0.043	0.090	0.001	0.001	N.D	0.000
Balteem El-Gedida canal	On	8.09	1.59[ab]	6.05[a]	0.024	0.090	0.003	0.012	N.D	0.006
	Off	8.15	1.93[ab]	7.73[abc]	0.426	0.086	0.001	0.001	N.D	0.001
Nyhaite Bahr Terra (Balteem) canal	On	8.22	1.62[ab]	6.40[a]	0.017	0.102	0.001	0.008	N.D	0.006
	Off	7.96	2.08[a]	9.64[a]	0.357	0.091	0.001	0.000	N.D	0.00
F. Test	C	N.S	**	N.S	N.S	N.S	N.S	N.S	-	N.S
	R	N.S	**	**	N.S	N.S	N.S	**	-	N.S
	C *x* R	N.S	N.S	N.S	N.S	N.S	N.S	N.S	-	N.S

Note: NS = not significant; N.D. = no data; C = canal; R = irrigation rotation; C × R = interaction.

3.3.2 AVAILABLE CONTENT OF HEAVY METALS AND MICRONUTRIENTS IN SOIL

The Table 3.4 indicates that the range of available content of heavy metals in soil after harvesting clover crop was: 2.26–4.17 of Mn, 0.72–1.41 of Zn, 0.19 –2.03 of

Cu, 0–0.25 of Ni, 0–0.126 of Cd and 0.054 to 0.216 ppm of Co in Beiala district. The highest content of available Mn, Cu, Cd and Co was found in the soil, which was irrigated from Marrzoka canal, which had accepted wastewater during the winter season. Meanwhile, the highest content of available Zn and Ni was found in soil, which was irrigated from Ebshan canal, which received wastewater from the human activity.

In the soil samples after harvesting of wheat crop (Table 3.4), the concentration of available elements ranged between 2.37–3.079 of Mn, 0.8–2.45 of Zn, 0.38–1.78 of Cu, 0.17–0.25 of Ni, 0–0.063 of Cd and 0.54–0.216 ppm of Co, respectively. The highest content of available Mn, Zn and Ni, Cu and Co was found in the soils, which were irrigated with water from El-Shorafaa, Bahr El-Nour, Ebshan and El-Sharkawia canals, respectively.

TABLE 3.4 DTPA Extractable of Heavy Metals (mg.kg^{-1}) in the Soil Samples After Harvesting of Clover and Wheat Crops, which were Adjacent to Irrigation Canals

Irrigation water source	After clover						After wheat					
	Mn	Zn	Cu	Ni	Cd	Co	Mn	Zn	Cu	Ni	Cd	Co
Beiala district (Kafr El-Sheikh Governorate)												
Bahr Beialla canal	2.26	0.96	0.19	0.00	0.063	0.054	2.78	0.96	0.38	0.17	0.063	0.162
Bahr El-Nour canal	3.06	0.93	0.32	0.17	0.063	0.00	2.37	2.45	0.57	0.25	0.00	0.127
Card El-Agamy canal	3.27	0.80	0.76	0.08	0.053	0.108	2.61	0.85	0.44	0.25	0.021	0.108
Drain No. 5	3.27	0.72	0.83	0.25	0.053	0.054	3.03	1.87	0.95	0.25	0.021	0.216
Ebshan canal	3.13	1.41	0.76	0.25	0.00	0.00	2.54	1.07	1.78	0.17	0.01	0.108
El-Sharkawia canal	3.65	1.25	1.14	0.08	0.074	0.0544	2.43	0.80	1.27	0.17	0.053	0.216
El-Shorafa canal	3.30	0.69	0.89	0.08	0.011	0.216	3.76	0.91	1.21	0.17	0.021	0.054
Fouda canal	3.48	1.07	1.59	0.08	0.042	0.162	3.48	0.80	0.95	0.17	0.00	0.216
Marzouka canal	4.17	1.33	2.03	0.17	0.126	0.324	2.61	0.83	1.27	0.25	0.042	0.108
El-Hamoul district (Kafr El-Sheikh Governorate)												

TABLE 3.4 *(Continued)*

Irrigation water source	After clover						After wheat					
	Mn	Zn	Cu	Ni	Cd	Co	Mn	Zn	Cu	Ni	Cd	Co
Bahr El-Banawan canal	4.52	1.52	1.78	0.00	0.011	0.108	2.33	0.8	0.63	0.25	0.00	0.054
Bahr El-Mansour canal	2.30	0.67	0.76	0.17	0.011	0.00	2.54	0.59	0.44	0.17	0.053	0.216
El-Ganabia El-Sabaa canal	3.20	0.64	0.76	0.08	0.032	0.00	2.61	0.96	0.06	0.08	0.053	0.00
El-Hallab canal	4.87	0.67	0.83	0.00	0.084	0.108	2.10	0.80	0.89	0.17	0.032	0.054
El-Kafr El-Sharki canal	4.45	1.12	1.59	0.25	0.042	0.054	3.48	1.01	0.32	0.17	0.053	0.108
El-Walda canal	2.61	0.69	0.32	0.17	0.053	0.054	2.78	0.75	1.14	0.08	0.032	0.054
Fom El-Khalieg canal	2.78	0.80	0.13	0.08	0.00	0.00	2.19	0.48	0.44	0.08	0.012	0.00
Kitchener Drain	4.43	0.85	0.44	0.08	0.021	0.108	2.16	2.80	1.40	0.25	0.084	0.054
Ragheb canal	5.04	1.07	1.46	0.17	0.042	0.054	2.37	0.85	0.63	0.17	0.042	0.162
Zouba canal	3.65	0.96	2.03	0.00	0.011	0.054	2.23	0.53	0.44	0.00	0.032	0.00
El-Borullus district (Kafr El-Sheikh Governorate)												
El-Ganabia El-Gharbia canal	3.90	1.47	1.59	0.25	0.00	0.170	2.61	1.65	0.87	0.25	0.021	0.162
El-Hellmya canal	4.52	0.61	0.76	0.25	0.00	0.187	6.43	1.044	2.86	0.08	0.011	0.00
El-Khashaa El-Gedida canal	2.68	0.75	0.95	0.08	0.00	0.162	3.09	2.05	0.94	0.25	0.00	0.160

TABLE 3.4 *(Continued)*

Irrigation water source	After clover						After wheat					
	Mn	Zn	Cu	Ni	Cd	Co	Mn	Zn	Cu	Ni	Cd	Co
El-Magazz canal	4.07	0.64	1.14	0.17	0.042	0.162	3.62	0.67	1.14	0.17	0.00	0.00
El-Nahda canal	4.00	0.56	0.63	0.17	0.032	0.054	3.23	0.69	0.63	0.080	0.032	0.054
Farha Terra 1 canal	5.11	1.89	2.16	0.08	0.042	0.054	3.20	1.07	0.70	0.17	0.021	0.108
Farha Terra 2 canal	4.34	0.67	0.89	0.17	0.021	0.162	3.8	1.41	0.32	0.25	0.011	0.108
Farha Terra 4 canal	5.57	0.85	1.90	0.25	0.032	0.162	3.76	0.59	0.32	0.25	0.011	0.054

In El-Hamoul district (Table 3.4), observations reveal that available content of all heavy metals in soils were found in all irrigated canals, after harvesting of clover and wheat crops. The highest contents of available Zn, Cu, Ni and Cd were found in soils which was irrigated from Kitchener drain after wheat, while the highest content of Mn and Co elements in the soils was found in case of El-Kafr El-Sharki and Bahr El-Mansour canals, respectively. Furthermore after harvesting of clover crop, the highest content of available Mn, Zn, Cu, Ni, Cd and Co in soils was found with irrigation water from Rajheeb, Bahr El-Banawaan, Zoubaa, El-Kafr El-Sharki and El-Hallab canals, respectively.

Regarding El-Borullus district, data in Table 3.4 revealed that after harvesting clover soil samples had highest concentration of available Mn, Zn, Cu and Cd, with irrigation from Farah Terra 1 canal. However, the highest concentration of Ni and Co were found in the soils adjacent to Farah Terra 4 and El-Helmia canals.

After harvesting of wheat crop, the highest content of available Mn and Cu were found with irrigation water from El-Helmia canal, while the highest values of Zn, Ni and Co were found in the soils with irrigation from El-Khashaa El-gedida canal. Also, the highest value of Cd element was found in the soil adjacent to Farah Terra 1.

We can conclude that available heavy metal concentrations in soils can be arranged in the ascending order: Borullus > El-Hamoul > Biela districts. This may due to because the water from all irrigated canals in El-Borullus district is a mixture of water source from the drainage, sewage and fresh waters, which contained more concentrations of heavy metals. Also, it was noticed that the concentration of the most available heavy metals was high in soils after harvesting of clover than in soils after harvesting of wheat. This may be due to variation of biological activity, which is accompanied by the cultivated plants that led to the increasing of organic acidity thus causing the increment of the availability of these elements.

These results are in agreement with those obtained by Abou Hussin et al. [3], Amer et al. [5], El-Henway [11] and Salt et al. [17]. They reported that use of the poor water quality and wastewater in irrigation increased the content of total and available heavy metals in soil.

3.3.3 CONCENTRATION OF HEAVY METALS AND MICRONUTRIENTS IN THE SHOOTS OF CLOVER

Table 3.5 indicates that the concentration (mg.kg^{-1} of dry matter plant) of heavy metals (Ni, Cd and Co) and micronutrients (Zn, Cu and Mn) varied from site to site, depending on the available concentration of these elements in soil samples (see Table 3.4), soil pH and the concentrations of these elements in irrigation water for the crop (see Tables 3.1–3.3). Data in Table 3.5 shows that Ni-concentration in shoot was higher than the safety limit of 8 ppm in all locations of the three districts [8], except in locations adjacent to Fouda and El-Shorafa canals (Biealla district) and Bahr El-Banawan canal (El-Hamoul district). Cd-concentration in shoot was higher than the safety limits (0.01–1.23 ppm) in most of the locations in the three districts [4]. Whereas, the highest concentration of Cd was 11.58 in locations adjacent to Bahr El-Banwan (El-Hamoul district) and 20.53 in locations adjacent to Farah terra 4 (El-Borullus district), respectively. Co-concentration in all locations was less than the safety limits (5–20 ppm) according to Cottenie et al. [9]. Zn, Mn and Cu concentrations in shoot for all locations were less than the safety limits (50 ppm of Zn, 100 ppm of Mn and 20 ppm of Cu) [4], except Cu-concentration was higher than the safety limit for locations adjacent to Garrd El-Agamy, Bahr Biealla canals (Biealla district); Farah terra 1, El-Khashaa El-Gedida and El-Nahdaa canals (El-Burullus district). These results are in harmony with those reported by Abd El-Naiem et al. [1] and El-Henawy [11].

We can conclude that the frequent utilization of agriculture drainage water or mixture of drainage water with different waste water for irrigating clover may cause an accumulation of some heavy metals (Ni and Cd), thus leading to detrimental effects on animals and humans.

TABLE 3.5 Concentrations of Heavy Metals in the Shoots of Clover Plants That Have Been Irrigated From Locations Adjacent to Irrigation Canals in Three Districts of Egypt

Irrigation water source	Mn	Zn	Cu	Ni	Cd	C
Beiala district (Kafr El-Sheikh Governorate)						
Bahr Beialla canal	37.8	29.3	23.5	8.33	0.00	7.31
Bahr El-Nour canal	28.1	34.7	11.8	25.0	0.00	5.00
Drain No. 5	31.3	40.0	16.7	8.33	0.00	1.07

TABLE 3.5 *(Continued)*

Irrigation water source	Mn	Zn	Cu	Ni	Cd	C
Ebshan canal	40.6	18.7	17.6	8.33	44.21	2.23
El-Sharkawia canal	28.1	32.0	17.6	8.33	1.05	0.00
El-Shorafa canal	34.4	37.3	0.00	0.00	8.42	2.11
Fouda canal	34.4	32.0	5.88	0.00	2.11	7.01
Gard El-Agamy canal	37.5	32.0	29.4	8.33	9.47	3.17
Marzouka canal	31.3	24.0	17.6	33.3	9.47	2.15
El-Hamoul district (Kafr El-Sheikh Governorate)						
Bahr El-Banawan canal	25.0	32.0	11.8	0.00	11.58	4.82
Bahr El-Mansour canal	31.3	16.0	5.88	8.33	8.42	0.00
El-Ganabia El-Sabaa canal	31.3	26.7	17.6	16.7	2.11	1.15
El-Hallab canal	28.1	29.3	0.00	8.33	0.00	5.40
El-Kafr El-Sharki canal	37.5	56.0	17.6	8.33	1.05	7.23
El-Walda canal	25.0	26.7	17.6	8.33	2.26	0.00
Fom El-Khalieg canal	28.1	44.0	5.88	8.33	1.05	0.00
Kitchener Drain	37.5	26.7	0.00	16.7	0.00	2.12
Ragheb canal	25.0	37.3	11.8	25.0	2.11	0.00
Zouba canal	25.0	10.7	0.00	8.33	0.00	5.40
El-Borullus district (Kafr El-Sheikh Governorate)						
El-Ganabia El-Gharbia canal	62.5	53.3	11.8	16.7	5.26	6.82
El-Hellmya canal	40.6	42.7	17.6	8.33	5.26	3.22
El-Khashaa El-Gedida canal	37.8	37.3	29.4	16.7	0.00	5.40
El-Magazz canal	31.3	40.0	11.8	8.33	3.16	7.72
El-Nahda canal	62.5	26.7	29.4	25.0	6.32	1.97
Farha Terra 1 canal	40.6	32.0	29.4	33.3	4.21	0.00
Farha Terra 2 canal	34.0	32.0	17.6	25.0	9.47	3.17
Farha Terra 4 canal	18.8	2.0	17.6	16.7	20.53	0.00

3.3.4 CONCENTRATIONS OF HEAVY METALS AND MICRONUTRIENTS IN THE STRAW AND GRAINS OF WHEAT PLANT

The Table 3.6 reveals that the concentration of heavy metals (Ni and Cd) and Cu-element in straw of wheat plant was higher than the values in the grains for most of the locations. The highest values of Cu (23.5 $mg.kg^{-1}$), Ni (50 $mg.kg^{-1}$) and Cd (6.32 $mg.kg^{-1}$) were found in locations adjacent to Marzouka and Bahr El-Nour canals (Beialla district) and Farah terra 2 (El-Borullus district) for Cu; Fouda and Marrzoka canals (Beiala district) and El-Banawan and El-Halabe canals (El-Hamoul district) for Ni and Kitchener drain for Cd.

On the other hand, concentrations of Mn, Zn and Co in grains of wheat were higher than the corresponding values of straw for all locations. However, the highest value of Mn (38.4 mg kg^{-1}), Zn (45.3 mg kg^{-1}) and Co (6.17 mg kg^{-1}) were found in Zobaa, Bahr El-Mansour and El-Magazz canals, respectively.

TABLE 3.6 Concentrations of Heavy Metals and Micronutrients in the Grains and Straw of Wheat Plant, which Had Been Irrigated From Adjacent Irrigation Canals, During 1999/2000

Irrigation water source	Grains						Straw					
	Mn	Zn	Cu	Ni	Cd	Co	Mn	Zn	Cu	Ni	Cd	Co
Beiala district (Kafr El-Sheikh Governorate)												
Bahr Beialla canal	31.3	34.7	11.7	0.00	1.05	2.67	28.1	10.7	11.8	41.7	4.21	0.22
Bahr El-Nour canal	31.3	24.0	11.7	25.0	3.16	2.18	15.6	2.67	23.5	25.0	2.11	0.41
Drain No. 5	28.1	37.3	5.88	0.00	2.11	5.12	28.1	10.7	5.88	16.7	4.02	0.77
Ebshan canal	28.1	24.3	11.7	25.0	0.00	1.17	18.8	10.7	23.5	41.7	2.11	0.12
El-Sharkawia canal	34.4	21.3	5.88	25.0	0.00	4.21	18.8	2.67	23.5	33.3	1.05	1.41
El-Shorafa canal	37.5	40.0	11.7	16.70	0.00	4.21	21.9	8.00	17.6	25.0	3.16	0.32
Fouda canal	28.1	24.0	5.88	16.70	1.05	0.00	18.8	10.7	17.6	50.0	1.05	0.00
Gard El-Aga-my canal	25.0	26.7	11.7	25.0	0.00	0.165	12.5	2.67	0.00	33.3	2.00	1.23
Marzouka canal	37.5	26.7	0.00	16.7	2.11	4.21	21.7	2.67	23.3	50.0	1.05	0.00
El-Hamoul district (Kafr El-Sheikh Governorate)												
Bahr El-Banawan canal	25.0	26.7	5.88	0.00	44.21	2.16	21.9	13.3	11.8	50.0	3.16	0.15
Balir El-Man-sour canal	28.1	45.3	17.6	8.33	0.00	3.21	21.9	10.7	0.00	8.33	5.26	1.22
El-Ganabia El-Sabaa canal	21.9	16.0	4.70	16.7	3.16	5.12	25.0	13.3	5.88	25.0	4.21	0.17
El-Hallab canal	21.9	16.0	0.00	0.00	0.00	3.98	18.8	10.7	17.6	50.0	0.00	0.00

TABLE 3.6 *(Continued)*

Irrigation water source	Grains						Straw					
	Mn	Zn	Cu	Ni	Cd	Co	Mn	Zn	Cu	Ni	Cd	Co
El-Kafr El-Sharki canal	15.6	29.3	5.88	8.33	0.00	0.00	18.8	13.3	11.8	8.33	2.11	0.23
El-Walda canal	31.3	24.0	0.00	8.33	0.00	4.16	16.6	5.33	0.00	0.00	0.00	1.11
Fom El-lChal-ieg canal	25.0	21.3	4.70	0.00	0.00	1.15	21.9	13.3	0.00	0.00	3.16	0.52
Kitchener Drain	25.0	42.7	0.00	8.33	0.00	3.00	18.8	10.7	11.8	8.33	6.32	0.12
Ragheb canal	34.4	32.0	5.88	0.00	0.00	0.00	25.0	2.67	17.6	8.33	0.00	0.00
Zouba canal	38.4	24.0	5.88	8.33	0.00	1.15	21.9	13.3	0.00	16.7	1.05	1.00
El-Borullus district (Kafr El-Sheikh Governorate)												
El-Ganabia El-Gharbia canal	25.0	32.0	23.5	25.0	2.11	3.00	18.8	10.7	11.8	16.7	2.11	0.17
El-Hellmya canal	31.3	26.7	0.00	8.33	0.00	4.85	28.1	16.0	11.8	41.7	3.16	1.17
El-Khashaa El-Gedida canal	31.3	34.7	0.00	0.00	2.11	4.27	25.0	6.67	11.8	0.00	1.05	0.52
El-Magazz canal	34.4	32.0	11.7	8.33	1.05	6.17	18.8	13.3	17.6	0.00	3.16	1.35
El-Nahda canal	28.1	40.0	17.6	33.3	1.16	0.00	15.6	10.67	5.88	33.3	3.25	1.37
Farha Terra 1 canal	31.4	21.3	0.00	8.33	2.11	3.98	15.6	13.3	17.6	41.7	4.21	0.00
Farha Terra 2 canal	34.4	40.0	17.6	16.70	3.16	6.17	31.3	13.3	23.5	0.00	4.21	1.17
Farha Terra 4 canal	34.4	21.3	11.7	3.33	3.16	0.00	12.5	8.0	17.6	33.3	4.21	1.07

Furthermore, data indicates that concentrations of Mn, Zn and Cu in both straw and grains of wheat were less than the safety limits reported by Cottenie et al. [9]. However, Cu-concentration in some locations was above safety levels and was below the phytotoxic levels. The Cd and Ni concentrations in both straw and grains were higher than permitted levels, according to Cottenie et al. [8]. This implies that frequent use of agriculture drainage water or mixture of drainage water and wastewater for irrigation of wheat plant caused an accumulation of some heavy metals (Ni and Cd), which in turn was harmful to animals and human. Similar results have been reported by Davis and Smith [10] and El-Henawy [11].

3.4 SUMMARY

An experimental survey was carried out to evaluate the irrigation water resources in three districts (Beiala, El-Hamoul and El-Borullus) of Kafr El-Shiekh Governorate of Egypt, during season 1999/2000. About 10 irrigation canals were selected in each district. Chemical analysis for heavy metal concentrations and micronutrients in water, soil and plant (wheat and clover crops) was performed.

The values of salinity index (EC) and Sodium adsorption ratio (SAR) of irrigation water in all three districts had highly significant differences among irrigation canals and irrigation rotations, except EC values in El-Hamoul district and SAR values in El-Borullus district had only significant differences between irrigation canals and irrigation rotation, respectively.

Data shows that the irrigation water had a medium quality class (C_2–S_1) for all irrigation canals in Beiala district, except drain No. 5, which had a lower quality class (C_3–S_2). El-Borullus district had a low quality class (C_3–S_1) and (C_3–S_2) for all canals. Irrigation water in El-Hamoul district had a medium quality class (C_2–S_1) in most of canals, while canals of Kitchener, Bahr El-Mansour, Fom El-Khalieg and El-Hallab had a low quality class (C_3–S_2).

There was no significant difference among heavy metal concentrations for either canals or rotations. However, Zn concentration in both Beiala and El-Hamoul districts as well as Cd concentration in El-Borullus and El-Hamoul districts had a significant difference. It is concluded that all concentrations of heavy metals were less than the permitted limits.

Heavy metals concentration in soil after cultivation of clover and wheat were in the order of Borullus > El-Hamoul > Beiala districts. Zn, Ni, Cd and Cu had the highest concentration in Beiala (Marzouka canal), El-Hamoul (Kitchener) and in El-Borullus (Tiera No. 4 and El-Gannabia El-Gharbia canals). Also the concentration of all heavy metals in soil was less than the hazard levels. Ni and Cd showed a high level concentration above the permitted level in both wheat and clover crops, in some areas adjacent to irrigation canals in three districts, which was harmful to animals and humans.

KEYWORDS

- clover
- crop production
- Egypt
- electrical conductivity, EC
- heavy metals
- Micronutrients
- Nile Delta
- pH
- salts
- sodium adsorption ratio, SAR
- Tanta University
- water pollution
- water quality
- wheat

REFERENCES

1. Abd El-Naim, M., Abu El-Enien, R., Fahmiry, S., Said, A. (1998). Risk Assessment of Soil and Water Pollution On Agriculture Crops In Egypt. *World Congress of Soil Science,* Montpellier – France, 20–26 August.
2. Abo-Soliman, M. S. M., Khalifa, M. R., El-Sabry, W. S., Sayed, K. H. (1992). Use of drainage water in irrigation at North of Nile Delta, its effects on soil salinity and wheat production. *J. Agri. Res. Tanta University, 18(2),* 425–435.
3. Abou-Hussien, E. A., El-Koumey, B. Y. (1997). Influence of Waste Water on Some Soils and Plant in Menofiya Governorate. Menofiya. *J. Agric. Res., 22(6),* 1733–1748.
4. Alloway, B. J. (1995). *Heavy Metals In Soils*. Blackie Academic professional: Imprint of Chapman & Hull, Eastern Cleddens Rood, UK.
5. Amer, A. A., Abd El-Wahab, S. A., Abou El-Soud, M. A. (1997). Effect of water quality on soil salinity and some crops production. *J. Agric. Science Mansoura University, 22(4),* 1287–1265.
6. Ayers, R. S., Wescot, D. W. (1985). *Water quality for agriculture*. Irrigation and Drainage Paper 29, FAO, Rome, 174 pages.
7. Chapman, H. D., Pratt, P. F. (1980). *Methods of analysis of soil, plants and waters*. University of Calif. *Division of Agric. Sci.,* 60–69.
8. Cottenie, A., Dhaese, A., Camerlynek, R. (1976). Plant quality response to uptake of polluting elements. Qualities Planetarium, Plant Food for Human's Nutrition, 26: 293.
9. Cottenie, A., Verloo, M., Velghe, G., Kiekens, L. (1982). Biological and analytical aspects of soil pollution. Laboratory of Analytical Agro. State University, Ghent -Belgium.
10. Davis, R. D., Smith, C. (1980). Crops as indicators of the significance of contamination of soil by heavy metals. Technical Report TR 140 by Water Research Center, Tevenage, U.K. 43 pages.

11. El-Henawy, A. S. (2000). Impact of available water sources at North Delta on soil and some field crops. M.Sc. Thesis, Faculty of Agriculture Kafr El-Sheikh, Tanta University Egypt.
12. El-Wakeel, A. F., El-Mowelhi, N. M. (1993). Reuse of drainage water on farm with reference to soil properties", crop production and heavy metals contents in plant Tissues and diary products. Egypt. *J. Agric. Res., 71(4),* 845–861.
13. Khalifa, M. R. (1990). Evaluation of suitability of drainage water for irrigation purpose, its effect on some soil properties of clay soils in North Delta. *J. Agricultural Research Tanta University, 16(3),* 573–586.
14. Khalifa, M. R., Rabie, A., Youssef, S. M., El-Henawy, A. S. (2003). Evaluation of available sources of irrigation water at North Delta and its effect on soil salt storage under some field crops. *J. Agric. Science.* Mansoura University, Special issue for Scientific Symposium on "Problems of soils and waters in Dakahlia and Damietta Governorates." March 18.
15. Klute, A. (1986). *Methods of Soil Analysis (Part 1).* American Society of Agronomy, Madison, Wisconsin, USA, 3rd edition.
16. Lindsay, W. L., Norvell W. A. (1978). Development of DTPA soil Test for zinc, iron, Manganese and copper. *Soil Sci. Soc. Amer. Proc., 42,* 421–428.
17. Salt, D. E., Blaylock, M., Kumar, N. P., Dushenkov, V., Ensley, B. D., Chet, I., Raskin, I. (1995). Phytoremediation; A novel strategy for removed of toxic metals from the environment using plants. *Biotechnology, 13(5),* 468–474.

CHAPTER 4

MAXIMUM WETTING DEPTH UNDER AN EMITTER

HAMMAMI MONCEF, DAGHARI HEDI, BALTI JELLOUL, and MAALEJ MOHAMED

CONTENTS

4.1 INTRODUCTION

Drip irrigation supplies water directly in the root zone of a plant. This method allows the effective wetted soil volume to be reduced thus reducing the evaporation and deep (water and nutrients) percolation losses. The prediction of wetted soil volume under an emitter is a must for water management, because an over water application results in loss of water and fertilizers beyond the root zone, particularly in sandy soils [11, 22]. Mickelakis et al. [25] have observed negligible, moderate and high deep percolation losses in a drip irrigated avocado orchard for three water levels application: 0.30.Epan, 0.60.Epan and 0.90.Epan. Levin et al. [24] have evaluated the deep percolation beyond 60 cm depth in a sandy soil at 26% of the amount of water supplied.

Maximum soil depth due wetting, during water infiltration from an emitter on the soil surface, can help to reduce water and nutrients losses. Therefore, analytical solutions of the axisymmetric water infiltration equation are often preferred but they are valid only for steady state flow or for short time infiltration where the gravity effects can be neglected [4, 9]. However, even for daily-irrigated field, such conditions are seldom met under real micro irrigation practices [12, 21, 30]. Recently, Revol et al. [31] proposed an approximate time-dependent solution for wetting front position for trickle point source infiltration. He assumes that the steady state moisture regime prevails behind the wetting front [10, 28].

Numerical models yield an accurate prediction of the wetted soil volume dimensions but are practical because of the complexity, cost and difficulty to reproduce the ponded area extension on soil surface [1, 23]. Several equations from numerical models (1;19) can be useful but these must be validated for real field situations.

Empirical models [33] have also been developed for prediction of dimensions of wetted bulb. Close relationships, between the vertical Z_f and horizontal W_f wetted soil volume dimensions, were inferred. From experimental data in a sandy soil, Keller and Bliesner [22] have developed graphical relationships between Z_f and W_f similar to those by Schwartzmass and Zur [33]. However, measurement of the horizontal wetted bulb dimension W_f is a difficult task, thus limiting of these models in practice.

This chapter discusses the research results to evaluate a new approach for predicting the wetting front depth $Z_f(t)$, during an axisymmetric water infiltration from a surface emitter. Based on the continuity and cumulative infiltration equations, our approach has two advantages: (i) maximum front depth $Z_f(t)$, beneath an emitter can be inferred by measuring radius $R_f(t)$ of soil wetted surface around an emitter; and (ii) our method is valid for transient and steady state infiltration flows.

4.2 THEORETICAL BASIS

Water infiltration into the soil from an emitter is a three-dimensional saturated–unsaturated flow with a moving boundary or wetting front, separating a wetted and a dry zone. This boundary movement results in a convective-diffusive type of flow in the soil. Therefore, the wetted bulb pattern should depend on both convective and diffusive effects of water flow. In an isotropic homogeneous medium, water infiltration from a surface emitter is an axisymmetric flow that can be described by the Richards' equation:

$$\partial q/\partial t = \partial/(r\partial r)\{rK(h)\partial h/\partial r\} + \partial/(\partial z)\{K(h)\ \partial h/\partial z\} - \partial K(h)/\partial z \qquad (1)$$

where: θ is the volumetric water content, L^3L^{-3}; tT is the time, T; h is the matric water head, L; r and *z* are the horizontal and the vertical coordinates, L, respectively; and K(h) is the soil hydraulic conductivity, LT^{-1}. Along the horizontal axis *r* on soil surface (Fig. 4.1), the wetting front (L) advancement results from a horizontal infiltration with cumulative value is I_h. Using the initial and boundary conditions (the Eqs. (2) and (3)), I_h is approximated [20, 27, 32, 36] as shown in Eq. (4):

$$t = 0,\ \ h(r, z, 0) = h_i \qquad (2)$$

$$t > 0,\ \ h(0,0, t) = 0 \qquad (3)$$

$$I_h = St^{1/2} \qquad (4)$$

In Eq. (4), S is the soil sorptivity, $LT^{-1/2}$. Because of the flow symmetry, the wetting front advancement along the vertical axis "*z*" results from a vertical water flow.

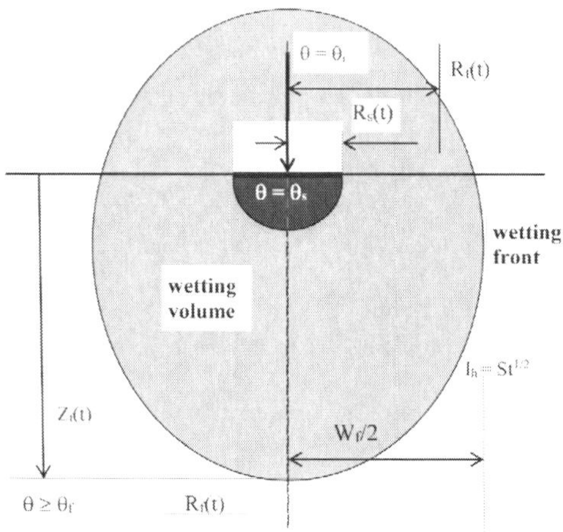

FIGURE 4.1 Typical pattern of the wetted soil volume (bulb).

With initial and boundary conditions in the Eqs. (2) and (3), Haverkamp et al. [17] and Smettem et al. [32] introduced the following solution, for the cumulative vertical infiltration I_v(L) for a disk infiltrometer on the soil surface:

$$I_v = St^{1/2} + \{K_i + (2 - \beta)/3(K_n - K_i)\}(t) \quad (5)$$

where: K_i and K_n are the soil hydraulic conductivity corresponding to initial and imposed pressure head; and β is a shape factor [32]. Green and AMPT assumptions ([4, 20] are:

$$R(r, z = 0, t) \leq R_f(t) \Rightarrow \theta(r, z = 0, t) = \theta_f$$

$$R(r, z = 0, t) > R_f(t) \Rightarrow \theta(r, z = 0, t) = \theta_i \quad (6)$$

$$Z(r = 0, z, t) \leq Z_f(t) \Rightarrow \theta(r = 0, z, t) = \theta_f$$

$$Z(r = 0, z, t) > Z_f(t) \Rightarrow \theta(r = 0, z, t) = \theta_i$$

where: the subscripts *i* and *f* refer to the initial and wetting front boundary (at time *t*) conditions, respectively. Though these assumptions were suggested in 1911, yet they remain useful because of the good approximation to provide for water redistribution into the soil [6, 7, 28, 31]. Philip [28] reported that the assumption of uniform water content is not realistic, but it can be admitted for a reduced frequently wetted soil volume (the case of a wetted bulb). Therefore, imposing a fixed pressure head on wetted soil volume, $h = h_f$ so that $K(h_f) = K_f = K_n$ and then applying the continuity equation to the amount of water infiltrated, I_h and I_v, we get:

$$I_h = (\theta_f - \theta_i)R_f(t) \quad (7)$$

$$I_v = (\theta_f - \theta_i)Z_f(t) \quad (8)$$

Substituting Eqs. (7) and (8) into Eqs. (4) and (5) gives:

$$St^{1/2} = (\theta_f - \theta_i)R_f(t) \quad (9)$$

$$St^{1/2} + \{K_i + [(2 - \beta)/3](K_f - K_i)\}t = (\theta_f - \theta_i)Z_f(t) \quad (10)$$

The sorptivity is an integral characteristic, which depends on the initial and the final soil water contents [20, 32, 38]. With assumptions in Eq. (6) and substituting Eq. (9) into Eq. (10) will yield:

$$Z_f(t) = R_f(t) + [\{K_i + [(2 - \beta)/3](K_{f\_}K_i)\}t]/[(\theta_{f-i})] \quad (11)$$

The shape factor β ranges between 0 and 1 [32]. Cindy et al. [7] reported that β should be constant for geometrically similar porous media. Haverkamp et al. [17] and Smettem et al. [32] used β = 0.56 and β = 0.60, respectively. In this study, choosing β = 1/2 and neglecting K_i as compared to K_f (for the experimental data used here, $K_i/K_f \approx 10^{-5}$), Eq. (11) reduces to:

$$Z_f(t) = R_f(t) + [(1/2K_f)t]/(\theta_{f-}\theta_i) \quad (12)$$

Equation (12) demonstrates that $Z_f(t)$ is an increasing monotonous function of $R_f(t)$, with the following properties:

$$\text{At } t = 0\text{: } R_f(t) = 0 \text{ and } Z_f(t) = 0$$

$$\text{At } t \to \infty\text{: } Z_f(t) \to \infty$$

$$\text{At } t > 0\text{: } dZ_f(t)/dR_f(t) \geq 0$$

$$\text{And at } \forall\, t > 0\text{: } Z_f(t) \geq R_f(t) \quad (13)$$

The Eq. (12) is comparable to a diffusive-convective water flow in the soil where the convective component increases with time. This implies that $R_f(t)$ reaches a limit whereas the term $[½.t.K_f/(\theta_f - \theta_i)]$ is an increasing monotonous function of time (t). Equation (12) enables to evaluate the effects of soil and infiltration parameters on $Z_f(t)$ and then on the evolution of bulb shape. In fact, with high K_f values (the case of sandy soils) Eq. (12) clearly demonstrates that $Z_f(t)$ will be higher and then the bulb shape will be deeper than in the heavy soils. It is also easy to verify that the drier the initial soil profile (the second term on the right hand side, the term $[K_f.t/2(\theta_f - \theta_i)]$ decreases as θ_i decreases), the slower the wetting front advancement.

The effects of discharge rate (Q) and water amount (V) are also implicitly evident. Indeed, in the same soil conditions and with the same volume of water (V = Q.t), an increase in the discharge rate Q results in a decrease in the infiltration time t, consequently, $\{K_f.t/2(\theta_f\text{-}\theta_i)\}$ decreases and $R_f(t)$ increases. However, a decrease in the discharge rate results in an increase in the infiltration time t. Therefore, the second term on the right hand side $\{K_f.t/2(\theta_f - \theta_i)\}$ increases and then $R_f(t)$ decreases. Brandt et al. [5], Levin et al. [24] and Akbar et al. [2] observed similar effects of discharge rate Q on the wetted soil volume beneath a point source. This behavior is attributed to the effects of discharge rate on the extension of ponded area at the soil surface. Dividing Eq. (12) by $R_f(t)$ gives:

$$Z_f(t)/R_f(t) = [1 + K_f.t/\{2R_f(t).(\theta_f - \theta_i)\}] \quad (14)$$

The quotient $[Z_f(t)/R_f(t)]$ is an increasing monotonous function of time, with the following properties:

$$t \to 0\text{: } Z_f(t)/R_f(t) \to 1 \quad (15a)$$

$$t \to \infty,\ Z_f(t)/R_f(t) \to \infty \quad (15b)$$

The properties in Eqs. (15a) and (15b) show that the wetted bulb shape changes the current infiltration: from an hemispherical to an elliptic shape. Also Eq. (15a) demonstrates that the shorter the infiltration time the more accurate are results of a hemispherical model.

To test the relevance of the proposed method in this chapter, laboratory experiments and numerical simulation of an axisymmetric water infiltration were performed for a silty soil. The recorded data for $Z_f(t)$ were compared with computed and predicted values, using the method in this chapter.

4.3 LABORATORY EXPERIMENTS

Infiltration experiments were carried out in the laboratory using a semicylindrical container: 1.20 m high and 1.50 m diameter. At its bottom, four drainage holes were made to prevent soil saturation. A glass plate, constituting the straight vertical face, enabled the observation of a wetting front advancement. The container was filled with a 0.10 m thin gravel and sand layer and then carefully filled upto 1.10 m height with silty soil (13% clay, 68% silt, 18% sand with a bulk density of 1.28). The soil surface was evened in order to favor an axisymmetric water distribution. Before each experiment, the container was left in the laboratory for 4 days in order to give enough time for soil ramming and pressure equilibrium. Water was delivered at the soil surface by a capillary tube connected to a constant level reservoir. The discharge flow rate of 1, 2 or 4 l/h was accurately fixed by adjusting the proper capillary length. However, the infiltration time (12 h) was kept the same in all cases. Hammami and Maalej [16] have described in detail about method and materials that were used in this section. The following parameters were measured:

1. The wetting front radius Rf(t) was measured, from the point source, on the soil surface.
2. The "maximum" wetting front depth was measured (through the transparent side), from the surface, along the symmetry axis.

4.4 NUMERICAL SOLUTION

Equation (1) was solved numerically using a finite difference approximation method for space operators in two dimensions and an altering–directions implicit method (ADI) for the time integration (code LOC.B1). These methods were used for their compatibility in this chapter [1, 3, 5] and their unconditional convergence and accuracy [18, 35, 36, 37].

4.4.1 SOIL CHARACTERISTICS

Volumetric soil water content (, determined gravimetrically) and the corresponding pressure head (h) values were sampled simultaneously. The current moisture redistribution was observed and was then adjusted to the model described Van Genuchten [34]:

$$\theta(h) = \theta_r + (\theta_s - {}_r)/\{1 + (\alpha h)^n\}^m \tag{16}$$

The relative hydraulic conductivity-pressure head relationship was according to the model by Mualem [26]:

$$K(h)/K_s = \{(\theta - \theta_r)/(\theta_{s-}\theta_r)\}^{1/2}\{1 - [1 - \{(\theta - \theta_r)/({}_s - \theta_r)\}^{1/m}]^m\}^2 \tag{17}$$

where: θ_s = saturated soil water content (L^3/L^3); θ_r = residual water content (L^3/L^3); K_s = saturated soil hydraulic conductivity (LT^{-1}); and K(h) = soil hydraulic conductivity at a pressure head h (LT^{-1}); and h = pressure head (L). The exponents *m* and *n* are empirical shape factors defines as: $m = [1 - (1/n)]$. For the numerical computing, authors used experimental data from Hammami and Maalej [16] that was obtained for the same soil: $\theta_s = 0.58$, $\theta_r = 0.25$, $K_s = 5.8$ cm/h, n = 2, m = ½, $\alpha = 0.025$ cm^{-1} and *h* is soil water head (mb). Compared to soils of analogous texture these values appear slightly overestimated, may be because they were determined on disturbed soil [15].

4.4.2 CHARACTERIZATION OF WETTING FRONT

The wetted soil volume (or bulb) around the point source is defined as the wetted soil volume for $\theta(r, z) \geq \theta_f$. Thus the wetting front is a boundary where $\theta(r, z) = \theta_f$. Physically it constitutes the lateral surface of the bulb volume, where the pressure head gradient is maximal which supplies the energy requirement for water transfer [8, 20]. The wetting front coordinates satisfy the following conditions:

$$\partial h/\partial r = \text{maximum}$$

$$\partial h/\partial z = \text{maximum}$$

The flow domain is divided into a network of equally spaced grids ($\Delta r = \Delta z = 2$ cm), where each point is designed by the subscripts *i* and *j*, referring to respectively the horizontal and vertical directions. The soil surface and the symmetry axis are respectively inserted between the lines (*i*, *j* = 1) and (*i*, *j* = 2) and the column (*i* = 1, *j*) and (*i* = 2, *j*). Therefore, referring to Fig. 4.2, we obtain:

$$R(t) = (i - 3/2)\Delta r, \quad 1 \leq i \leq I$$

$$R_f(t) = (i_{f-}3/2)\Delta r, \quad 1 \leq i \leq I_f$$

$$R_s(t) = (i_{s-}3/2)\Delta r, \quad 1 \leq i \leq I_s$$

$$Z(t) = (j - 3/2)\Delta z, \ \ 1 \leq j \leq J$$

$$Z_f(t) = (j_f - 3/2)\Delta z, \ \ 1 \leq j \leq J_f$$

where: $\{R_s(t), Z_s(t)\}$ are coordinates of the last node being saturated; and $\{R_f(t), Z_f(t)\}$ are those of the last node being wetted. Because of the derivative space approximation, conditions in Eq. (16) can involve an overestimation of the wetting front coordinates (equal to Δr for $R_f(t)$ and Δz for $Z_f(t)$, respectively).

4.4.3 INITIAL AND BOUNDARY CONDITIONS

The initial condition is: For t = 0, $h(r, z, 0) = h_i$, where: h_i corresponds to the average of the pressure head profile sampled before infiltration (the Eqs. (16) and (17)). Other boundary conditions are:

$$t > 0, z > 0: r \rightarrow \infty, h(r, z, t) = h(r, z, t = 0) = h_i$$

$$t > 0, r > 0: z \rightarrow \infty, h(r, z, t) = h(r, z, t = 0) = h_i$$

$$t > 0, z > 0: r = 0, \partial h/\partial r = 0 \tag{18}$$

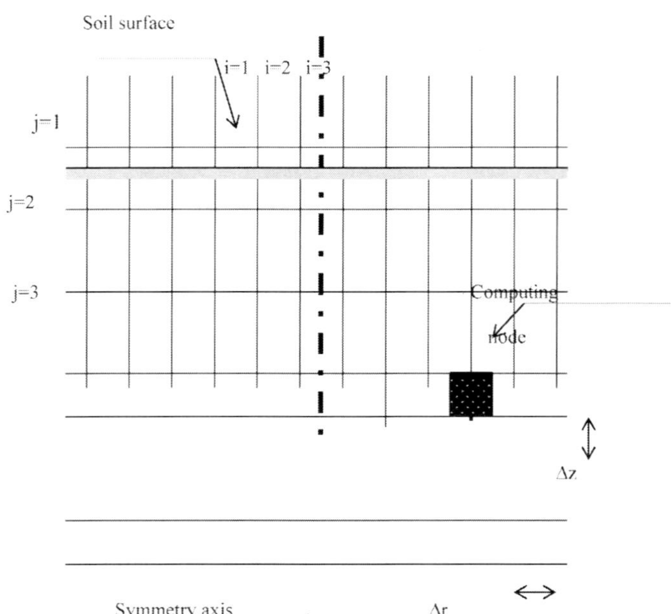

FIGURE 4.2 Schematic representation of the computing domain.

When water is supplied at a constant flow rate from a point source on the soil surface, a circular ponded zone appears. The radius $R_s(t)$ of this zone increases with time then reaches a limit R_{sMax}. Therefore, two conditions were imposed:

1. **$z = 0$, $r(t) \leq R_s(t)$:** The difficulty, with this condition, arises from the drop discontinuity, particularly at slow discharge rates. Consequently, an alternate saturated-unsaturated sequence may occur at the boundary of the ponded area. So we assumed that: $h(i, j = 1, t) < 0$. This condition may occur only at the limit of the last grid (on the soil surface) being saturated where the vertical flux q is given by:

$$q_{(i,\,j\,=\,1,\,t)} = Q_{residual} = \frac{Q_0 - E - \sum_{i=2}^{is-1} q(i).S(i)}{S(is)} \tag{19}$$

where: Q_0 = emitter discharge rate (L^3T^{-1}); q_i = vertical flow at node (i, j=1) on the soil surface (LT^{-1}); $S_{(i)}$ = surface of node (i, j=1) (L^2); $S_{(is)}$ = area corresponding to last node ($i=i_s$, j=1) being saturated (L^2); and E = evaporation rate (L^3T^{-1}). Because of the localized wetted area around the emitter, the total amount of water evaporated was 5.2% and 1.7% of the water supplied, respectively for q = 4 l/h and q = 1 l/h. Philip [27, 38, 29] and Revol [30] noted that for a current infiltration from a point source on the soil surface, evaporation is negligible compared to the total amount of water supplied. Therefore, Eq. (19) reduces to:

$$q_{(i,\,j\,=\,1,\,t)} = Q_{residual} = \frac{Q_0 - \sum_{i=2}^{is-1} q(i).S(i)}{S(is)} \tag{20}$$

After computing h(i,1, t) value, two cases were considered:

a. if $h(i,1, t) < h_s$, we start the next vertical (i+1) imposing the condition:

$$q_{(i+1,\,j\,=\,1,\,t)} = 0$$

b. if $h(i,1, t) \geq h_s$, we repeat computing the same vertical (i, j=1) but imposing Dirichlet condition: $h_{(i,1,t)} = h_s$ and then $i_s = i$

2. **$z = 0$, $r(t) > R_s(t)$:** Away from the ponded zone, the vertical infiltration from the surface is null, the evaporation was neglected, so the condition imposed was: $q_{(i,\,j\,=\,1,\,t)} = 0$. Wetting front coordinates $R_f(t)$ and $Z_f(t)$ have been numerically computed. Using conditions in the Eqs. (16) and (17), these are approximated below::

$$\{h(i+1, j = 1, n) - h(i, j = 1, n)\}/\Delta r = \text{maximum}$$

$$\{h(i = 1, j+1, n) - h(i = 1, j, n)\}/\Delta z = \text{maximum} \tag{21}$$

At any time step, $t = n.\Delta t$, we check:
if $\{h(i+1, j = 1, n) - h(i, j = 1, n)\}/\Delta r > \{h(i, j = 1, n) - h(i-1, j = 1, n)\}/\Delta r$,
then $i_f = i+1$, if not $i_f = i$
if $\{h(i = 1, j+1, n) - h(i = 1, j, n)\}/\Delta z > \{h(i = 1, j, n) - h(i = 1, j-1, n)\}/\Delta z$,

then $j_f = j+1$, if not $j_f = j$ (22)

Internodal hydraulic conductivities were calculated by arithmetic averages:

$$K(h_{(i\pm1/2),j}) = \{K(h_{i,j}) + K(h_{(i\pm1/2),j})\}/2$$

$$K(h_{i,(j\pm1/2)}) = \{K(h_{i,j}) + K(h_{i,(j\pm1/2)})\}/2 \quad (23)$$

4.5 RESULTS AND DISCUSSION

The experimental values of soil moisture $\theta(h)$ with the corresponding pressure head (h) are plotted, for three discharge rates of an emitter in Fig. 4.3. Parameters α, n and θ_r were determined using Van Genuchten [34] procedure, which often provides good agreement, because all experimental data were used for curve fitting. Wetting front dimensions $R_f(t)$ and $Z_f(t)$, as a function of time, are given in Fig. 3.4, which indicates experimental and computed values.

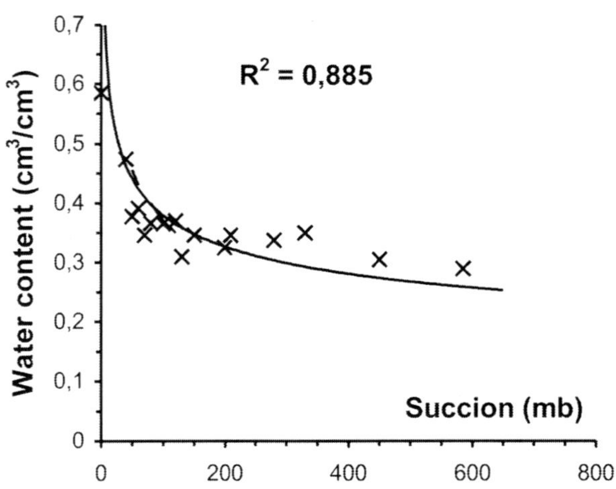

FIGURE 4.3 Soil water retention curve, showing the experimental data (crosses) and the fitted data.

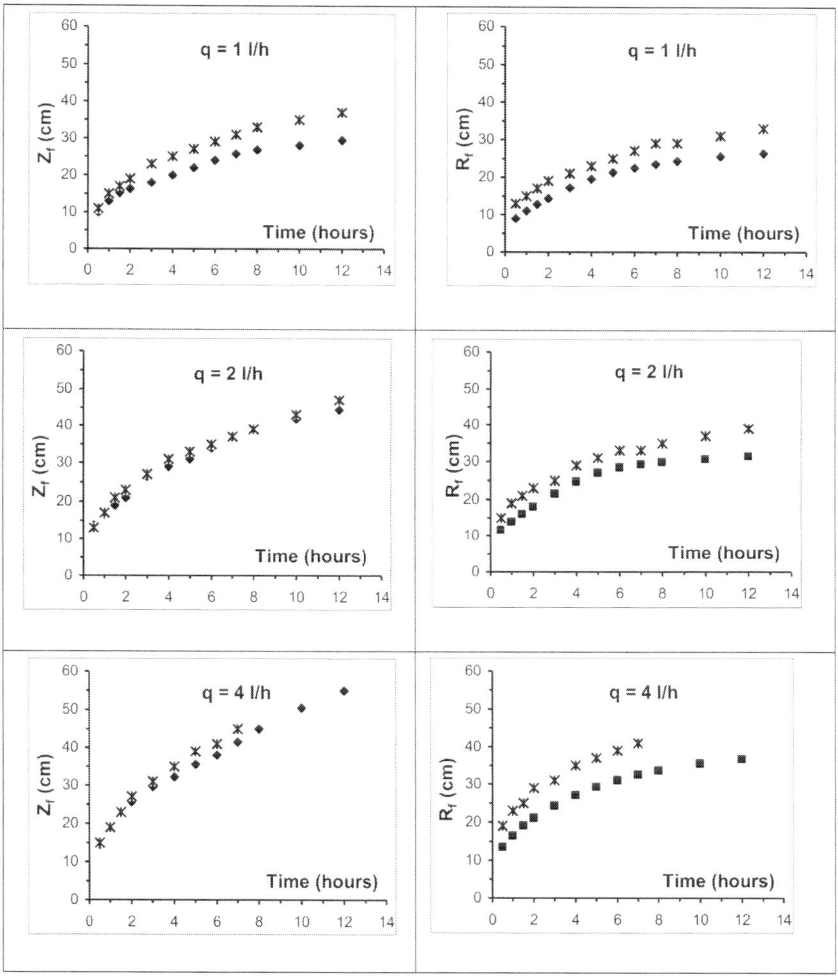

FIGURE 4.4 Relationships among $R_f(t)$ and $Z_f(t)$ and elapsed time (t), for three discharge rates of an emitter (q). Measured values are shown with solid squares and the computed values are shown by crosses.

In all cases, $R_f(t)$ and $Z_f(t)$ curves are similar to those reported by other researchers [1, 19]. The shape of the curve reveals that the wetting front advancement rates (horizontal rate $dR_f(t)/dt$ and vertical rate $dZ_f(t)/dt$) are decreasing monotonous functions of time. This behavior is explained by the fact that the same amount of water must wet by an increasing soil volume, which would result in a decrease in the wetting front advancement rates, under a constant flux source. Computed $R_f(t)$ values (Fig. 4.4) are always higher than experimental values with a relative error deviation, $E_R = (R_{fm} - R_{fn})/R_{fn}$, ranging between 8% and 30% for 12 h infiltration

(Fig. 4.5a). The reader may note the slightly E_R higher values mainly at the beginning and at the end of infiltration. In the beginning, these discrepancies can result from the finite difference instability for the first computing time steps. But at the end of infiltration, these discrepancies can result from lack of precision in computing the extension of a ponded area due to inaccurate definition of the soil surface boundary conditions. The complexity of boundary conditions is a major source error in modeling the water infiltration under an emitter [1, 5, 23].

The computed values of $Z_f(t)$ are also higher than the measured values but with a relative deviation $E_Z = (Z_{fm} - Z_{fn})/Z_{fn}$ much lower than E_R, ranging between +5% and –5% for q = 2 l/h and q = 4 l/h (Fig. 4.5b), but it is slightly higher for q = 1 l/h. The numerical model enables computing the wetting front depth better than the wetted area extension on the soil surface.

Data points corresponding to the couples ($R_f(t)$, $Z_f(t)$) are plotted in Fig. 4.6. These points fall into curves similar to those plotted by Keller [22]. The curves clearly show that $Z_f(t)$ is an increasing monotonous function of $R_f(t)$, with a coefficient of determination of $R^2 > 0.99$; and satisfies the boundary conditions in Eq. (13) for Eq. (12).

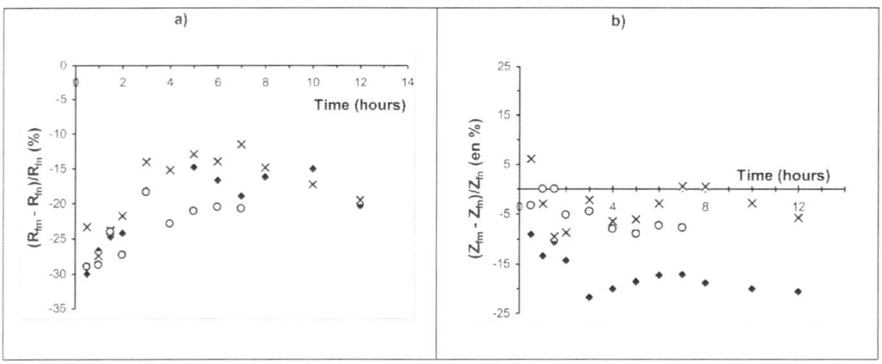

FIGURE 4.5 Relative deviations (a) E_R and (b) E_Z as a function of elapsed time, for three discharge rates: Solid squares, q = 1 lph; Crosses, q = 2 lph; and circles, q = 4 lph.

Using measured values of $R_f(t)$ and Eq. (12), the corresponding $Z_f(t)$ values were calculated. θ_i and θ_f values that were found with a soil–water retention curve (Fig. 4.3): $\theta_i = \theta(h_i) = 0.27$ and $\theta_f = \theta(h_f) = 0.40$: where, h_i corresponds to the average of the matric water head profile sampled in the infiltration at beginning ($\theta_i = 0.27$); and h_f corresponds to the inflection points ordinate of the matric water head profile sampled current redistribution (Eq. 16).

The $Z_f(t)$ predicted values versus $R_f(t)$ are plotted with computed and measured data in Fig. 4.6. In general, computed and predicted (using Eq. (12)) values are comparable, although the relative deviation, ($E = (Z_{fc} - Z_{fn})/Z_{fn}$, between computed

Z_{fn} and predicted Z_{fc} data), is relatively important, giving a values of 23% and 10% (Fig. 4.7a). These discrepancies can be explained by following three assumptions:

1. Lack of precision in estimating soil hydraulic characteristics and measurement errors. Haverkamp [17] reported that a relative difference of only 2% in surface water content caused a relative difference of 24% in the wetting front position;
2. The model of Haverkamp et al. [18] (Eq. (5) in their model) under-estimated the cumulative infiltration with relative error ranging from 9% to 21% for the silty soils [38];
3. An overestimation of the R_f computed values compared to measured ones shown in Figs. 4.5a and 4.5b, where the relative deviation E_R, between wetting front radius measured on the soil surface R_{fm} and computed R_{fn}, remains higher than the 10% for entire time range.

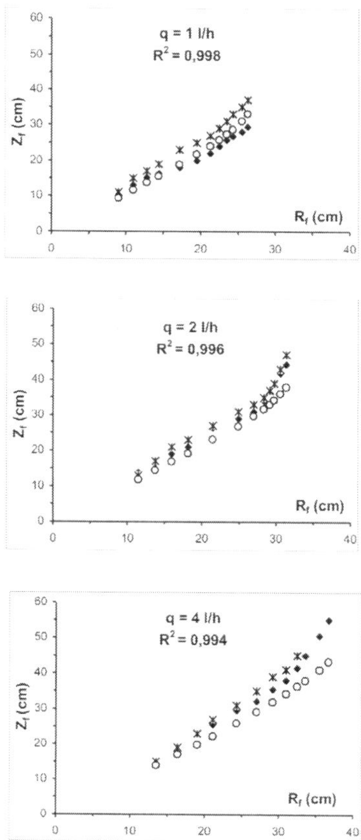

FIGURE 4.6 Relationships among $R_f(t)$ and $Z_f(t)$, for three discharge rates of an emitter (q): Solid squares, q = 1 lph (top); Crosses, q = 2 lph (center); and circles, q = 4 lph (bottom).

In order to take into consideration of above assumption #3, authors calculated once again $Z_f(t)$ values using, in Eq. (12), the R_{fn} computed data instead of measured data R_{fm}. The rectified Z_{fr} values are compared to the computed Z_{fn} data in Fig. 4.7b. The corresponding relative deviation, ($E' = (Z_{fr} - Z_{fn})/Z_{fn}$), falls within 2 h of infiltration, to less than 5%. The high values of E' in the beginning can be explained by the numerical overestimated data due to the finite difference method instability for the first computing time steps.

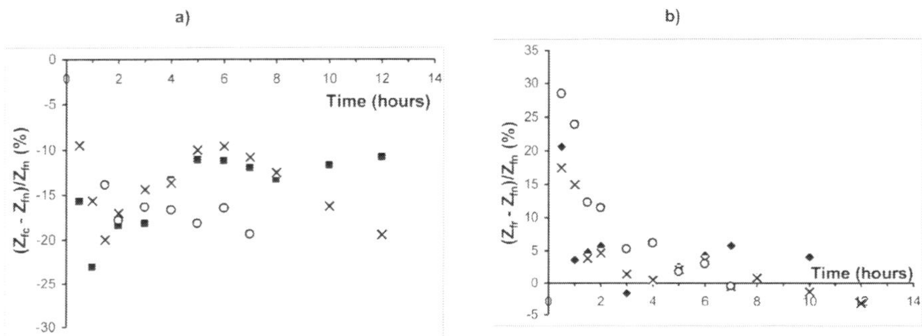

FIGURE 4.7 Comparison of rectified Zfr values with the computed Zfn data for three discharge rates of an emitter (q): Solid squares, q = 1 lph; Crosses, q = 2 lph; and circles, q = 4 lph.

4.6 SUMMARY

High amounts of water application always induce deep-water percolation and fertilizers seepage, under trickle-irrigation. Predicting the wetting front depth under an emitter can be a reliable method to prevent such a practice and can enable irrigation manager to reduce water and nutrients losses, Using experimental results, continuity and dynamic equations, a simple method was developed for determining the maximum wetting front depth $Z_f(t)$, during water infiltration from an emitter on the soil surface. Knowing soil hydraulic conductivity K_f, initial θ_i and wetting front θ_f water content, the method in this chapter enables to calculate the wetting front depth by only measuring its radius $R_f(t)$ on the soil surface. Thus, the proposed approach can help so that that the wetting depth remains just equal to that of the rooted soil.

Because of its simplicity, the method in this chapter can be practically useful to reduce deep percolation in trickle irrigation management. This study was performed during an axisymmetric water infiltration into bare, homogeneous and isotropic soil. Further research is needed to test the suitability of the approach for heterogeneous, anisotropic and cropped soils.

KEYWORDS

- **emitter**
- **hydraulic conductivity**
- **infiltration**
- **point source**
- **silty soil**
- **soil moisture**
- **soil water**
- **Trickle irrigation**
- **Tunisia**
- **unsaturated soil**
- **water distribution**
- **wetting front**
- **wetting front radius**

REFERENCES

1. Ababou, R. (1981). Modélization des transfers hydriques dans le sol en irrigation. Thèse Docteur-Ingénieur, Institut Polytechnique de Grenoble (217 p).
2. Akbar, Ali Khan, Muluneh Yitayew, Warrick A.W. (1996). Field evaluation of water and solute distribution from a point source. *Journal of Irrigation and Drainage Engineering,* 221–227 pages.
3. Balti, J. (1993). Etude numérique et expérimentale de l'irrigation par la méthode des jarres. Thèse de doctorat de spécialité en physique de la faculté des sciences de Tunis (115 p).
4. Ben-Asher, J., Charach, Ch., Zemel, A. (1986). Infiltration and water extraction from trickle irrigation source: The effective hemisphere model. Soil Science Soc. Am. J., 50, 882–887.
5. Brandt, A., Bresler, E., Diner, N., Ben-Asher, I., Heller, J., Goldberg D. (1971). Infiltration from a trickle Source: 1. Mathematical models. *Soil Science Soc. Am. Proc., 35,* 675–682
6. By-Fred, L. O., Bahram, S. (1997). Green and AMPT infiltration with redistribution. *Journal of Irrigation and Drainage Engineering, 123,* 386–393.
7. Cindy, S. K., James R. Hunt, (1996). Prediction of Wetting Front Movement During One-Dimensional Infiltration Into Soil. *Water Resources Res., 32,* 55–64.
8. Clement, T. P., Wise, W. R., Molz, F. J. (1994). A physically based, two-dimensional, finite-difference algorithm for modeling variably saturated flow. *Journal of Hydrology, 161,* 71–90.
9. Clothier, B. E., Scotter D. R. (1982). Constant flux infiltration from a hemispherical cavity. *Soil Science Soc. Am. J., 46,* 696–700
10. Clothier, B. E. (1984). Solute travel times during trickle irrigation. Water Resour. Res., 20, 1848–1852.
11. Darusman, A. H. K., Stone, L. R., F. R. Lamm, 1997. Water flux below the root zone vs. Drip-Line spacing in drip-irrigated corn. *Soil Science Soc. Am. J., 61,* 1755 –1760.

12. Eugenio, C. F., Dani, O. (1996). A parametric model for two-dimensional water uptake intensity by corn roots under drip irrigation. *Soil Science Soc. Am. J., 60,* 1039–1049.
13. Eugenio, C. F., Dani, O. (1997). Applicability of analytical solutions for flow from point sources to drip irrigation management. *Soil Science Soc. Am. J., 61,* 1331–1341.
14. Gardner, W. R., Hillel, D., Benyamini, Y. (1970). Post irrigation movement of soil water: I. Redistribution. *Water Resour. Res., 6(3),* 851–861.
15. Hammami, M., Maalej, M. (1999). Prediction of soil moisture depth under an emitter (*Prédiction de la profondeur de sol humidifiée sous goutteur*). *Rev. Science Eau., 12(2),* 273–284.
16. Hammam, I. M. (2001). Novel approach to determine the soil wetting volume under an emitter (*Nouvelle approche pour déterminer le volume de sol humidifié par un goutteur*). Thèse de Doctorat d'Etat en Sciences physiques, Université de Tunis (214 p).
17. Haverkamp, R., Vauclin, M., Touma, J., Wierenga, P. J., Vachaud, G. (1977). A comparison of numerical simulation models for one-dimensional infiltration. *Soil Science Soc. Am. J., 41,* 285–293.
18. Haverkamp, R., Ross, P. J., Smettem, K. R. J., Parlange, J. Y. (1994). Three-dimensional analysis of infiltration from the disc infiltrometer, II: Physical based infiltration equation. *Water Resour. Res., 30,* 2931–2935.
19. Healy, W., Warrick A. W. (1988). A generalized solution to infiltration from surface soil point source. *Soil Science Soc. Am. J., 52,* 1245–1251.
20. Hillel, D. (1988). *Principles Soil Physical Processes* (L'eau et le sol: principles et processes physiques). Translated by Loi W. De Backer. Louvain-la-neuve Académia (288 p).
21. Jury, W. A., Earl, K. D. (1977). Water movement in bare and cropped soil under isolated trickle emitter, I: Analysis of bare soil experiments. *Soil Science Soc. Am. J., 41,* 852–856.
22. Keller, J., Bliesner, R. D. (1990). Sprinkler and Trickle Irrigation. An AVI Book – Van Nostrand Reinhold. New York (640 p).
23. Lafolie, F., Guennelon, R., Van Genuchten, M. T. (1989). Analysis of water flow under trickle irrigation, I: Theory and numerical solution. *Soil Science Soc. Am. J., 53,* 1310–1318.
24. Levin, I., Van Rooyen, P. C., Van Rooyen, F. C. (1979). The effect of discharge rate and intermittent water application by point-source irrigation on the soil moisture distribution pattern. *Soil Science Soc. Am. J., 43,* 8–16.
25. Michelakis, N., Vougioucaloum, E., Clapaki, G. (1993). Water use, wetted soil volume, root distribution and yield of avocado under drip-irrigation. *Agricultural Water Management, 24,* 119–131.
26. Mualem, Y. (1976). A new model for predicting the hydraulic conductivity of unsaturated porous media. *Water Resour. Res., 12,* 513–522.
27. Philip, J. R. (1957). Theory of infiltration, IV: Sorptivity and algebraic infiltration equations. *Soil Sci., 84,* 257–264.
28. Philip, J. R. (1984). Travel times from buried and surface infiltration point sources. *Water Resour. Res., 20*, 990–994.
29. Philip, J. R. (1989). The scattering analog for infiltration in porous media. Review of Geophysics, 27, 431–448.
30. Revol, P. (1994). Hydrodynamic characterization of soil infiltration for two dimensions modeling for drip irrigation (*Caractérization hydrodynamique des sols par infiltration bidimensionnelle axisymétrique et modélization simplifiée de la microirrigation*). Thèse université de Grenoble (221 p).
31. Revol, P., Clothier, B. E., Mailhol, J. C., Vachaud, G., Vauclin, M. (1997). Infiltration from a surface point and drip irrigation, II: An approximate time-dependent solution for wet-front position. *Water Resour. Res., 33*, 1869–1874.

32. Smettem, K. R. J., Ross, P. J., Haverkamp, R., Parlange, J. Y. (1995). Three-dimensional analysis of infiltration from the disk infiltrometer, III: Parameter estimation using a double-disk tension infiltrometer. *Water Resour. Res., 31*, 2491–2495.
33. Schwatzmass, M., Zur, B. (1985). Emitter spacing and geometry of wetted soil volume. *Journal of Irrigation and Drainage Engineering, 112,* 242–253.
34. Van Genuchten, M. T. (1980). A closed-form equation for predicting the hydraulic conductivity of unsaturated soils. *Soil Science Soc. Am. J., 44,* 892–897.
35. Vauclin, M., Haverkamp, R., Vachaud, G. (1979). Numerical analysis of nonlinear diffusion equation: Application for infiltration in nonsaturated soils (*Résolution numérique d'une équation de diffusion non linéaire: application à l'infiltration de l'eau dans le sol non saturé*). Presses Universitaires de Grenoble (183 p).
36. Vauclin, M. (1988). Flow of water and air in soils: Theoretical and experimental aspects. Seytoux NATO ASI Series, 53–87
37. Vauclin, M. (1989). Flow of water and air in soils: Theoretical and experimental aspects. In: H.J. Morel Seytoux (ed.), *Unsaturated Flow in Hydrologic Modeling: Theory and Practice.* Kluwer Academic Publishers. 53–91.
38. Zhang, R. (1997). Infiltration Models for the Disk Infiltrometer. *Soil Science Soc. Am. J., 61,* 1597–1603.

CHAPTER 5

WETTED ZONE BEHAVIOR UNDER MICRO IRRIGATED CROPS

HAMMAMI MONCEF, DAGHARI HÉDI, and HATIRA ABDESSATAR

CONTENTS

Modified and printed from Hammami Moncef, Daghari Hédi and Hatira Abdessatar, 2011. An empirical concise approach to predict the maximal wetted soil depth under trickle irrigated crops. *American-Eurasian J. Agric. Environ. Sci., 11(3),* 334–340, idosi@idosi.org. Open Source Access. Authors are thankful to the European Union for the financial support through project: "*QUALIWATER: Diagnosis and control of salinity and nitrate pollution in Mediterranean irrigated agriculture (Project N°: INCO-CT-2005-015031).*"

5.1 INTRODUCTION

Trickle irrigation has widely extended throughout the world. Indeed, the irrigated area under trickle irrigation has increased by 330% during the 1990s. During 2000, more than 3 million hectares were expected to be trickle irrigated worldwide [4]. In most arid countries where water resource is limiting factor, using trickle irrigation to sustain irrigated agriculture is a must. In fact, this system enables to increase the crop yield and to reduce water losses up to 50% as compared to furrow or basin irrigation [3, 22].

The principal mission of the trickle irrigation is to supply water directly in the rhizosphere and then to keep the rooted soil volume within prescribed humidity thresholds. Consequently:

- The wetted area on the soil surface is to be reduced. Thus, water losses by evaporation from surface are significantly reduced.
- The wetted soil volume of onion shape is limited to beneath emitters. Thus, deep percolation and nutrient losses are substantially reduced.

To achieve maximum profits from these opportunities, the water distribution network and trickle irrigation management must be designed so that the wetted soil volume is matched with the rooting zone. To achieve this objective, the shape and the dimensions of the wetted soil volume behavior is to be known [4, 5].

Several analytical and numerical models for predicting water infiltration into the soil have been proposed. Because of the computational simplicity, the general insights and the direct link among the inputs and outputs, the analytical solutions are useful tools for design of trickle irrigation network and management. But most of these solutions remain valid only for steady state flow, in homogeneous and uniform soil conditions [20, 24].

Many numerical models have been proposed to simulate soil water redistribution pattern beneath point and/or linear surface sources [1, 2, 17]. Although these models are powerful in solving complexity of nonlinear soil problems, yet they are less practical because of their complexity and the saturated zone's extension on the surface remain difficult to be accurately reproduced. Moreover, only few of these models allow for water uptake by plant. In 1974, Keller and Karmelli [13] presented a table linking the soil texture (coarse, medium or fine), the emitter spacing and the emitter discharge rate to the wetted soil fraction (P) induced by 40 mm water depth. Empirical expressions have been adjusted [6, 12, 23] to allow reproducing bulb's extension. Hammami et al. [11] proposed a compact physical based approach for predicting the wetted soil depth $Z_f(t)$, beneath an emitter on the soil surface. Comparison with measured and theoretical results revealed that this approach is more reliable [11]. Because of the simplicity and feeless, some of these models remain useful.

This chapter proposes a new empirical approach that enables to predict the maximum wetted soil depth $Z_f(t)$ under trickle irrigated sweet melon and tomato.

5.2 MATERIAL AND METHODS

5.2.1 CLIMATIC DATA AND FIELD TRIAL SITE

The trials were carried out at two private plots in Kalaât Landalous district located in the north-eastern region (latitude: $37°02' \leq \alpha \leq 37°06'$ N; longitude: $10°05' \leq \varphi \leq 10°10'$ E and $0 \leq$ AMSL ≤ 5 m) of Tunisia. It is one of the widest (2905 ha) irrigated land in the country.

Environmental conditions are favorable for trickle irrigation management (shortage of water resources, the fertile soil depth did not exceed 1 m in the major parts of district, orchards and vegetables are the most irrigated summer crops). More than 85% of the average annual rainfall (497 mm) occurs between October and April (Table 5.1).

Because of the acute imbalance between annual precipitation and potential evapotranspiration (1344 mm), irrigating crops in summer is a must. The average temperature ranges from 11°C (January) to 27°C (August) (Table 1). The soil texture is a loamy-clay loam. Tables 5.2 and 5.3 show a quite uniform textured soil profile with relatively high bulk densities D_b for plots with sweet melon and tomato. Medjerda River is the main water source with a salinity ranging between 1 g/L (in winter) and 2.5 g/L (in summer).

5.2.2 MEASUREMENTS

The soil samples were taken by an auger–hole method at three random locations in each plot to determine the physical characteristics such as: particles size partition, bulk density [8], saturated soil water content and hydraulic conductivity [16].

The data were taken on two private trickle irrigated fields. In the first plot, the tomato seedlings were transplanted on 24th March of 2009. In the second plot, the melon seedlings were transplanted on 4th April of 2009. For both plots, each crop row was irrigated by a single lateral equipped with in-line emitters at 30 cm apart.

Emitter discharges (Q) were monitored using valves that were installed on the laterals upstream. Identical experimental devices were used. However, in the tomato plot, trials were performed with two different emitter discharge rates and three initial water suction (H_i, mb) values. Contrary in the melon plot, irrigation measurements were taken with three different emitter discharge rates, but the average initial water suction (H_i, mb) was similar. Soil water suction in each plot was measured with the sensors (Fig. 5.2) that were installed as shown in Fig. 5.1.

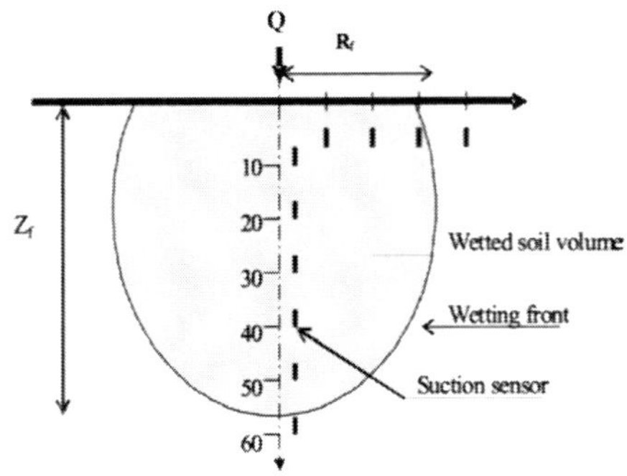

FIGURE 5.1 Schematic description of the wetted cross-section and the location of sensors to measure soil water suction at the experimental site (Fig. 5.2). R_f = radius of the wetted surface, Z_f = depth of the wetted bulb and Q = discharge rate from the emitter.

FIGURE 5.2 Measurement of wetted front advance: visual and tensiometric (sensors).

TABLE 5.1 Climatic Data: Average Monthly Temperature T (°C), Rainfall P (mm) and Potential Evapotranspiration ETP (mm) Values

	Values, mm												
	Sep	**Oct**	**Nov**	**Dec**	**Jan**	**Feb**	**Mar**	**Apr**	**May**	**Jun**	**Jul**	**Aug**	**Total**
T	23.8	20.0	15.7	12.3	11.0	11.2	13.0	14.7	19.5	23.0	26.0	27.0	—
P	46.7	35.6	67.7	86.5	75.0	63.7	38.0	40.4	24.1	13.4	2.8	3.4	497
ETP	133	115	84	69	71	80	98	106	126	136	161	165	1344

TABLE 5.2 Soil Characteristics in the Tomato Plot: Soil Texture, Bulk Density (D_b), Saturated Soil Moisture (θ_s) and Saturated Hydraulic Conductivity (K_s)

Soil depth	Particle size distribution			Soil properties			
	Sand	Loam	Clay	Texture class	D_b	θ_s	K_s
cm	%	%	%	—	$g.m^{-3}$	%	$cm.h^{-1}$
0–20	38.5	40.5	21.0	Loam	1.47	0.45	2.40
20–40	48.0	34.0	18.0	Loam	1.50	0.44	1.65
40–60	32.5	46.5	21.0	Loam	1.48	0.46	1.20

Each observation is an average of three soil samples.

TABLE 5.3 Soil characteristics in the Melon Plot: Soil Texture, Bulk Density (D_b), Saturated Soil Moisture (θ_s) and Saturated Hydraulic Conductivity (K_s)

Soil depth	Particle size distribution			Soil properties			
	Sand	Loam	Clay	Texture class	D_b	θ_s	K_s
cm	%	%	%	—	$gm.cm^{-3}$	%	$cm.h^{-1}$
0–20	22.0	43.5	33.5	Clay loam	1.48	0.46	2.10
20–40	20.0	40.0	38.0	Clay loam	1.51	0.45	1.50
40–60	20.0	41.0	39.0	Clay loam	1.50	0.43	1.52

Each observation is an average of three soil samples.

Each value of Q is the average of four observations for two adjacent emitters at the beginning and the end of each irrigation event. However, H_i value corresponds to the average of suction readings made just before irrigation on five sensors placed at 10, 20, 30, 40 and 50 cm depth (Fig. 5.1). Supplied water depths (Ds) were calculated as follows:

$$D_s = [Z_r(\theta_c - \theta_i)] \quad (1)$$

where: D_s = supplied water depth (mm); Z_r = rooted soil depth (mm); θ_i and θ_c are initial and at field capacity soil moisture (determined using soil water suction sensors). Irrigations were initiated as soon as soil water suction reached the previously fixed H_i value (= 200, 400 and 600 mb). The following variables were recorded:

- The average width of the wetted area, R_f (t) (cm), was measured visually on the soil surface at elapsed times [19, 20, 21]. Each R_f(t) value is an average of three observations on three consecutive emitters at each trial site (Fig. 2).

- The maximum depth of the wetted bulb, $Z_f(t)$, at 5 cm parallel to the symmetrical axis, determined using the soil water suction sensors: The wetting front depth was recorded once a water suction reduction was observed on the tensiometer placed at the same point [22].

5.3 RESULTS AND DISCUSSION

5.3.1 CLIMATE AND SOIL CHARACTERISTICS

The long period (1970–2010) climatic data in Table 5.1 reflect an acute imbalance between precipitation and potential evapotranspiration especially for summer crops (vegetables and orchards).

The Tables 5.2 and 5.3 indicate a homogeneous loamy textured soil in the tomato plot and homogeneous clay loam textured soil in the melon plot. The relatively higher saturated hydraulic conductivity of the topsoil layer results from the frequent soil cropping activities.

5.3.2 HORIZONTAL AND VERTICAL WETTING FRONT ADVANCES

Recorded $R_f(t)$ and $Z_f(t)$ values for the elapsed time are plotted in Figs. 5.3 and 5.4. These curves are similar to those reported by several researchers [1, 6, 11]. In fact, the higher emitter discharge rates result in faster horizontal wetting front advance. The effect of such flow rates is not so clear on the vertical wetting front velocity (advance). On the other hand, it seems that the drier initial soil moisture conditions result in slower wetting front advance. This behavior is due to the fact, that under constant flux source with initial drier soil profile, the same amount of water should wet an increasing volume of soil pores, which would result in a decrease in the wetting front advance rate.

Experimental $Z_f(t)$ values as function of the corresponding $R_f(t)$ data are plotted in Figs. 5.5 and 5.6. In all cases, $Z_f(t)$ is strongly correlated with $R_f(t)$ ($r > 0.92$). The corresponding ($R_f(t)$, $Z_f(t)$) data observations are scattered on an exponential shaped curves identical to those reported by Keller and Bliesner [14] and by Hammami et al. [11]. This exponential form is as follows:

$$Z_f = a.[exp\ (b.R_f)] \tag{2}$$

where: Z_f is the maximum wetted soil depth (cm); R_f refers to the width of wetted strip (cm) measured on soil surface and a and b are exponential regression coefficients. The values of these nonlinear regression coefficients were determined using nonlinear regression analysis. As a rule of thumb, the following boundary conditions must be satisfied:

$$R_f \rightarrow 0, Z_f \rightarrow 0 \quad (3a)$$

$$R_f \rightarrow R_{Max}, Z_f \rightarrow Z_{Max} \quad (3b)$$

where: at the end of irrigation, Z_{Max} is the maximum wetting front depth; and R_{Max} is the maximum width of the wetted area on the soil surface. Then, substituting the regression constants, a and b, in Eq. (2) and rearranging yields:

$$Zf = (ZMax).exp^{[(\mathrm{Rf}- \mathrm{RMax})\div Rf]} \quad (4)$$

It is clear that Eq. (4) satisfies the physical boundary conditions (Eqs. 3a and 3b). The fitting parameters (Z_{Max} *and* R_{Max}) must be adjusted for in-situ cropping conditions. The Z_{Max} value is previously fixed equal to the maximum rooted depth and R_{Max} is fixed equal to the shaded width or the canopy lateral spread. However, these parameters are strongly dependent on soil properties and irrigation conditions. In fact, in the same textured soil and initial water content, increased R_{Max} value results with higher emitters' flow rates. However, with the same emitter flow irrigation times, lower R_{Max} and higher Z_{Max} values appear in the coarse textured soil.

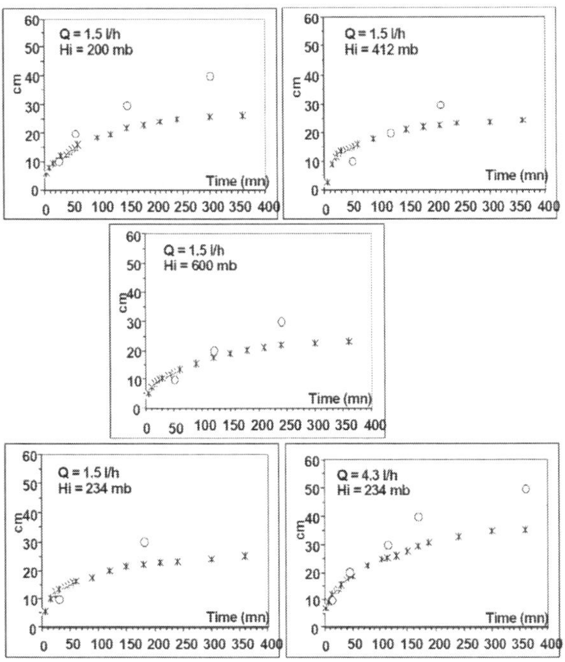

FIGURE 5.3 Tomato plot: vertical (data shown by circles) and horizontal (data shown by crosses) wetting front advances versus elapsed time for two emitters' discharge rates (Q) and varying initial water suctions (H_i).

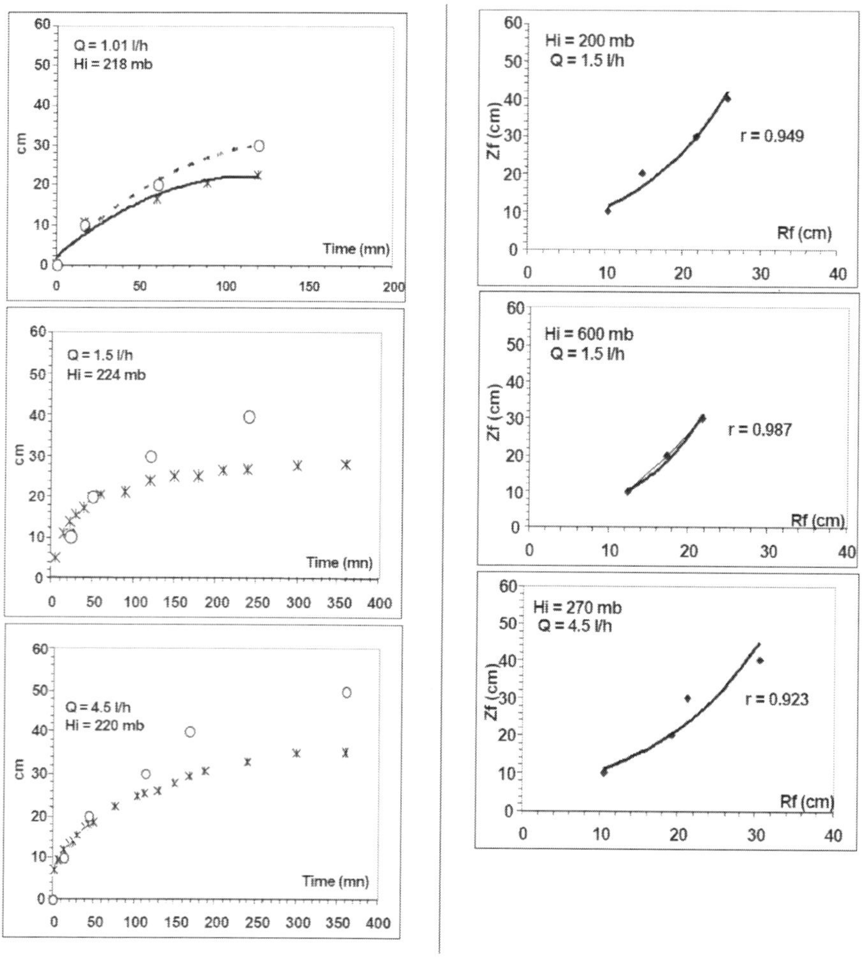

FIGURE 5.4 Sweet melon plot: vertical (data shown by circles) and horizontal (data shown by crosses) wetting front advances for three emitters' discharge rates (Q) and initial water suctions (H_i).

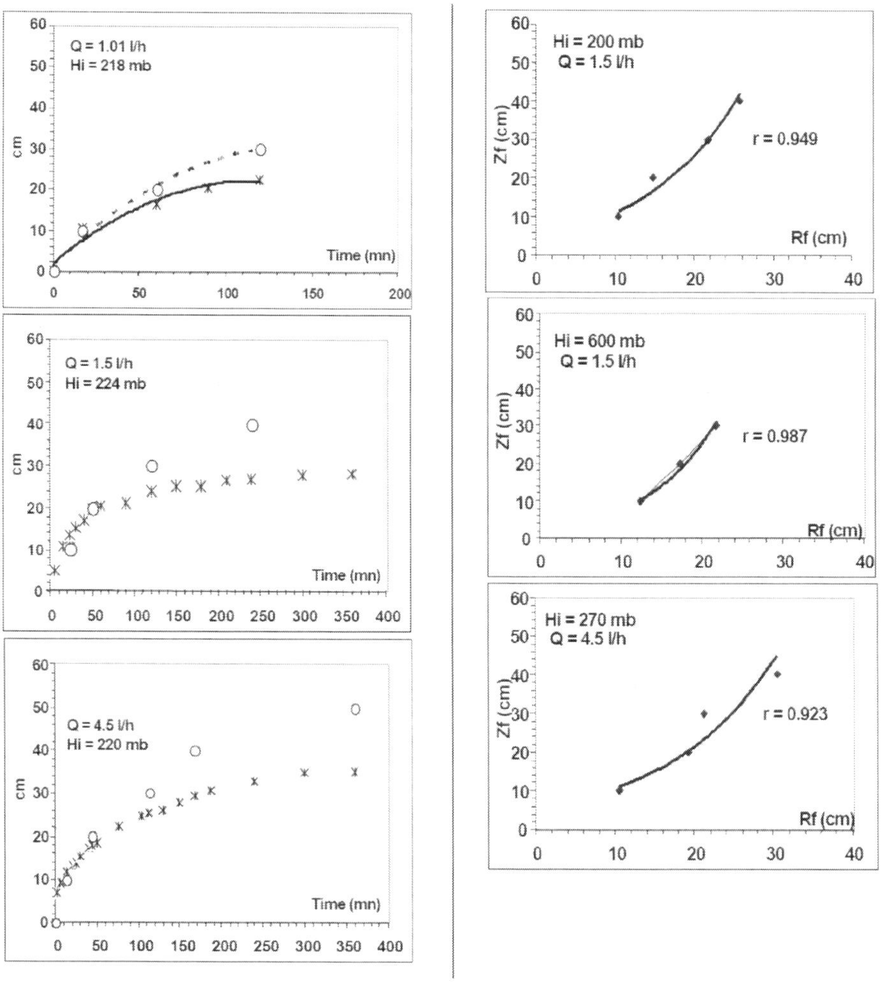

FIGURE 5.5 Tomato plot: Z_f as a function of R_f for two emitters' discharge rates (Q) and three initial water suctions (H_i).

Then using the same R_{Max} value in Eq. (4), the resulted wetting front (Z_f) will be deeper in coarse textured soils than in fine textured soils. These results agree those reported by several investigators [1, 5, 6, 11, 17]. The parameters in Eq. (4) are based on the experimental data for the two-cropped plots, distinguished emitter discharge rates and different initial water contents. It satisfies the physical boundary conditions and is in agreement with the published results on this topic.

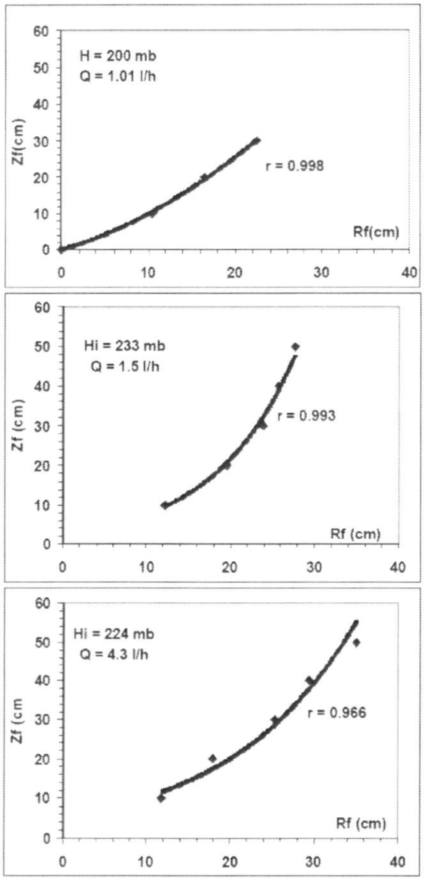

FIGURE 5.6 Melon plot: Z_f as a function of R_f for three emitter discharge rates (Q) and varying initial water suctions (H_i).

Then Eq. (4) can be a practical helpful tool to predict the wetting front depth under trickle irrigated crops, although it is valid only for the infiltration phase.

5.4 CONCLUSIONS

Using horizontal and vertical wetting front advance data, an empirical equation for predicting the maximum wetted soil depth was obtained for trickle-irrigated crops. The proposed equation was established using data recorded on two cropped plots, with different emitters' discharge rates and distinguished initial soil water contents. Based on the measurements of surface wetted area width, the proposed equation enables to compute the corresponding depth of wetting front. The fitting parameters (Z_{Max} and R_{Max}) values are strongly dependent on the local soil and irrigation management conditions. Thus these must be always adjusted for in-situ conditions. Because of its simplicity, this approach can be helpful tool for deep percolation and fertilizer-leaching control in trickle irrigated crops. But further trials are needed to test the relevance of the proposed approach for wide range of trickle-irrigated crops though it remains valid for only the watering phase.

5.5 SUMMARY

An easy, empirical and reliable new approach for predicting the wetted soil depth for the trickle-irrigated crops is proposed. The approach was adjusted using field measurements of the maximum wetting front depth $Z_f(t)$ and lateral spread $R_f(t)$ in both tomato and sweet melon plots. Within each plot, measurements were made for different initial water contents and three emitter discharge rates. For all cases, results showed that $Z_f(t)$ is strongly correlated ($r > 0.92$) with $R_f(t)$. An empirical exponential relationship was inferred. Knowing the lateral wetting front spread (in-situ conditions), the proposed approach enables to predict the correspondent maximum wetted soil depth. The only two empirical parameters were easily fitted to the in-situ measurements. Because of its simplicity, the proposed approach is a practical tool for trickle irrigation management, deep-water percolation and fertilizers leaching control.

KEYWORDS

- **deep water percolation**
- **emitter**
- **fertigation**
- **irrigation**
- **leaching**
- **Maximum wetted depth**
- **melon**
- **point source**
- **tomato**
- **trickle irrigation**
- **Tunisia**
- **wetted strip width**
- **wetting front**

REFRENCES

1. Ababou, R. (1981). Modeling of hydraulic conductivity of irrigated soil (Modélization des transfers hydriques dans le sol en irrogation). PhD Dissertation in Engineering. Polytechnic Institute of Grenoble.
2. Brandt, A., Bresler, E., Diner, N., Ben-Asher, I., Heller, J., Godelber, D. (1971). Infiltration from a trickle source: I – Mathematical models. *Soil Science Soc. Am. Proc., 35*, 675–682.
3. Clyder, R., Timothy, K., Carlos, N. (1989). Comparison of subsurface Trickle and furrow irrigation on plastic-mulched and bare soil for tomato production. *J. Am. Hor. Sci., 114,* 40–43.
4. Coelho, F. E., Or, D. (1997). Applicability of analytical solutions for flow from point sources to drip irrigation management. *Soil Science Soc. Am. J., 61*, 1331–1341.
5. Elmaloglou, S, Diamantopoulos, E. (2007). Wetting front advance patterns and losses by deep percolation under the root zone as influenced by pulsed drip irrigation. *Agric. Water Management, 90,* 160–163.
6. Elmaloglou, S., Malamos. N. (2006). A methodology for determining the surface and vertical components of the wetting front under a surface point source, with root water uptake and evaporation. *Irrig. and Drain, 55,* 99–111.
7. Fernandez-Galvez, J., Simmonds, L. P. (2006). Monitoring and modeling the three-dimensional flow of water under drip irrigation. *Agric. Water Management, 83,* 197–208.
8. Gee, G. W, Bauder, J. W. (1986). Particle-size analysis. In: *Methods of Soil Analysis – Part 1.* Agronomy Monograph 9. American Society of Agronomy, Madison, WI, 383–409.
9. Hammami, M. (2001). New approach to determine the soil wetting volume under an emitter (*Nouvelle approche pour déterminer le volume de sol humidifié par un goutteur*). PhD Dissertation for the Faculty of Physical Sciences, Tunisia.
10. Hammami, M, Maalej, M. (1999). Prediction of soil wetting depth under an emitter (*Prédiction de la profondeur de sol humidifiée sous goutteur*). *Rev. Science Eau., 12(2),* 273–284.
11. Hammami, M., Daghari, H. Balti, J. Maalej, M. (2002). Approach for predicting the wetting front depth beneath a surface point source: Theory and numerical aspect. *Irrig. Drain, 51,* 347–360.
12. Healy, W, Warrick, A. W. (1988). A generalized solution to infiltration from surface soil point source. *Soil Science Soc. Am. J., 52,* 1245–1251.
13. Jiusheng, L., Jiangjun, Z., Minjie, R. (2004). Wetting patterns and nitrogen distributions as affected by fertigation strategies from a surface point source. *Agric. Water Management, 67,* 89–104.
14. Keller, J., Bliesner Ron, D. (1990). Sprinkler and Trickle irrigation. An AVI Book Van Nostrand Reinhold, New York.
15. Keller, J., Karmelli, D. (1974). Trickle irrigation design parameters. *Transactions of the ASAE, 17(4),* 678–684.
16. Klute, A., Dirksen, C. (1986). Hydraulic conductivity and diffusivity – laboratory methods, In: *Methods of soil analysis, Part 1.* Agronomy Monograph 9. American Society of Agronomy, Madison, WI.
17. Lafolie, F., Guennelon, R., Van Genuchten, Th. M. (1989). Analysis of water flow under trickle irrigation: I. Theory and numerical solution. *Soil Science Soc. Am. J., 53,* 1310–1318.
18. Lubana Singh, P. P., Narda, N. K. (1998). Soil water dynamics model for trickle irrigated tomatoes. *Agric. Water Management, 37,* 145–161.
19. Michelakis, M., Vougioucalou, E., Clapaki, G. (1993). Water use, wetted soil volume, root distribution and yield of avocado under drip-irrigation. *Agric. Water Management, 24,* 119–131.
20. Philip, J. R. (1985). Steady absorption from spheroid cavities. *Soil Science Soc. Am. J., 49,* 828–830.

21. Raats, P. A. C. (1971). Steady infiltration from point sources cavities and basins. *Soil Science Soc. Am. J., 35*, 689–694.
22. Satpute, G., Bendales, S., Kausal, A. (1992). Water requirement of tomato crop under drip and furrow irrigation. *KPV Res. J., 16,* 83–87.
23. Schwar Tzmass, M, B. Zur (1985). Emitter spacing and geometry of wetted soil volume. *Journal of Irrigation and Drainage Engineering (ASCE), 112,* 242–253.
24. Warrick, A. W. (1974). Time-dependent linearized size infiltration. I. Point sources. *Soil Science Soc. Am. J., 38,* 384–386.

CHAPTER 6

DESIGN OF BURIED MICRO IRRIGATION LATERALS BASED ON SOIL WATER RETENTION

HAMMAMI MONCEF, ZAYANI KHEMAIES, and HÉDI BEN ALI

CONTENTS

Printed from Hammami Moncef, Zayani Khemaies and Hédi Ben Ali. *Required lateral inlet pressure head for automated SDI management*. International Journal of Agronomy Volume 2013, 6 pages. Open Access article, http://www.hindawi.com.

6.1 INTRODUCTION

In subsurface drip irrigation (SDI), water emits from the buried drippers into the soil and spreads out in the rhizosphere due to capillary and gravity forces [20, 25]. Thus, SDI system permits direct application of water to the wetted soil volume and maintaining dry the non-rooted topsoil. This pattern has advantages such as minimizing soil evaporation, deep percolation, weeds growth and thus affects evapoconcentration phenomenon. The SDI improves the water application uniformity, increases the laterals and emitters longevity, reduces the occurrence of soil-borne diseases and infestation of weeds. Several field trials have revealed relevant profits due to adequate management of SDI for crop production. Nevertheless, the appropriate depth of buried laterals remains debatable [10, 14, 21, 28]. Comparing evaporation from surface and subsurface drip irrigation systems, Evett et al. [7] reported a saving of 51 mm and 81 mm irrigation depth with drip laterals buried at 15 cm and 30 cm, respectively. Neelam and Rajput [20] recorded maximum onion yield (25.7 t per ha) with drip laterals buried at 10 cm depth. They reported maximum drainage with drip laterals at 30 cm depth. Several investigators have analyzed the effects of soil properties on the discharge of SDI emitters and water distribution uniformity [1, 17, 23]. The analytical method by Sinobas et al. [25] predicted reasonably well the soil water suction and the pressure head distribution in the laterals and SDI units [26]. The water oozes out from the buried emitters due to inlet lateral pressure head and the soil water suction. Therefore, the emitter discharge is high at the beginning of irrigation due to dry root zone. Gradually, as the soil pore space in the vicinity of the dripper outlet is filled with water, a positive pressure head develops, which may cause a decrease in dripper discharge [24]. If the discharge is greater than the soil infiltration capacity, the resulting overpressure near the nozzle tends to reduce the flow rate [17, 30].

6.2 BASICS OF ANALYTICAL METHOD

The pressurized irrigation systems are customarily designed so that the mean pressure head throughout the pipe is equal to the nominal pressure head. On the other hand, irrigation management is based on the replenishment of the soil holding capacity. Hence, the soil moisture should range between predetermined and minimum allowable soil moisture. It is assumed that the average pressure head is equal to the emitter operating pressure head. The emitter discharge equation is defined below [15]:

$$q = KH^{x} \tag{1}$$

where: q [L^3T^{-1}] and H [L] are emitter discharge and the emitter pressure head; K [$L^{3-x}T^{-1}$] and x are nonlinear regression coefficients. Equation (1) is valid for a pressure head ≥ 5.0 m. It is worth pointing out that most long-path turbulent flow and pressure-compensating emitters require an operating pressure head fulfilling this condition. For buried emitters, the emitter pressure head is lumped with the water suction near the outlets, as shown below:

$$H = h_e - h_i \tag{2}$$

where: h_e and h_i refer to the pressure heads [L] at the inner and outer of the emitter, respectively. For emitters in surface drip irrigation, h_i is the atmospheric pressure. Conversely, for buried emitters, h_i is a spatial-temporal variable dependent on the prevailing soil water content. Hence, we will consider the sigmoid retention curve of Van Genuchten [27] given below:

$$\theta = \{\theta_r + (\theta_s - \theta_r)[1 + (\alpha h)^n]^m\} \tag{3}$$

where: θ [L^3L^{-3}] and h [L] refer to the volumetric water content and to the soil suction head, respectively. The residual water contents are denoted as θ_r, α [L^{-1}]. The constants n and m are nonlinear regression coefficients that are found by fitting the curve to the scattered data (θ, h) according to Eq. (3); and θ_s refers to the saturated soil water content. The dimensionless parameters n and m are expressed by the Mualem [19] as shown below:

$$m = [1 - (1/n)] \tag{4}$$

The soil capillary capacity C [L^{-1}] is derived straightforwardly by differentiating Eq. (3) with respect to the suction head h as follows:

$$C = d\theta/dh = -\{mn\alpha\,(\theta_s - \theta_r)\,(\alpha h^{\,n-1})[1 + (\alpha h)^n]^{m+1}\} \tag{5}$$

Equation (5) shows that additional increase in the suction head produces an additional release of water from the soil. Besides, the value of C is the highest if the second derivative of the soil moisture content with respect to the suction head is zero. Under these conditions, the crops absorb the maximum water from the root zone for the same additional energy increment. Further analysis indicates that the coordinates of the inflection point of the retention curve as well as the maximum capillary capacity are as follows:

$$h_{op} = -m^{\,1/n}/\alpha$$

$$\theta_{op} = \theta_r + (\theta_s - \theta_r)/(1 + m)^m$$

$$C_{max} = nm^{\,m+1}\alpha(\theta s - \theta r)/(1 + m)^{m+1} \tag{6}$$

where: h_{op}, θ_{op} and C_{max} refer to the optimal water suction, optimal soil water content and maximum capillary capacity, respectively. Therefore, the design of SDI systems should ascertain a suction head at the emitter outlet that matches the optimal water status within the root zone. Combining Eqs. (1) and (2) yields the following:

$$q = K(h_e - h_{op})^x \tag{7}$$

Equations (6) and (7) reveal the dependence of the emitter discharge on the pressure heads at the inner and outer tips of the nozzle. In as much as the soil is more or less dry at the beginning of the irrigation, the discharge decreases with the elapsed time. Incidentally as the soil becomes wetter, the soil pressure head increases and the emitter discharge stabilizes to a minimum value. Gil et al. [9] found that the decrease of the flow rate is steeper in loamy than in sandy soils. Yao et al. [30] recorded that the wetted soil volume in medium loam and sandy loam is virtually invariant as the inlet pressure head was increased from 60 to 150 cm. This increase of pressure head may lead to the backpressure development. Yao et al. [30] recommended that the emitter discharge should be matched to the soil conditions so that backpressure occurrence is avoided. According to Ben-Gal et al. [2] and Lazarovitch et al. [17], one of the main issues with SDI systems is the soil saturation. This phenomenon induces temporary asphyxia of crops and may stop the emitter discharge even though the moistened bulb is not yet spatially well extended. Based on equations (1) and (2), the emitter discharge is null whenever the outlet pressure head (h_i) matches the predetermined inlet one (h_e). Afterwards, the redistribution process provides drier rooted soil profiles. Subsequently, the pressure near the emitter (h_i) decreases until the pressure differential between the outlet tips overtakes a minimum level.

The threshold value Δh_{min} is required for the emitter operation. The Δh_{min} is dependent on the structural form, dimension and material of the emitter pathway. For any emitter model, Δh_{min} may be inferred from the emitter discharge-pressure head relationship provided by the manufacturer. Thus, the next irrigation is automatically triggered, once the following inequality is fulfilled:

$$[h_e - h_i] = [h_e - h_{op}] \geq \Delta h_{min} \tag{8}$$

Therefore, the required minimum pressure head at the emitter inlet h^*_{min} should comply with:

$$h^*_{min} \geq [h_{op} + \Delta h_{min}] \ (9)$$

It is emphasized that the suction head at the vicinity of the emitter cannot be maintained constant and equal to h_{op}. Unavoidable fluctuations of the suction head are expected owing to evapotranspiration and water redistribution processes. For the sake of convenience, the suction head in the root zone should be circumscribed within a prescribed interval $[(h_{op} + \Delta h_{op})$ and $(h_{op} - \Delta h_{op})]$. Therefore, the minimum required emitter inlet pressure head h^{min}_{req} is given by:

$$h^{min}_{req} = [h_{op} - \Delta h_{op} + \Delta h_{min}] \tag{10a}$$

where: the maximum required emitter inlet pressure head h^{max}_{req} is given by:

$$h^{max}_{req} = [h_{op} + \Delta h_{op} + \Delta h_{min}] \quad (10b)$$

The magnitude of the interval $[h_{op} \pm \Delta h_{op}]$ should account for the sensitivity of the crop to the water stress. As a matter of fact, for tomato crop, the reduction of the water requirement by 20% resulted in 20% increase in yield [6]. However, the decrease of the onion water requirement by 20% resulted only 2% decrease in yield [21]. It should be highlighted that these yield reductions are more or less significant according to the physiological stages.

TABLE 6.1 Tolerable Soil Pressure Head Variations For Selected Crops

Crop	Pressure range, cm		Reference
	Upper limit	Lower limit	
Grape	- 2	- 1000	[12]
Grass	- 25	- 800	[3]
Soybean	- 25	- 800	[3]
Spring wheat	- 25	- 1000	[18]
Tomato	- 2	- 800	[8, 12]

6.3 REQUIRED LATERAL PRESSURE HEAD

For a buried lateral equipped with N identical emitters, the inlet discharge Q will vary within the following limits:

$$Nq_{min} \le Q \le Nq_{max} \quad (11)$$

where: q_{max} and q_{min} are the maximum and minimum emitter average discharge, respectively. For design purpose, only the maximum average emitter discharge is considered. Therefore, the lateral inner diameter is designed to allow the conveyance of the upper bound of the discharge. Consequently, the minimum pressure head required at the upstream end of nontapered flat lateral is:

$$h_{Lm} = [Z_d + J_L + \Delta h_{min} + h_{op} - \Delta h_{op}] \quad (12a)$$

where: the maximum pressure head required at the upstream end of the lateral is:

$$h_{LM} = [Z_d + J_L + \Delta h_{min} + h_{op} + \Delta h_{op}] \quad (12b)$$

where: Z_d[L] and J_L[L] are emitter burial depth and head loss along the lateral, respectively. By convention, the gravitational potential Z_d is computed negatively downwards.

According to the aforementioned basics, the design procedure of SDI systems should lead to the automation of micro irrigation. Indeed, the irrigation events are triggered, whenever the mean pressure head within the root zone is reduced to the minimum prescribed value of $(h_{op} - \Delta h_{op})$. The events are automatically ended, once the pressure head within the root zone exceeds the maximum value of $(h_{op} + \Delta h_{op})$. From theoretical standpoint, a self-regulation of the flow rate by soil water properties and moisture conditions should prevail. Moreover, the variations in emitter discharge due to the head losses are offset by soil pressure head gradients. Accordingly, the irrigation events as well as the uniformity of the flow rates are controlled by the soil suction head at the depth of burial of emitters. These results agree with Gil et al. [9] who indicated higher variability in the flow rates with surface emitters than with the buried emitters.

Tolerable variations in the soil pressure head for some crops are summarized in Table 6.1. It is worth pointing out that the abovementioned approach remains valid regardless of the used soil water-retention relationship.

6.3.1 DESIGN STEPS

Step 1: Carry out simultaneous in situ field measurements of soil moisture and suction heads.

Step 2: Fit the experimental dataset (θ, h) in accordance with the appropriate soil water-retention curve (for example: Eq. (5)).

Step 3: Derive twice the moisture content with respect to the suction head and infer h_{op}.

Step 4: Select the proper interval of the soil suction head Δh_{op} for a particular crop (for example, data provided in Table 6.1).

Step 5: For the emitter type under consideration, calculate the minimum inlet pressure head h^*_{min} using Eq. (9).

Step 6: Calculate the minimum and maximum required emitter inlet pressure heads using Eqs. (10a) and (10b), respectively.

Step 7: Using Eq. (11), calculate the required lateral inlet discharge.

Step 8: Determine the minimum and maximum required lateral inlet pressure heads, using Eqs. (12a) and (12b), respectively.

6.4 EXAMPLE

Determine the minimum and maximum required lateral inlet pressure heads, for the following data:

Length of polyethylene nontapered flat pipe: 100 m.

In-line emitter spacing: 40 cm equally spaced.

Emitter depth, according to Patel and Rajput [21]: 15 cm
Crop: Tomato
Soil texture: homogeneous sandy soil.

6.4.1 PROCEDURE

Step 1: Simultaneous in situ measurements of the soil moisture and suction heads were performed [11] on three randomized locations during water redistribution. In each soil profile, suction heads were measured using three tensiometers installed at 10, 30 and 50 cm soil depth. Soil cores sampled at the same depths were used to determine gravimetrically the corresponding soil moisture. For each depth, the average of the three observations was considered.

Step 2: Experimental data were fitted in accordance with Van Genuchten [27] model [11]. Scattered and fitted data are shown in Fig. 6.1. The inferred fitting parameters (θ_r, m, n and α) are summarized in Table 6.2.

TABLE 6.2 Fitting Parameters for Van Genuchten's Equation for the Sandy Soil

θ_s (cm^3/cm^3)	θ_r (cm^3/cm^3)	α (cm^{-1})	n	R^2
0.38	0.02	0.05	1.70	0.991

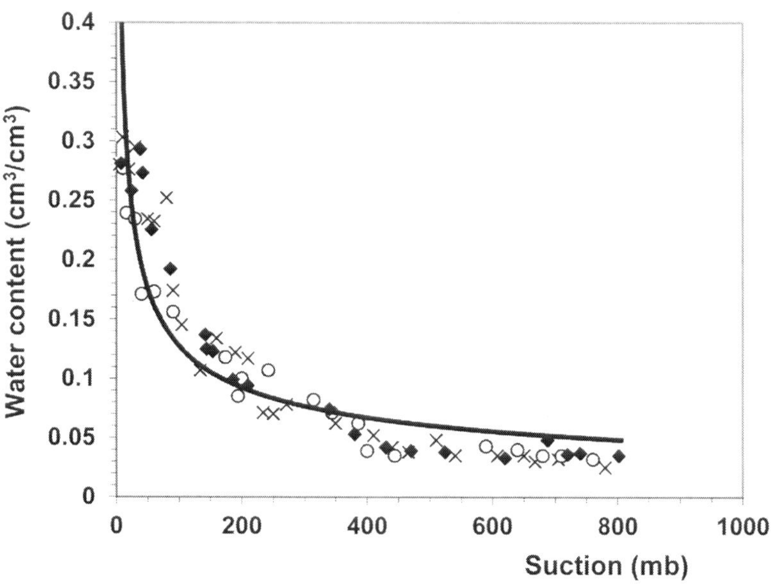

FIGURE 6.1 Soil water retention curve and measured data at different depths. Legend: 10 cm = xx; 30 cm = oo; and 50 cm = ■■.

Step 3: Using Eq. (6), the optimum suction head h_{op} is approximately −12 cm. This value is within the optimal range of the suction head for tomato crop (*see* Table 6.2, [8, 12]). To prevent asphyxia risk or relative water stress at upper (−2 cm) and lower (−800 cm), tolerable pressure heads, $\Delta h_{op} = 400$ cm is acceptable.

Step 4: Therefore, the prescribed soil pressure head limits for tomato crop are determined as follows:

$$h_{op} - \Delta h_{op} \approx -12 - 400 = -412 \text{ cm and}$$

$$h_{op} + \Delta h_{op} \approx -12 + 400 = 388 \text{ cm} \tag{13}$$

In order to avoid eventual backpressure development, the suction head should be maintained within [−412 and 0.0] cm.

Step 5: A trapezoidal labyrinth long-path emitter is used with a minimal differential operating pressure head of $\Delta h_{min} = 500$ cm. The discharge-pressure head relationship of these emitters is shown below [22]:

$$q = [0.752(h_e - h_i)^{0.478}] \tag{14}$$

where: q = emitter discharge (l/h), h_e = emitter inlet pressure head (m) and h_i = the emitter outlet pressure head (m).

Step 6: Using equations (10a) and (10b), the required emitter inlet pressure h_{req} should comply with:

$$[(-12 - 400 + 500) = 88] \leq h_{req} \text{ (cm)} \leq [(0 + 500) = 500] \tag{15}$$

To maintain an optimal suction head within the root zone (−12 cm) and to compensate the minimum differential operating pressure head ($\Delta h_{min} = 500$ cm), the optimal required emitter inlet pressure should be $h_{oreq} = (-12 + 500) = 488$ cm. Compared with the pressure heads customarily required for on-surface drippers (approximately 1000 cm), the obtained value underlines an outstanding energy saving with SDI systems. Therefore, according to Eq. (14), the corresponding emitter discharge q is given as:

$$\{0.752[0.88 - 0.00]0.478 = 0.707\} \leq q \text{ (l/h)} \leq \{0.752[5.00 - (-4.12)]\ 0.478 = 2.163\} \tag{16}$$

As long as the lowest differential pressure head (0.88 m) is less than the minimum differential operating pressure head ($\Delta h_{min} = 500$ cm), the emitter discharge vary within the interval [0.00, 2.163]. Nevertheless, the optimal required emitter discharge matching the optimal soil suction head q_{op} will be:

$$q_{op} = \{0.752[4.88 - (-0.12)]0.478\} \approx 1.623 \text{ l/h} \tag{17}$$

Step 7: The number of emitters along the lateral equals 100 m/0.4 m = 250. According to (11), the optimal required discharge at the lateral inlet tip is:

$$Q_{op} = 250 \times 1.623 = 405.75 \text{ l/h} \tag{18}$$

The head loss gradient j may be estimated by Watters and Keller's formula [29] as follows:

$$j = \alpha Q^{\beta} D^{-\gamma} \tag{19}$$

where: Q and D are the discharge and the lateral inside diameter, respectively. For j (m/m), Q(l/h) and D(mm), the parameters in Eq. (19) are β = 1.75, γ = 4.75 and α = 14.709598υ, where: υ (m^2 s^{-1}) is the kinematic viscosity of water. At 20°C, α is equal to 0.4655. Considering Eq. (19) and an inner diameter of 16 mm, the head loss throughout the lateral J_L is given [29]:

$$J_L = \alpha Q^{\beta}_{max} D^{-\gamma}/(1+\beta)\text{L}$$

$$\text{J}_L = 0.4655(405.75)^{1.75}(16.0)^{-4.75}100/(1+1.75) = 1.184 \text{ m} \tag{20}$$

This value is doubled if we take into consideration the head losses due to emitters' connection as computed by Juana et al. [13] method.

Step 8: Using Eqs. (12a) and (12b) and accounting for emitters' connection head losses, the required pressure head at the inlet tip of the lateral will be:

$$\{(-15+2\times118.4+500-12-400)=309.8\} \leq h_L \text{ (cm)} \leq \{(-15+2\times118.4+500+0)=721.8\} \tag{21}$$

In the same way, the optimal required pressure head h_{Lo} at the lateral inlet will be:

$$h_{Lo} = (-15 + 2 \times 118.4 - 12 + 500) = 709.8 \text{ cm} \tag{22}$$

Therefore, it is possible to ensure a complete automation of the SDI system via the installation of an overhead basin with a constant water level.

6.5 CONCLUSIONS

Besides savings in water, energy and labor-input, the SDI system offers the opportunity to fully automate the micro irrigation and to include best management practices in agriculture. In fact, the adequate control of variation of soil moisture in the vicinity of emitters is a milestone in the management of subsurface drip irrigation. The rationale is that the flow rate of buried drippers is a function of pressure head at the soil depth of subsurface drip lines. Therefore, the temporal variation of the flow rate is dependent on soil water redistribution and water uptake by roots. The design procedure developed in this chapter provides appropriate emitter discharge and inlet lateral pressure head that fit the water uptake by plant roots. Knowing soil retention curve and water uptake, the procedure provides guidelines to design SDI laterals. The main objective of the design is to ascertain optimal suction head within the installation depth of emitters so that irrigation events are automatically controlled based on the soil moisture variations. The case study showed that soil moisture can be circumscribed within an interval suitable for plant growth. This approach can be a helpful tool for the optimum design of SDI system and the best irrigation management. However, it is worthwhile to note that the current approach completely overlooks the effects of burial drippers on clogging.

6.6 SUMMARY

Subsurface drip irrigation is based on small and frequent water application near the root zone. Since emitter lines are buried in the SDI, the emitter discharge is dependent on the soil moisture status in the vicinity of the emitters. This chapter includes design of subsurface laterals based on the soil water-retention characteristics and water uptake by the roots. The approach in this chapter permits systematic triggering and cut-off of irrigation events based on fixed water suctions in the rhizosphere. Therefore, the soil moisture is maintained at an optimal threshold value to ensure the best plant growth. The method in this chapter is a helpful tool for the optimum design of the SDI system and appropriate water management. Knowing the soil water-retention curve, the appropriate water suction for the plant growth and the emitter discharge-pressure head relationships were developed. The method by authors allows the computation of the required hydraulics of the laterals (e.g., inlet pressure head, inside diameter, etc.). An illustrative example is presented for the design of SDI laterals in tomato.

KEYWORDS

- **Buried dripper**
- **Clogging**
- **Deep percolation**
- **Drip line**
- **Dripper**
- **Evapoconcentration**
- **Hydrus-2D**

- **Irrigation design**
- **Lateral**
- **Micro irrigation**
- **Plant growth**
- **Pressure head**
- **Soil evaporation**
- **Soil moisture**
- **Soil water**
- **Subsurface drip irrigation**
- **Suction head**
- **Tomato**
- **Water distribution**
- **Water management**
- **Water use efficiency**

REFERENCES

1. Ayars, J. E., Phene, C. J., Hutmacher, R. B. (1999). Subsurface drip irrigation of row crops: a review of 15 years of research at the Water Management Research Laboratory. *Agricultural Water Management*, *42(1)*, 1–27.
2. Ben-Gal, A., Lazorovitch, N., Shani, U. (2004). Subsurface drip irrigation in gravel-filled cavities. *Vadose Zone Journal*, *3(4)*, 1407–1413.
3. Clemente, R. S., De Jong, R. H., Hayhoe, N., Reynolds, W. D., Hares, M. (1994). Testing and comparison of three unsaturated soil water flow models. *Agricultural Water Management*, *25(2)*, 135–152.
4. Clothier, B. E., Green, S. R. (1997). Roots: the big movers of water and chemical in soil. *Soil Science*, *162(8)*, 534–543.
5. Colaizzi, P. D., Schneider, A. D., Evett, S. R., Howell, T. A. (2004). Comparison of SDI, LEPA and spray irrigation performance for grain sorghum. *Transactions of the American Society of Agricultural Engineers*, *47(5)*, 1477–1492.
6. Doorenbos, J., Kassem, A. H., Bentverlsen, C. L. M. (1987). Yield Response to water. FAO Paper 33.
7. Evett, S. R., Howell, T. A., Schneider, A. D. (1995). Energy and water balances for surface and subsurface drip irrigated corn. *Proceedings of the 5th International Micro irrigation Congress*, Orlando, Florida, USA, April, pages 135–140.
8. G¨arden¨as, A. I., Hopmans, J. W., Hanson, B. R., Simunek, J. (2005). Two-dimensional modeling of nitrate leaching for various fertigation scenarios under microirrigation. *Agricultural Water Management*, *74(3)*, 219–242.

9. Gil, M., Sinobas, L. R., Juana, L., Sanchez, R., Losada, A. (2008). Emitter discharge variability of subsurface drip irrigation in uniform soils: effect on water-application uniformity. *Irrigation Science, 26(6),* 451–458.
10. Grabow, G. L., Huffman, R. L., Evans, R. O., Jordan, D. L., Nuti, R. C. (2006). Water distribution from a subsurface drip irrigation system and drip line spacing effect on cotton yield and water use efficiency in a coastal plain soil. *Transactions of the ASABE, 49(6),* 1823–1835.
11. Hammami, M., Zayani, K. (2009). Effect of trickle irrigation strategies on tomato yield and roots distribution. *World Journal of Agricultural Sciences, 5,* 847–855.
12. Hanson, B. R., Simunek, J., Hopmans, J. W. (2006). Evaluation of urea-ammonium-nitrate fertigation with drip irrigation using numerical modeling" *Agricultural Water Management, 86(1),* 102–113.
13. Juana, L., Sinobas, L. R., Losada, A. (2002). Determining minor head losses in drip irrigation laterals. I: methodology. *Journal of Irrigation and Drainage Engineering, 128(6),* 376–384.
14. Kandelous, M. M., Simunek, J. (2010). Numerical simulations of water movement in a subsurface drip irrigation system under field and laboratory conditions using HYDRUS-2D. *Agricultural Water Management, 97(7),* 1070–1076.
15. Khemaies, Z., Moncef, H. (2009). Design of level ground laterals in trickle irrigation systems. *Journal of Irrigation and Drainage Engineering, 135(5),* 620–625.
16. Lazarovitch, N. (2008). *The effect of soil water potential, hydraulic properties and source characteristic on the discharge of a subsurface source', [Ph.D. thesis],* Faculty of Agriculture of the Hebrew University of Jerusalem.
17. Lazarovitch, N., Simunek, J., Shani, U. (2005). System-dependent boundary condition for water flow from subsurface source. *Soil Science Society of America Journal, 69(1),* 46–50.
18. Li, K. Y., De Jong, R. J. Boisvert, B. (2001). An exponential root water-uptake model with water stress compensation. *Journal of Hydrology, 252(1–4),* 189–204.
19. Mualem, Y. (1976). A new model for predicting the hydraulic conductivity of unsaturated porous media" *Water Resources Research*, 12(3), 513–522.
20. Neelam, P., Rajput, T. B. S. (2008). Dynamics and modeling of soil water under subsurface drip irrigated onion. *Agricultural Water Management, 95(12),* 1335–1349.
21. Patel, N., Rajput, T. B. S. (2009). Effect of subsurface drip irrigation on onion yield. *Irrigation Science, 27(2),* 97–108.
22. Qingsong, W., Yusheng, S., Wenchu, D., Gang, L., Shuhuai, H. (2006). Study on hydraulic performance of drip emitters by computational fluid dynamics" *Agricultural Water Management, 84(1–2),* 130–136.
23. Safi, B., Neyshabouri, M. R., Nazemi, A. H., Massiha, S., Mirlatifi, S. M. (2007). Water application uniformity of a subsurface drip irrigation system at various operating pressures and tape lengths. *Turkish Journal of Agriculture and Forestry, 31(5),* 275–285.
24. Shani, U., Xue, S., Gordin-Katz, R., Warrick, A. W. (1996). Soil limiting flow from subsurface emitters. I: Pressure Measurements. *Journal of Irrigation and Drainage Engineering, 122(5),* 291–295.
25. Sinobas, L. R., Gil, M., Juana, L., Sanchez, R. (2009). Water distribution in laterals and units of subsurface drip irrigation, I: Simulation. *Journal of Irrigation and Drainage Engineering, 135(6),* 721–728.
26. Sinobas, L. R., Gil, M., Juana, L., Sanchez, R. (2009). Water distribution in laterals and units of subsurface drip irrigation. II: Field Evaluation. *Journal of Irrigation and Drainage Engineering, 135(6),* 729–738.
27. Van Genuchten, M. T. (1980). A closed-form equation for predicting the hydraulic conductivity of unsaturated soils. *Soil Science Society of America Journal, 44(5),* 892–898.
28. Vories, E. D., Tacker, P. L., Lancaster, S. W., Glover, R. E. (2009). Subsurface drip irrigation of corn in the United States Mid-South. *Agricultural Water Management, 96(6),* 912–916.

29. Watters, G. Z., Keller, J. (1978). Trickle irrigation tubing hydraulics. Tech. Rep. 78–2015, ASCE, Reston, VA, USA.
30. Yao, W. W., Ma, X. Y., Li, J., Parkes, M. (2011). Simulation of point source wetting pattern of subsurface drip irrigation. *Irrigation Science, 29(4),* 331–339.

CHAPTER 7

ECONOMIC RETURNS FOR DRIP IRRIGATED TOMATO

AJAI SINGH

CONTENTS

7.1 INTRODUCTION

Efficient water use is necessary for sustainable crop production and drip irrigation proved to efficiently provide irrigation water and nutrients to the roots of plants, while maintaining high yield. Because not all the soil surface is wetted under drip irrigation, less water is required for irrigation. Modern drip irrigation has become the most valued innovation in agriculture. Higher water application efficiencies are achieved with drip irrigation due to reduced soil evaporation, less surface runoff and minimum deep percolation. The Government of India has been considering rapid promotion of use of plastics in agriculture and micro irrigation as a major step in improving overall horticultural crop yields and water use efficiency. The micro irrigation has gained considerable growth in the country due to financial assistance provided by the centrally sponsored subsidy scheme. Presently, drip irrigation has the greatest potential where (i) water and labor are expensive or scarce; (ii) water is of marginal quality viz., saline; (iii) soils are sandy, rocky or difficult to level; (iv) steep slopes and undulated topography; and (v) high value crops are produced. The principal crops under drip irrigation are commercial field crops (sugarcane, cotton, tobacco etc.), horticultural crops – fruit and orchard crops, vegetables, flowers, spices and condiments, bulb and tuber crops, plantation crops and silviculture/forestry plantations. This method of irrigation continues to be important in the protected agriculture viz., greenhouses, shade nets, shallow and walking tunnels etc., for production of vegetables and flowers. Drip irrigation is also used for landscapes, parks, highways, commercial developments and residences. Undoubtedly, the area under drip irrigation will continue to increase rapidly as the amount of water available to agriculture declines and the demands for urban and industrial use increase. Drip irrigation is also one of the techniques that enable growers to overcome salinity problems that currently affect 8.0 million ha area in India. As this area increases, so too will the use of Drip irrigation to maintain crop production. In addition, because growers are looking to reduce cost of production but at the same time improve crop quality, the improved efficiency provided from drip irrigation technology will become increasingly important.

This chapter discusses research findings on the application of drip in vegetable production, economic returns and some important issues to design of the drip irrigation system.

7.2 QUANTITATIVE APPROACH OF WETTING PATTERNS

Due to the manner in which water is applied by a drip irrigation system, only a portion of the soil surface and root zone is wetted unlike surface and sprinkler irrigation systems. Water flowing from the emitter is distributed in the soil by gravity and capillary forces creating the contour lines similar to onion shape. The exact shape of the wetted volume and moisture distribution depends on the soil texture, initial soil moisture and to some degree, on the rate of water application. The water savings in drip irrigation are due to reductions in deep percolation, surface runoff and evaporation from the soil. In the line-source type of drip irrigation system, where the emitters are spaced very closely, individual onion patterns creates a continuous moisture zone. The knowledge about the wetting patterns under emitters is essential in selecting the appropriate spacing of the emitters. Emitter spacing and emitter flow rates must match to the wetting characteristics of the soil and the amount and timing of water to be supplied to meet the crop needs. Under drip irrigation, the ponding zone that develops around the emitter is strongly related to both the application rate and the soil properties. The water application rate is one of the factors, which determine the soil moisture regime around the emitter and the related root distribution and plant water uptake patterns [3, 4].

Drip irrigation systems generally consist of emitters that have discharge varying from 2.0 to 8.0 lph. In semiarid climates, crop water use during the summer can be 6 to 8 mm/d with water supplied two or three times a week. When the water application is exactly equal to the plant water need, then also, part of the water may not be used by the plant and it would most likely leach below the root zone. Therefore, lowering the emitter discharge to as close as possible to the plant water uptake rate can improve irrigation efficiency. Recently, microdrip irrigation systems have been developed that provide emitter discharges of 0.5 lph. These systems have been studied most intensively in greenhouses [7] and preliminary results have shown that farmers were able to reduce water consumption of tomato plant by 38%, to increase yield by 14 to 26% and to reduce leaching fraction by 10 to 40%. In a recent study on sweet corn under field conditions, it was shown that microdrip irrigation may improve yield, reduce drainage flux and affect the water content distribution within the root zone, especially through an increased drying of the 0.60 to 0.90 m soil layer compared with conventional drip irrigation [2].

The microdrip technology still raises some problems concerning the uniformity of application and the steadiness of the discharge. However, soil moisture regimes similar to those resulting from continual low water application rates can be achieved by means of pulsed drip irrigation. Infiltration experiments on a sandy loam soil showed that the water content distribution and the rate of wetting front advance under a pulsed water application were similar to water applied in a continuous manner and those temporal fluctuations in flux and in soil water content exponentially damped with depth for periodic pulses applied at the soil surface. Consequently, pulsed irrigation using conventional drip emitters could be one way of creating the water regime observed with continual low application rates while bypassing technical problems associated with microdrip emitters.

The relationships between water application rates, soil properties and the resulting water distribution for conventional emitters (2.0 lph) are well documented. The wetting patterns during application generally consist of two zones: (i) a saturated zone close to the emitter and (ii) a zone where the water content decreases toward the wetting front. Increasing the emission rate generally results in an increase in the wetted soil diameter and a decrease in the wetted depth [1, 8]. In microdrip irrigation, field observations seem to indicate that there is no saturated zone and that the wetted soil volume is greater compared to that for conventional emitter discharges [7]. The relationship between the water application rate and the resulting water content distribution is complex because it is a three-dimensional outcome related to soil properties and crop uptake characteristics. Therefore, a quantitative representation of the flow processes by means of a simulation model can be beneficial in studying the effects of emitter discharge on the water regime of drip irrigated crops.

Many attempts have been made to determine water movement and wetting pattern under drip emitters using mathematical and numerical models. The Richards equation, formulated by Lorenzo A. Richards in 1931, describes the movement of water in unsaturated soils. It is a nonlinear partial differential equation, which is often difficult to approximate. Partial differential equations are types of differential equations, which formulate a relation involving unknown functions of several independent variables and their partial derivatives with respect to those variables. Ordinary differential equations usually model dynamical systems whereas partial differential equations are used to model multidimensional systems. Darcy's law was developed for saturated flow in porous media. The Richards' equation is based solely on Darcy's law and the continuity equation for the water movement in unsaturated soils. Therefore, it is strongly physically based, generally applicable and can be used for fundamental research and scenario analysis. The Richards equation can be stated in the following form:

$$\frac{\partial \theta}{\partial t} = \frac{\partial}{\partial z}\left[K(\theta)\left(\frac{\partial \psi}{\partial z}+1\right)\right] \tag{1}$$

where: K = hydraulic conductivity, ψ = pressure head, z = elevation above a vertical datum, θ = water content, and t = time.

Under drip irrigation, it has been discussed above that only a portion of the horizontal and cross sectional area of the soil is wetted. The percentage-wetted area as compared with the entire field covered with crops, depends on the volume and rate of discharge at each emitter, spacing of emitter and the type of soil being irrigated. For widely spaced crops, the percentage-wetted area should be less than 67% in order to keep the area between the rows relatively dry for cultural practices. Low value of percentage-wetted area also reduces the loss of water due to evaporation and involves less cost. For closely spaced crops such as vegetables with rows and laterals spaced less than 1.8 m, percentage wetted area often approaches 100% [6]. Several efforts have been made to estimate the dimensions of the wetted volume of soil under an emitter. The wetted soil volume depends upon the hydraulic conductivity of the soil, discharge of the emitter and amount of water available in the soil [8]. The following empirical equations have been developed to estimate the wetted depth and width. The equations were derived using three-dimensional cylindrical flow geometry and results were verified from plane flow model.

$$Z = 29.2\left(V_w\right)^{0.63}\left(\frac{K}{q}\right)^{0.45} \quad (2)$$

$$w = 0.031\left(V_w\right)^{0.22}\left(\frac{K}{q}\right)^{-0.17} \quad (3)$$

By combining the Eqs. (1) and (2), one can find out the relationship between depth of wetting front (*Z*) and width of wetted soil volume (*w*). The relationship can be expressed as follows.

$$w = 0.0094\left(Z\right)^{0.35} q^{0.33} K^{-0.33} \quad (4)$$

where: Z = depth of wetting front in m, w = wetted width or diameter of wetted soil in m, V_w = volume of water applied in liters, K = saturated hydraulic conductivity of soil in m/s, q = discharge of emitter in lph.

7.3 COMPUTATION OF CROP WATER REQUIREMENT WITH LIMITED WETTING

In drip irrigation systems, only part of the soil surface is wetted and for widely spaced crops, crop canopy coverage is also limited. It is not appropriate to consider the soil evaporation from the entire soil surface under drip irrigation systems. A correction factor (K_r) was introduced to take into the account of percentage of crop canopy coverage of cultivated land [12]. The relationship is expressed below:

$$ET_{CROPCor} = K_r \times ET_{CROP} \quad (5)$$

where: ET_{CROCor} is corrected crop water requirement, K_r is correction factor and ET_{crop} is crop water requirement without considering limited area wetting. The following formula was developed to estimate crop evapotranspiration or crop water requirement for limited wetted areas [6].

$$ET_{CROPCr} = ET_{CROP}\left[0.1\times\sqrt{P_d}\right] \quad (6)$$

where: P_d is percentage of crop canopy coverage.

7.3.1 EXAMPLE

A drip irrigated tomato crop is at development stage and the relative humidity is 70% with wind velocity of 2 m/c. Crop canopy coverage is 60% and K_c value for development stage is 0.75. Assume ET_0 as 5.2 mm/day and correction factor as 0.92. Compare the evapotranspiration estimations without area wetting [12] and for limited area wetting [6].

7.3.1.1 PROCEDURE

1. Let us estimate the crop evapotranspiration when limited area concept is not being applied, Eq. (5).
   $ET_{CROP} = K_c \times ET_0 = 0.75\ x\ 5.2 = 3.9$ mm/day
2. When limited area concept is being applied, according to Vermeirn and Jobling [12]
   $ET_{CROPCor} = K_r \times ET_{CROP} = 0.92 \times 3.9 = 3.59$ mm/day
3. Finally, according to Keller and Bliesner [6], Eq. (6):
   $ET_{CROPCr} = ET_{CROP}\left[0.1 \times \sqrt{P_d}\right] = 3.9\ (0.1 \times \sqrt{60}) = 3.02$ mm/day

7.4 ECONOMIC ANALYSIS

Investment on installation of drip irrigation system must be economically viable and justified. If the project gives economic surplus, we can say the project is economically viable. As drip irrigation involves considerable cost, the economic appraisal of the installation of drip system must take full account of all the cost and benefit likely to be accrued from the crops to be grown and its byproduct. Economic analysis is carried out to determine whether the returns from the project will be able to justify the investment or not. Drip irrigation project cost will include all the expenditure made on procurement, installation, operation and cultivation cost. The annual cost of a project includes both fixed and variable costs. The benefits of irrigation through drip irrigation are many such as better crop survival, earlier fruit production, more yields, efficient distribution of nutrients, less plant stress, reduced yield variability and improved fruit quality. We will present herein the methodology applied to evaluate the economics of irrigation. Growers/farmers operating drip irrigation must identify the drip irrigation investment, operating cost or yield response. Before taking up analysis part, it is essential to understand the few basic terms and method of economic analysis.

7.4.1 DEPRECIATION

It is well known fact that if you a buy a product today, its value after 3–4 years will not be the same. This means that the product has been depreciated due to use over the years. Depreciation is the value reduction of any asset due to physical use over the time periods. The annual depreciation is calculated from the following formula.

$$Annual\ depreciation = \frac{Primary\ cost - Salvage\ value}{Useful\ life\ in\ years} \tag{7}$$

7.4.2 SALVAGE VALUE

Salvage value is an estimate of the remaining value of an investment at the end of its useful life. It is desirable to determine the present worth of a future value of a product, which is called discounting, for any analysis of a project, which is going to be operative for a long time. The present worth of a future value of a product at the end of n years at an interest rate of i can be computed by using the following expression.

$$PW = F\left(\frac{1}{(1+i)^n}\right) \tag{8}$$

where: PW is the present worth of the future income value and F is the future values of the income.

7.4.3 ESCALATION COST

The rate of escalation can be incorporated in the analysis of present worth and annual cost. If e is the annual rate of escalation, the present worth value, which incorporates the effect of escalation in the cost, can be estimated by the following formula:

$$PW(e) = PW \times \left(\frac{(1+e)^n}{(1+i)^n}\right) \tag{9}$$

Keller and Blisener [6] considered the interest rates (i), the expected life of investment 'n' and an estimate of the expected annual rate of escalation in the calculation of annual energy cost. The present worth of the escalating energy factor and the equivalent annualized cost of escalating energy factor were computed by the following equations:

$$PW(e) = \left[\frac{(1+e)^n - (1+i)^n}{(1+e)-(1+i)}\right] \cdot \left[\frac{1}{(1+i)^n}\right] \tag{10}$$

and equivalent annualized cost of escalating energy (EAE) factor at annual rate i is calculated as:

$$EAE(e) = \left[\frac{(1+e)^n - (1+i)^n}{(1+e)-(1+i)}\right] \cdot \left[\frac{i}{(1+i)^n - 1}\right] \tag{11}$$

where: *PW(e)* = present worth factor of escalating energy cost taking into account the time values of money over the life cycle; and *EAE(e)* = equivalent annualized cost factor of escalating energy taking into account the value of money over life cycle.

7.4.4 CAPITAL RECOVERY FACTOR

A Capital Recovery Factor (CRF) converts a present value into a stream of equal annual payments over a specified time at a given discount rate (interest rate). If P is the present value of a product, then amount of each level payment to be made at the end of each of *n* periods can be determined by multiplying it with *CRF*. The standard capital recovery factor CRF is computed by:

$$CRF = \frac{i(1+i)^n}{(1+i)^n - 1} \tag{12}$$

7.4.5 DISCOUNTING METHODOLOGY

While analyzing the drip irrigation project, when we convert all the cost and benefits to a common time base, we call it discounted cash flow technique. In short, all the costs and benefits are compared on the basis of a common time scale, though this may occur at different time periods. Every investment project will have cost and benefit. Based on simple knowledge of total cost and total benefit, we can measure three indicators of economic evaluation such as Net Present Value (NPV), Benefit-Cost Ratio (BCR) and Internal Rate of Return (IRR). We will make it clear with a solved example in this chapter.

7.4.6 NET PRESENT VALUE

This is a single value representing the difference between the sum of the projected discounted cash inflows and outflows attributable to a capital investment, using a discount rate that properly reflects the relevant risks of those cash flows. Using NPV as indicator, we convert all the cost and benefit of any year into present year. If the value of NPV is positive, the project benefit has more than the cost and then the project is feasible and can be taken up for implementation. It may be interpreted as the present worth of the income generated by the investment. The NPV can be calculated as follows:

$$\sum_{t=0}^{n} \frac{B_t}{(1+i)^t} = \sum_{t=0}^{n} \frac{C_t}{(1+i)^t} \tag{13}$$

where: B_t = benefit at time t, C_t = cost at time t, i = discount rate and n = number of years.

7.4.7 BENEFIT COST RATIO

The discounted measure of the project worth can be expressed by benefit-cost ratio. This is the ratio implies the return per rupee investment. The benefits and costs of any year are converted into the equivalent basis (i.e. present year) to find out the benefit cost ratio. If BCR value is bigger than 1, benefit has more than the cost, then the project is feasible. The B/C ratio can be worked out by using the following formula:

$$BCR = \frac{\sum_{t=0}^{n} \frac{B_t}{(1+i)^t}}{\sum_{t=0}^{n} \frac{C_t}{(1+i)^t}} \tag{14}$$

7.4.8 INTERNAL RATE OF RETURN

The average annual percentage return is expected from a project, where the sum of the discounted cash inflows over the life of the project is equal to the sum of the discounted cash outflows. Therefore, the IRR represents the discount rate that results in a zero NPV of cash flows. In this method also, we convert all the costs as well as benefits of any year into equivalent basis (i.e. present year). IRR can be compared with the existing bank interest rate to judge the economic feasibility of the project. IRR is calculated from the following principle:

$$\sum_{t=0}^{n} \frac{B_t}{(1+i)^t} = \sum_{t=0}^{n} \frac{C_t}{(1+i)^t} \tag{15}$$

The cost of the drip irrigation system includes all of the fixed costs, operation costs, maintenance costs and all the costs incurred to the project. The benefit of the investment will include, income from production, any form of by-product, salvage value etc.

7.5 ECONOMIC RETURNS FOR TOMATO: AN EXAMPLE

7.5.1 EXAMPLE

Drip irrigated tomato crop is to be cultivated on an area of one hectare. Determine the NPV and BCR. The cash flow pattern is given in Table 1. Assume an interest rate as 12% and life of the drip system as 10 years. The operation cost includes the maintenance cost of drip systems and cost of tomato cultivation, which may increase over the period of time. However, the return, that is, cash inflow has been assumed as constant with a tomato yield of tomato of 30 t/ha.

TABLE 7.1 Cash Flow Pattern

Year	Fixed cost	Operation and maintenance	Cash inflow	Year	Fixed cost	Operation and maintenance	Cash inflow
1	180,000	40,000	120,000	6	–	50,000	120,000
2	–	40,000	120,000	7	–	50,000	120,000
3	–	40,000	120,000	8	–	60,000	120,000
4	–	45,000	120,000	9	–	60,000	120,000
5	–	45,000	120,000	10	–	60,000	120,000

7.5.2 PROCEDURE

First we will calculate the discount rate, which is also called the discount factor with the following formula

$$\text{Discount factor} = \left(\frac{1}{(1+i)^n} \right) \tag{16}$$

where: n is 1, 2, 3, …., 10.

$$\text{Cash outflow} = \text{Fixed cost} + \text{Operation and maintenance} \tag{17}$$

$$\text{Cash flow} = \text{Cash inflow} - \text{Cash outflow} \tag{18}$$

$$\text{Discounted cash flow} = \text{Cash flow} \times \text{Discount factor} \tag{19}$$

$$\text{Discounted cash outflow} = \text{Cash outflow} \times \text{Discount factor} \tag{20}$$

$$\text{Discounted cash inflow} = \text{Cash inflow} \times \text{Discount factor} \tag{21}$$

The results are presented in a Table 7.2.

TABLE 7.2 Procedure to Calculate the Discounted Cash Inflow

Year	Fixed cost	Operation and maintenance	Cash outflow	Cash inflow	Cash flow	Discount factor	Discounted cash flow	Discounted cash outflow	Discounted cash inflow
	Table 7.1	Table 1	Eq. (17)	Table 7.1	Eq. (18)	Eq. (16)	Eq. (19)	Eq. (20)	Eq. (21)
1	180,000	40,000	220,000	120,000	–100,000	0.8929	–89,290	196,438	107,148
2		40,000	40,000	120,000	80,000	0.7972	63,776	31,888	95,664
3		40,000	40,000	120,000	80,000	0.7118	56,944	28,472	85,416
4		45,000	45,000	120,000	75,000	0.6355	47,662	28,597	76,260
5		45,000	45,000	120,000	75,000	0.5674	42,555	25,533	68,088
6		50,000	50,000	120,000	70,000	0.5066	35,462	25,330	60,792
7		50,000	50,000	120,000	70,000	0.4523	31,661	22,615	54,276
8		60,000	60,000	120,000	60,000	0.4039	24,234	24,234	48,468
9		60,000	60,000	120,000	60,000	0.3606	21,636	21,636	43,272
10		60,000	60,000	120,000	60,000	0.322	19,320	1,9320	38,640
Total			670,000	1,200,000			253,960	424,063	678,024

Net Present Value (NPV) = sum of discounted cash flow for the periods of 10 years = Rs. 2,53,960 > 0.

Since NPV is greater than zero, investment on drip irrigation is economically feasible. Now calculate benefit cost ratio (BCR).

Discounted benefit cost ratio = (Discounted cash inflow)/(Discounted cash outflow) = (678,024)/(424,063) = 1.60

Now what will happen to BCR if discounted cash flow technique is not followed in economic analysis?

Undiscounted cash flow technique = (Cash inflow)/(Cash outflow) = (1,200,000)/(670,000) = 1.80

BCR is higher in case of undiscounted cash flow technique, which may not be appropriate technique of project appraisal and feasibility study. In India, this system is gaining popularity among fruit growers and in water scarce area but a substantial area is being covered annually under vegetables crops. One of the major concerns raised by farmers about the drip system is its economic viability.

In one of the research study undertaken by the author [9, 11], the economic viability of drip irrigation system for growing capsicum crop based on discounted cash flow technique (net present worth and benefit cost ratio) was investigated. Eight irrigation treatments were laid under drip with and without plastic mulch. The irrigation levels were 1, 0.8 and 0.6 of the crop evapotranspiration. The pan evaporation method was used for estimation of reference evapotranspiration and Water Balance Approach was used for irrigation scheduling. The average amount of water supplied under treatment VD (100% irrigation requirement supplied with drip) was found to be 415 mm for whole growing season of the crop. Similarly the amount of water was 332 mm and 249 mm for the treatment 0.8 VD (80% irrigation requirement supplied with drip) and 0.6 VD (60% irrigation requirement supplied with drip), respectively. Highest yield was recorded in case of treatment VD + PM (100% irrigation requirement supplied with drip plus plastic mulch) followed by VD. Yield under treatments 0.8 VD, 0.6 VD, 0.8 VD+PM and 0.6 VD+PM were significant while treatments VD, VF and VF + PM were at par with the treatment VD+PM. Net Present Worth (NPW) was found to be positive for all the treatments. The highest NPW was Rs. 309,734.90 in treatment VD and lowest was Rs. 144,172.24 in case of 0.6 VD+PM. The yield per mm of water used was 35 (high value) in both the treatments VD and VD + PM. However, the yield per mm of water used was lowest and was 18.07 in VF and 19 in VF +PM, respectively.

In another study on tomato crop, author investigated impact of drip irrigation. Highest yield of 34.3 t/ha was recorded in case of treatment VD+PM (100% irrigation requirement supplied with drip plus plastic mulch) followed by VD [10]. In the 0.8 VD, the yield per mm of water used was 53 with a BCR of 1.13, which was the best among all the treatments considering BCR, yield obtained and yield per mm of water used. The income from produce was estimated using prevailing average market price of Rs. 2000/t and the results are shown in Table 7.3.

TABLE 7.3 Economic Analysis of Drip Irrigated Tomato for Two Years Data for An Area of One ha.

Item	VD	0.8 VD	0.6 VD	VF	VD+PM	0.8 VD+PM	0.6 VD+PM	VF+PM
	Indian rupees, Rs.							
Fixed system cost	135,811	135,811	135,811	0	183,411	183,411	183,411	47,600
Annualized cost	31,514	31,514	31,514	0	41,082	41,082	41,082	13,784

TABLE 7.3 *(Continued)*

Item	VD	0.8 VD	0.6 VD	VF	VD+PM	0.8 VD+PM	0.6 VD+PM	VF+PM
	Indian rupees, Rs.							
Cost of cultivation	26,000	26,000	26,000	30,000	26,000	26,000	26,000	30,000
Yield of produce (t/ha)	34	32.5	26	28	34.3	32.7	26.85	29.3
Income from produce	68,000	65,000	52,000	56,000	68,600	65,400	53,700	58,00
Water used (mm)	760	608	456	1064	760	608	456	1064
Yield per mm of water	45	53	57	18.07	45	54	59	28
Benefit cost ratio	1.18	1.13	0.90	1.86	1.02	0.97	0.80	1.33

Coefficient of Determination at 5% = 0.96

7.6 SUMMARY

Water resources must be used and developed keeping in view the needs of people. People involved with water management in agriculture comprise a diverse group of subsistence, emerging and commercial farmers and permanent and seasonal laborers, with their dependents and functionaries of State Department of Agriculture, India. All these water users or managers are the target groups of the research findings. In any strategic research plan for agricultural water management, the point of departure of applied research is the real-life problems experienced by water users/managers for irrigated and rain-fed crop production.

There are several reasons for sustainable water saving through drip irrigation method. First, since water is supplied only at root zone of the crop, the evaporation and distribution losses are completely reduced. Second, water is supplied only to the crop, whereas land is irrigated in surface irrigation method, which consumes obviously more water. Third, uneven land surface consumes enormous water under surface irrigation method; this problem does not arise in drip irrigation method where water is supplied through pressurized system. Fourth, controllability of irrigation is easier under drip irrigation method, which helps the farmers to conserve the water. The other issue of drip irrigation is related to its economic viability, as farmers are often reluctant to adopt this technology fearing that the technology may not be economically viable. This chapter discusses economic feasibility analysis for drip-irrigated tomato.

KEYWORDS

- **benefit cost ratio**
- **capsicum**
- **cash inflow**
- **cash outflow**
- **drip irrigation**
- **economic feasibility**
- **evaporation**
- **evapotranspiration**
- **tomato**
- **vegetables**

REFERENCES

1. Ah Koon, P. D., Gregory, P. J., Bell, J. P. (1990). Influence of drip irrigation emission rate on distribution and drainage of water beneath a sugar cane and a fallow plot. *Agric. Water Manage, 17,* 267–282.
2. Assouline, S., Cohen, S., Meerbach, D., Harodi, T., Rosner, M. (2002). Microdrip irrigation of field crops: Effect on yield, water uptake and drainage in sweet corn. *Soil Science Soc. Am. J., 66*, 228–235.
3. Coelho E. F., Or, D. (1996). A parametric model for two-dimensional water uptake by corn roots under drip irrigation. *Soil Science Soc. Am. J., 60,* 1039–1049.
4. Coelho, E. F., Or, D. (1999). Root distribution and water uptake patterns of corn under surface and subsurface drip irrigation. *Plant Soil, 206*, 123–136.
5. Howell, T. A., Hiler, E. A. (1974). Designing trickle irrigation laterals for uniformity. *Journal of Irrigation and Drainage Div. ASCE, 100(IR4), 443*–445.
6. Keller, J., Ron D. Bliesner (Eds.) (1990). *Sprinkler and Trickle Irrigation.* Van Nostrand Reinhold, New York.
7. Koenig, E. (1997). Methods of microirrigation with very small discharges and particularly low application rates (In Hebrew). *Water Irrig., 365*, 32–38.
8. Schartzman, M., Zur, B. (1986). Emitter spacing and geometry of the wetted soil volume. *J. Irrig. Drain. Eng., 112,* 242–253.
9. Singh, Ajai, (2008). Economic feasibility of growing capsicum crop under drip irrigation in West Bengal, India. *J. Irrigation Drainage System, Springer Science, 22,* 179–188.
10. Singh, Ajai, (2007). Yield response of drip irrigated tomato to different levels of irrigation. *J. Interacademicia, 11(2),* 200–207.

11. Tiwari, K. N., Ajai Singh, Mal, P. K. (2003). Effect of drip irrigation on yield of cabbage (Brassica oleracea L. var. capitata) under mulch and nonmulch conditions. *Journal of Agricultural Water Management,* The Netherlands, 58, 19–28.
12. Vermeiren, L., Jobling, G. A. (1984). *Localized Irrigation.* FAO Irrigation and Drainage Paper No. 36, Rome.

CHAPTER 8

TOMATO ROOT DISTRIBUTION UNDER DRIP IRRIGATION: TUNISIA

HAMMAMI MONCEF and ZAYANI KHEMAIES

CONTENTS

Printed from Hammami Moncef and Zayani Khemaies, 2009. Effect of Trickle Irrigation Startegies on Tomato Yield and on Roots' Distribution. World Journal of Agricultural Sciences, 5(S): 847–855. http://www.idosi.org/wjas/wjas5%28S%29.htm, Open Access article.

8.1 INTRODUCTION

Crop water uptake and subsequent root distribution/depth within the soil profile are essential parameters for irrigation scheduling, fertigation management, soil tillage operation, for designing on farm irrigation and drainage networks, adequate irrigation and/or optimal depth of buried drains. Moreover, a thorough understanding of root distribution is required to determine the appropriate location of soil moisture sensors. These sensors should be installed at representative soil depth where root density and activities are the highest. Indeed, the optimal plant growth is linked to the water status within the soil profile and to the size of the moistened bulb [11]. The desired moistened bulb should encompass the rooted soil volume [10, 20].

Root distribution remains among the most challenging inputs influencing the reliability of simulation models of water uptake by roots [9, 15, 21]. Large variation in the root distribution has been observed in space and time [16, 23]. This variability depends especially on changes in soil air/water balance in the root zone and is affected by agricultural operations, physical and chemical soil conditions and water supply. Salgado and Cautin [16] reported that avocado trees in fine soils had 25% more roots than in coarse sandy soils. They also reported that drip irrigation produced about 30% more roots than microsprinkler. Zotarelli et al. [23] indicate that root distribution of tomato is essentially governed by the development stage, soil moisture and nutrient availability. According to Keller and Bliesner [11] and Dalvi et al. [5], more root penetration and further production may be obtained if an acceptable water stress is imposed. For Mickelakis et al. [14], the issue, how a reduced moistened soil volume affects the rooting system and improves the crop yield, is still inconclusive. Brent and Steve [3] and Steve et al. [19] highlighted the need for a better understanding of the response of roots to variable soil conditions. A great deal of research is still needed to evaluate effects of irrigation and root distribution on water use efficiency [23]. Misunderstanding about effects of root expansion on soil properties and water status may lead to ineffective management of on farm water management.

This chapter discusses the research results to: monitor the root development in trickle irrigated tomato during the growing season; assess the effects of four irrigation strategies on tomato yield and the rooting system; and compare the root distribution along and across the lateral sides.

8.2 MATERIALS AND METHODS

Field trials were carried out at the experimental farm of the Agricultural Center for Professional Studies of El-Alia (latitude 37°10'N, longitude 10°01'E and AMSL = 48 m). The farm is located in the north-eastern region of Tunisia in one of the widest trickle irrigated area. Environmental conditions at the site are favorable for trickle irrigation (shortage of water resources, 55% of cropped lands are sandy soils, orchards and vegetables cover more than 66% of the cultivated area). It should be stressed that more than 88% of the average rainfall (625 mm/year) falls during October through April. Because of the acute imbalance between annual precipitation and potential evapotranspiration (1330 mm/year), summer crops must be irrigated. The average temperature ranges from 11 °C (January) to 28 °C (August). The soil texture is sandy at the experimental site. Table 8.1 shows a quite uniform textured soil profile with relatively high soil bulk densities.

TABLE 8.1 Particle Size Distribution and Soil Bulk Density For Three Soil Depths At the Experimental Site

Soil depth	Sand	Silt	Clay	Bulk density
cm	%	%	%	$g.cm^{-3}$
0–20	87.5	4.7	7.8	1.67
20–40	88.7	3.9	7.4	1.68
40–60	94.0	2.5	3.5	1.70

A rectangular field plot (25 × 60 m) was divided into 16 blocks. Each block (5 × 14 m) consisted of five crop rows. Measurements of water pressure head, root length and density and crop yield were recorded on three center rows. Four representative plants were randomly chosen in each center row (3 × 4 plants per block). Each crop row was irrigated by a single lateral with emitters (2 lph) that were spaced 30 cm. A control valve was installed at the entrance of each sub main to allow the irrigation of blocks separately (Fig. 8.1). In April, tomato seedlings were transplanted 40 cm apart along the side of the laterals that were 120 cm apart from one another.

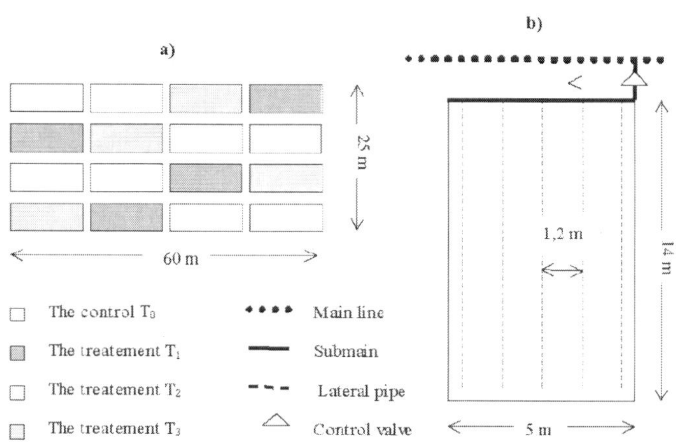

FIGURE 8.1 Experimental plots (a) and layout of drip system (b) within each plot.

To assess the effects of water application on the crop yield and on the root system, four irrigation treatments were used, namely:

T0: Control treatment: Irrigation depth was based on Class A pan evaporation (E_p, mm per day). The crop evapotranspiration ET_c (mm per day) was estimated according to Doorenbos and Pruitt [6]:

$$ET_c = K_c K_p (E_p) \quad (1)$$

where: K_c = crop coefficient, K_p = Pan coefficient and E_p = Class A pan evaporation, mm/day.

T1: The irrigation was initiated when mean suction head within the rooted soil volume dropped to 250 mb.

T2: The irrigation was initiated when mean suction head within the rooted soil volume dropped to 400 mb.

T3: The irrigation was initiated when mean suction head within the rooted soil volume dropped to 550 mb.

The aforementioned suction heads (negative pressure) were based on the soil moisture conditions required for root water uptake. Elmaloglou and Malamos [7] indicate that root water uptake is optimum as long as the suction head was within −300 mb to −25 mb. For tomato crop, optimal interval for the suction head is −800 mb to −2 mb [2, 8].

The irrigation events were based on the readings of four tensiometers that were installed at: A (R = 0 cm, Z = 10 cm), B (R = 0 cm, Z = 30 cm), C (R = 15 cm, Z = 10 cm) and D (R = 15 cm, Z = 30 cm), where: R = lateral distance from the plant stem, Z = soil depth, respectively. Tensiometers at locations A and C were used to initiate the irrigation during the vegetative growth stage (10–30 days after transplantation). And tensiometers at B and D were used to trigger the irrigation during the second bloom and the harvesting stages. It is obvious that the wetting front should be deeper than or equal to the rooting depth [4]. During the first ten days, all blocks received the same amount of irrigation to ensure good seedling transplantation. Afterwards, the irrigation treatments were applied. The gross water volume V_g [L^3] in each irrigation was estimated as below:

$$V_g = N q_m t \quad (2)$$

where: N = number of emitters per block, q_m = the mean emitter discharge [L^3/T] and t = the irrigation duration [T], respectively. Because of clogging hazards, the mean emitter discharge q_m was measured at the initial, at the mid and at the end of the irrigation season as recommended by Merriam and Keller [13]. Neglecting evaporation during the infiltration phase, the irrigation time required to replenish the vadoze zone up to the desired soil water content θ_f is given by:

$$T_i = [(\theta_f - \theta_i) A_w Z_r] \div [N q_m] \quad (3)$$

where: θ_i = the initial soil water content [L^3L^{-3}], θ_f = the final soil water content [L^3L^{-3}], Z = the rooted soil depth [L] and A_w = the area [L^2] of the wetted strip on the soil surface. For the sake of simplicity, the area of the wetted strip is computed as follows:

$$A_w = L_r S_w \tag{4}$$

where: L_r = the lateral length and S_w = the average width of the wetted strip. The S_w was the average of three measurements made at upstream, center and downstream of the row. It should be emphasized that soil water contents were estimated from the soil moisture retention curve that was drawn for the site using tensiometer readings (Fig. 8.2). In equation (3), it is assumed that θ_f and θ_i were homogeneous within the sampled soil volume V_s. This assumption holds true as long as V_s is small and the steady state condition prevails.

Split-plots in randomized blocks with four replications (four representative plants) were used to evaluate the effects of irrigation strategies on crop yield and distribution of root length density (RLD). For this purpose, the following parameters were determined:

- the distribution of root length density.
- the number of fruits
- the average weight of each fruit.

Roots samples were taken 20, 55 and 90 days after transplantation (DAT). The soil sample size was 10 × 10 × 10 cm³. The soil samples were taken at locations: R(0, 20 cm) and Z(0, 30 cm) at 20 DAT, R(0, 35 cm) and Z(0, 60 cm) at 55 and 90 DAT, respectively. Roots within each core sample were immediately cleaned, tagged, packed up into plastic bags and were immediately transferred to the laboratory and then counted and measured on a plate illuminated by a light.

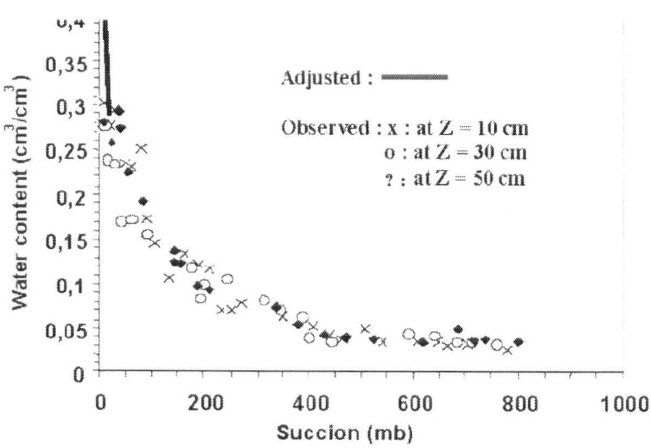

FIGURE 8.2 Soil moisture retention curve for the experimental site.

8.3 RESULTS AND DISCUSSION

The soil texture was based on the mechanical analysis of soil at (0 cm, 20 cm), (20 cm, 40 cm) and (40 cm, 60 cm). Table 8.1 shows the predominance of the sand fraction in the soil profile. High values of the bulk density are attributable to the soil texture (Table 8.1). The soil profile is virtually homogeneous since the bulk densities are similar at all soil depths and the retention curves are almost identical (Fig. 8.2). The equation by Van Genuchten [22] was fitted to observed data:

$$\theta = \theta_r + \{[\theta_s - \theta_r] \div [1 + (\alpha h)^n]^m\} \quad (5)$$

where: θ_s = saturated soil water content, θ_r = residual soil moisture condition, θ = soil moisture condition and α, n and m are constants given in Table 8.2, respectively. Using a nonlinear procedure for optimization, scattered data were fitted in Eq. (5). The results are summarized in Table 8.2. The coefficient of determination (R^2) of 0.99 indicates a good agreement between actual and fitted data (Fig. 8.2).

TABLE 8.2 Fitting Parameters for Van Genuchten Relationship

θ_s	θ_r	α	n	m	R^2
Cm^3/cm^3	cm^3/cm^3	cm^{-1}	–	–	–
0.38	0.02	0.05	1.70	0.41	0.99

The RLD data were measured along six vertical planes and were used to have a better understanding of the rooting system. The vertical planes were:

V1 = the vertical confounded with the plant stem;
V2 = the vertical at 10 cm from the plant stem and perpendicular to the crop row;
V3 = the vertical at 20 cm from the plant stem and perpendicular to the crop row;
V4 = the vertical at 30 cm from the plant stem and perpendicular to the crop row;
V5 = the vertical on the crop row at 10 cm from the plant stem; and
V6 = the vertical on the crop row at 20 cm from the plant stem.

The univariate ANOVA was used to evaluate the effects of the treatments on the average of RLD values (cm/cm^3). Twenty days after transplantation (DAT) and regardless of the treatment, the rooting depth along V1 was confined within 10 cm of the topsoil in the vicinity of the plant stem. The average RLD value was approximately equal to 0.40 cm/cm^3. This similarity is attributable to the fact that during the first 10 DAT, the same quantity of irrigation depth was provided in all blocks. Afterwards, significant differences in the rooting depths and RLD values were recorded among the treatments. Table 8.3 shows that the rooting depth at 55 DAT reached 22, 20, 25 and 30 cm in T0, Tl, T2, and T3, respectively. The rooting depth at 90 DAT was culminated at 40, 42, 60 and 62 cm in T0, T1, T2, and T3, respectively. These results are in agreement with those reported by Lubana and Narda [12], who stated that the rooting depth of tomato crop seldom exceeds 60 cm in field conditions.

TABLE 8.3 Root Length Density (RLD, cm/cm^3) Distribution Along V1 by Treatment and For Three Growing Stages

Soil layer (cm)	Days after transplanting, DAT											
	20	55	90	20	55	90	20	55	90	20	55	90
	T0			T1			T2			T3		
0–10	0.42a	0.813b	0.855b	0.373a	0.840b	0.935b	0.418a	0.835b	0.953b	0.403a	0.820b	0.990b
10–20	–	o.125a1	0.210b1	–	0.123a1	0.360cl	–	0.150a1b1	0.400c1	–	0.200b1	0.427c1
20–30	–	–	0.080a2b2	–	–	0.050a2	–	–	0.150b2	–	0.043a2	0.155b2
30–40	–	–	0.040a3b3	–	–	0.020a3	–	–	0.090b3	–	–	0.100b3
40–50	–	–	–	–	–	–	–	–	0.050a4	–	–	0.060a4
50–60	–	–	–	–	–	–	–	–	0.015a5	–	–	0.040a5

The values with the same letter are not significantly different at 5% level.

TABLE 8.4 Root Length Density (RLD, cm/cm^3) Distribution Along V2 by Treatment and For Three Growing Stages

Soil layer (cm)	Days after transplanting, DAT											
	20	55	90	20	55	90	20	55	90	20	55	90
	T0			T1			T2			T3		
0–10	0.090	0.140	0.455	0.085	0.140	0.453	0.085	0.172	0.525	0.105	0.190	0.548
	a	ab	a	a	ab	c	a	ab	c	ab	b	c
10–20	–	0.060	0.130	–	0.050	0.140	–	0.068	0.143	–	0.100	0.175
		a1b1	blcl		a1	blcl		albl	blcl		a1b1	cl
20–30	–	–	0.060	–	–	0.035	–	0.020	0.120	–	0.043	0.125
			a2b2			a2		a2	b2c2		a2	c2
30–40	–	–	0.030	–	–	0.010	–	–	0.058	–	–	0.068
			a3			a3			a3			a3
40–50	–	–	–	–	–	–	–	–	0.015	–	–	0.030
									a4			a4
50–60	–	–	–	–	–	–	–	–	–	–	–	–

The values with the same letter are not significantly different at 5% level.

RLD at 20 DAT within the upper soil layer was as high as 0.40 cm/cm^3 for all the treatments. RLD at 90 DAT reached 0.855, 0.935, 0.953 and 0.99 cm/cm^3 in T0, T1, T2 and T3, respectively. RLD at 20 DAT was zero cm/cm^3 in the second and in the underlying soil layers ($Z \geq 20$ cm), in all treatments. Conversely, RLD at 90 DAT within the second soil layer were 0.210, 0.360, 0.400 and 0.427 cm/cm^3 in T0, T1, T2 and T3, respectively. Significantly nonzero RLD values (0.05 and 0.06 cm/cm^3) in the third layer were observed only in treatments T2 and T3. Therefore, it is clear that the less frequent irrigations resulted in deeper and denser roots. These results agree with those of Keller and Bliesner [11] and Dalvi et al. [5], who suggest that more root expansion would occur if acceptable water stress is imposed. For apple trees, Sokalska et al. [18] observed higher number of roots in the surrounding dry soil than in the moistened bulb. Besides, the these results agree with those of Zotarelli et al. [23], who recorded RLD at 20 DAT between 0.25 cm/cm^3 and 1.0 cm/cm^3 in the top 10 cm of the soil profile. According to Gardenas et al. [8], the rooting depth of fertigation tomato seldom exceeds 25 cm. In the same context, Blaine et al. [2] considered a rooting depth of 20 cm as sufficiently convenient for managing irrigation of tomato crop. Therefore, it appears that tomato roots tend to be deeper and denser during the growing season. This behavior is of great significance as the crop is less frequently irrigated. Thus, it may be concluded that as tomato plants grow, their rooting system expands into the soil profile in order to fulfill the needs of increasing water and nutrients.

Nevertheless, the overwhelming majority of observed root lengths (more than 80%) were still confined within 40 cm of the topsoil even in the treatment T3. These results agree with those of Mickelakis et al. [14], Gardenas et al. [8] and Zotarelli et al. [23]. This behavior is attributed to soil resistance to penetration of roots and availability of nutrients and essentially soil pore density in this layer for the trickle irrigated fields. This may explain why reduced and frequent irrigations are more efficient than high and less frequent irrigations. It is obvious that excess of water in the root zone induces air shortage and harmful effects due to plant asphyxiation. Conversely, tomato crop cannot survive amid excessive water restrictions. Along the vertical V2, the same trend was observed. The effect of the treatment on the rooting depth and RLD was significant especially during the third growing stage. Maximum rooting depths were 42, 40, 50 and 54 cm in T0, T1, T2 and T3, respectively. It is worth emphasizing that RLD differences between growing stages on one hand and the treatments on the other hand were slightly less important than those observed along the vertical V1. As a matter of fact, Table 4 shows that RLD increased from 0.09, 0.085, 0.085 and 0.105 cm/cm^3 up to 0.455, 0.453, 0.525 and 0.548 in T0, T1, T2 and T3 treatments, respectively.

Table 8.5 shows that effects of the treatments on RLD distribution along the vertical V3 were also significant. On the contrary, the effects of the treatments on RLD distribution were not significant along the vertical V4. Table 8.6 shows that RLD values within the upper two layers were zero cm/cm^3 at 20 and 55 DAT. RLD values at 90 DAT ranged between 0.02 and 0.06 cm/cm^3 in all treatments.

TABLE 8.5 Root Length Density (cm/cm³) Distribution Along V3 by Treatment and For Three Growing Stages

Soil layer (cm)	Days after transplantation of tomato seedlings, DAT											
	20	55	90	20	55	90	20	55	90	20	55	90
	T0			T1			T2			T3		
0–10	–	0.0225a	0.060ab	–	0.045ab	0.065ab	–	0.045ab	0.093b	–	0.065ab	0.105b
10–20	–	0.010a1	0.038albl	–	0.020a1	0.050a1b1	–	o.025a1	0.085a1b1	–	0.080a1b1	0.105b1
20–30	–	–	0.015a2	–	–	0.035a2	–	–	0.075a2	–	0.020a2	0.080a2
30–40	–	–	–	–	–	–	–	–	0.040a3	–	–	0.060a3
40–50	–	–	–	–	–	–	–	–	–	–	–	0.0300
50–60	–	–	–	–	–	–	–	–	–	–	–	–

The values with the same letter are not significantly different at 5% level.

TABLE 8.6 Root Length Density (cm/cm³) Distribution Along V4 by Treatment and For Three Growing Stages

Soil layer (cm)	Days after transplantation of tomato seedlings, DAT											
	20	55	90	20	55	90	20	55	90	20	55	90
	T0			T1			T2			T3		
0–10	–	–	0.030a	–	–	0.033a	–	–	0.045a	–	–	0.053a
10–20	–	–	0.033a1	–	–	0.040a1	–	–	0.040a1	–	–	0.060a1
20–30	–	–	–	–	–	0.020a2	–	–	0.025a2	–	–	0.050a2
30–40	–	–	–	–	–	–	–	–	–	–	–	0.020a
40–50	–	–	–	–	–	–	–	–	–	–	–	–
50–60	–	–	–	–	–	–	–	–	–	–	–	–

The values with the same letter are not significantly different at 5% level.

TABLE 8.7 Root Length Density (cm/cm³) Distribution Along V5 by Treatment and For Three Growing Stages

Soil layer (cm)	Days after transplantation of tomato seedlings, DAT											
	20	55	90	20	55	90	20	55	90	20	55	90
	T0			T1			T2			T3		
0–10	0.100ab	0.132ab	0.423c	0.115ab	0.140ab	0.437c	0.095ab	0.167b	0.490c	0.085a	0.165b	0.500c
10–20	–	0.050a1	0.147bl	–	0.050a1	0.140bl	–	0.050a11	0.152bl	–	0.075a1	0.160bl
20–30	–	–	0.050a2b22	–	–	0.035a2b2	–	–	0.080b2	–	0.023a2	0.090b2
30–40	–	–	0.015a3	–	–	0.010a3	–	–	0.035a3	–	–	0.063a3
40–50	–	–	–	–	–	–	–	–	–	–	–	0.025
50–60	–	–	–	–	–	–	–	–	–	–	–	–

The values with the same letter are not significantly different at 5% level.

TABLE 8.8 Root Length Density (cm/cm³) Distribution Along V6 by Treatment and For Three Growing Stages

Soil layer (cm)	Days after transplantation of tomato seedlings, DAT											
	20	55	90	20	55	90	20	55	90	20	55	90
	T0			T1			T2			T3		
0–10	–	0.040a	0.061a	–	0.047a	0.075’	–	0.051a	0.090a	–	0.061a	0.095a
10–20	–	0.020a1	0.047a1	–	0.020a1	0.045a1	–	0.050a1	0.052a1	–	0.048a1	0.075a1
20–30	–	–	0.015a2	–	–	0.010a2	–	–	0.045a2	–	0.020a2	0.057a2
30–40	–	–	–	–	–	–	–	–	0.020a3	–	–	0.015a3
40–50	–	–	–	–	–	–	–	–	–	–	–	–
50–60	–	–	–	–	–	–	–	–	–	–	–	–

The values with the same letter are not significantly different at 5% level.

According to Tables 8.4 and 8.7, the comparison between on rows and within the same row, RLD distribution shows that approximately the same rooting depth was obtained regardless of the sampling direction. For the same distance from the stem, RLD values were similar throughout the growing season. The same behavior might be observed when comparing results of Table 8.5 with those of Table 8.8. Nevertheless, the spread of the rooting system was as large as 35 cm across the row, except one data along the row <20 cm. It should be noticed that these results are almost the same for all the treatments. It might be attributed to the relatively high soil moisture contents along the row and to tomato crop density (40 cm × 120 cm). It should be taken into consideration that regular spacing between emitters (30 cm) generated a continuous moistened strip along the row. These results are helpful for managing tomato fertigation and monitoring nutrient leaching.

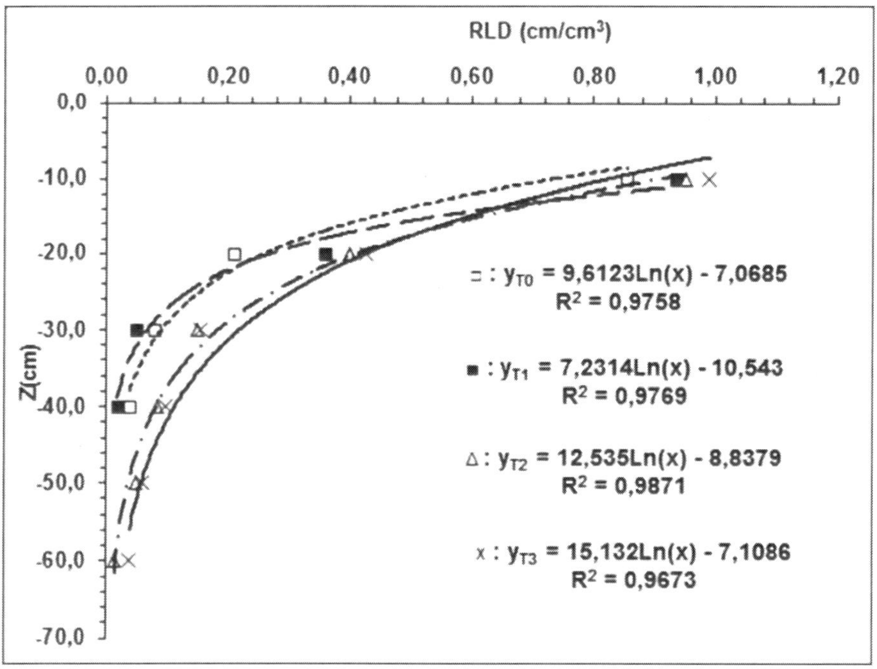

FIGURE 8.3 Variations of Tomato RLD in the Soil Profile (Along Vl at 90 DAT) in Four Irrigation Treatments.

The RLD values at 90 DAT and along the vertical VI are plotted in Fig. 8.3. The logarithmic equation was fitted to the scattered data for all the treatments:

$$Z = a \log_e (RLD) + b \quad (6)$$

where: Z (cm) is the rooting depth measured downwards, RLD = root length density and the fitting parameters a and b are found with regression analysis using appropriate boundary conditions:

$$\lim_{Z\to\infty} RLD(Z) = 0 \quad \text{and} \quad \lim_{Z\to\infty} RLD(Z) = e^{-b/a}$$

The parameters (a, b) are specific to the cropping conditions. Thus, Eq. (6) is useful for modeling water uptake by roots and managing fertigation in trickle irrigation. It is worthwhile to observe that sampling of roots is cumbersome, tedious and destructive process for the soil and plant. Therefore, empirical equations are often used to predict the rooting depth. Models described by Lubana and Narda [12] and Shashi and Sandhya [17] can be used for this purpose. The good-fit between measured and fitted rooting values can be evaluated by the following equations [1]:

$$I_a = 1 - \frac{\sum_{i=1}^{N}(Pi-Oi)^2}{\sum_{i=1}^{N}(|Pi'|+|Oi'|)^2} \tag{7}$$

$$RMSE = \sqrt{\frac{\sum_{i=1}^{N}(P_i-O_i)^2}{N}} \text{ and,} \tag{8}$$

where: P_i and O_i refer to the predicted and observed rooting depths, respectively. In Eq. (7), $P_i' = (P_i - O_m)$ and $O_i' = (O_i - O_m)$, with O_m and N are the average rooting depth and the number of observations, respectively.

It should be stressed that I_a is a valuation of the discrepancy between observed and predicted values. Likewise, the root mean square error (RMSE) reflects the more or less agreement between observed and predicted rooting depths. Apart from the treatment T1, Table 8.9 shows that I_a is higher than 0.86, which indicates a good agreement between actual and predicted data. With the treatments T2 and T3, Lubana and Narda [12] model provides lower RMSE value than the model by Shashi and Sandhya [17]. Consequently, model by Lubana and Narda [12] model can be used to reproduce the temporal distribution of tomato rooting system. The relatively high values of $RMSE_L$ recorded with the treatments T0 and T1 reflect a mitigated development of the rooting system as compared to the predicted data.

TABLE 8.9 Evaluation Models of Lubana and Narda [12] and Shashi and Sandhya [17] For Observed Data

RMSE	Irrigation treatments			
	T0	T1	T2	T3
$RMSE_L$ (cm)	13.29	18.94	9.51	7.70
$RMSE_S$ (cm)	10.97	12.70	11.96	12.5
I_{aL}	0.86	0.64	0.94	0.96
I_{aS}	0.89	0.86	0.90	0.89

The suffixes: L = Lubana model and S = Shashi model, respectively.

TABLE 8.10 Average Fruit Weight, Number of Fruits and Mean Yield of Tomato For Four Irrigation Treatments

Parameter	Units	Irrigation treatments			
		T0	T1	T2	T3
Number of fruits	—	39.00a	40.00a	43.25a	38.75a
Mean weight	g	64.63a’	79.00b’	79.53b’	61.40a’
Yield	Kg/plant	2.52 s”	3.16b"	3.44b"	2.38b”

The values with the same letter are not significantly different at 5% level.

Notwithstanding the nonsignificant difference between the treatments T1 and T2 from yield standpoint of view, data summarized in Table 8.10 indicate that the average fruit weight was significantly affected by the irrigation strategies and was increased with irrigation frequency. In turn, the average fruit number was not significantly affected. These results confirm the effects of water stress on crop yield [6]. Furthermore, these results conclude that an increase of irrigation frequency acts to reduce the gap between water application and plant needs. To reap the best from the previous results, it is advised to manage trickle irrigation of tomato in El Alia perimeter in accordance with treatment T2. As a matter of fact, T2 can save up to 23% of irrigation water and can boost the tomato yield by more than 40% with respect to the actual yield with common practices in Tunisia.

8.4 CONCLUSIONS

Sustainable irrigation should lead to timely application of appropriate gross water depths to meet crop demand, holding promise of increased yield and quality. The rooting system of trickle irrigated tomato crop was sampled at three growing stages for four irrigation treatments. Experimental results show similar trend in rooted soil volume distribution irrespective of the applied treatments during the initial growing stage. As plants thrive, the rooting system expands in all the directions. Twenty days after transplantation, the effects of irrigation strategies on the rooted soil volume was not significant. In turn, it was significant at 55 DAT and more significant at 90 DAT. During these stages, records indicate that the less frequent irrigation caused deeper and denser rooting system. Besides, experimental results show expansion of larger roots across the row than along the row. These results are useful for fertigation management of tomato crop where the irrigation at 55 DAT saved 23% of water and boosted the crop yield by more than 40% compared to the control treatment.

8.5 SUMMARY

Crop water uptake and subsequent root distribution are basic for soil tillage operation, water transfer modeling, fertigation and management of drainage networks. To assess the impact of irrigation scheduling on root length density, four watering strategies were used for trickle irrigated tomato in Tunisia. Irrigation was started as soon as the pressure head within the rooted zone was lowered to −250 mb, −400 mb and −550 mb for the treatments T1, T2 and T3, respectively. The control treatment T0 was based on a Class A pan evaporimeter. Experimental results illustrated virtually similar rooted soil volumes for all the treatments during the first growing stage. The rooting depth remained confined within the 10 cm of the topsoil with approximately similar root length densities (roughly 0.40 cm/cm^3). Afterwards, roots expanded in all the directions during the growing season. Moreover, it is concluded that infrequent irrigation generated deeper and denser roots, thus better root anchorage. Likewise, the root expansion along the row was lower than that across the row. These results are helpful for farm water management.

KEYWORDS

- **class A pan**
- **days after transplantation**
- **irrigation frequency**
- **irrigation management**
- **irrigation strategies**
- **root length density, RLD**
- **soil moisture curve**
- **suction head**
- **tomato**
- **trickle irrigation**
- **Tunisia**

REFERENCES

1. Ayman Suleiman, A. M., Soler, C. T. Hoogenboom, G. (2007). Evaluation of FAO-56 crop coefficient procedures for deficit irrigation management of cotton in a humid climate. *Agric. Water Manage, 9,* 33–42.
2. Blaine, R. H., Similnek, J., Hopmans, J. W. (2006). Evaluation of urea-ammonium-nitrate fertigation with drip irrigation using numerical modeling. *Agric. Water Manage., 86,* 102–113.
3. Brent, Clothier, E., Steve, R. S. G. (1994). Root zone processes and the efficient use of irrigation water. *Agric. Water Manage., 25,* 1–12.
4. Chopart, J. L., Vauclin, M. (1990). Water balance estimation model: Field test and sensitivity analysis. *Soil Science Soc. Am. J., 54,* 1377–1384.
5. Dalvi, V. B., Tiwari, K. N., Pawade, M. N., Phirke, P. S. (1999). Response surface analysis of tomato production under micro irrigation. *Agric. Water Manage, 41,* 11–19.
6. Doorenbos, J., Pruitt, W. O. (1977). *Guidelines For Predicting Crop Water Requirements.* FAO Irrigation and Drainage paper No. 24, Rome.
7. Elmaloglou, S. T., Malamos, N. (2006). A methodology for determining the surface and vertical components of the wetting front under a surface point source, with root water uptake and evaporation. *Irrig. and Drain, 55,* 99–111.
8. Gardenas, A. I., Hopmas, J. W., Hanson, B. R., Simunek, J. (2005). Two-dimensional modeling of nitrate leaching for various fertigation scenarios under microirrigation. *Agric. Water Manage., 74,* 219–242.
9. Gardner, W. R. (1991). Modeling water uptake by roots. *Irrig. Sci., 12,* 109–114.
10. Hammami, M., Dagliari, H., Balli, J., Maalej, M. (2002). Approach for predicting the wetting front depth beneath a surface point source: Theory and Numerical Aspect. *Irrig. Drain. Eng., 51,* 347–360.
11. Keller, J., Bliesner, R. D. (1990). *Sprinkler and Trickle Irrigation.* New York, Van Nostrand Reinhold.
12. Lubana, P. P. S., Narda, N. K. (1998). Soil water dynamics model trickle irrigated tomatoes. *Agric. Water Manage., 37,* 145–161.
13. Merriam, J. L., Keller, J. (1978). *Farm Irrigation System Evaluation.* Utah State University. Pages 217.
14. Michelakis, N., Vougioucalou, E., Clapaki, G. (1993). Water use, wetted soil volume, root distribution and yield of avocado under drip-irrigation. *Agric. Water Manage., 24,* 119–131.
15. Protopopas, A., Bras, R. L. (1987). A model for water uptake and development of root system. *Soil Sci., 144,* 352–366.
16. Salgado, E., Cautin, R. (2008). Avocado root distribution in fine and coarse-textured soils under drip and microsprinkler irrigation. *Agric. Water Manage., 95,* 817–824.
17. Shashi, M., Sandhya, R. (1999). Modeling water uptake plant roots. *Journal of Irrigation and Drainage Engineering, 124(3),* 158–165.
18. Sokalska, D. I., Haman, D. Z., Szewczuk, A., Sobota, J., Deren, D. (2009). Spatial root distribution of mature apple trees under drip irrigation system. *Agric. Water Manage., 96,* 23–34.
19. Steve, R. G., Kirkham, M. B., Clothier, B. E. (2006). Root uptake and transpiration: From measurements and models to sustainable irrigation. *Agric. Water Manage., 86,* 165–174.
20. Tliabet, M., Zayani, K. (2008). Wetting patterns under trickle source in a loamy sand soil of south Tunisia. American-Eurasian. *J. Agric. and Environ. Sci., 3(1),* 38–42.
21. Todd, Skaggs, H., Martinus, Th., Genuchten, Van, M. T., Peter Shouse, J., James Poss, A. (2006). Macroscopic approaches to root water uptake as a function of water and salinity stress. *Agricultural Water Management, 86,* 140–149.
22. Van, Genuchten, M. T. (1980). A closed-form equation for predicting the hydraulic conductivity of unsaturated soils. *Soil Science Soc. Am. J., 44,* 892–897.

23. Zotarelli, L., Scholberg, J. M., Dukes, M. D., Carpena, R. M., Icerman, J. (2009). Tomato yield, biomass accumulation, root distribution and irrigation water use efficiency on a sandy soil, as affected by nitrogen rate and irrigation scheduling. *Agric. Water Manage., 96,* 23–34.

CHAPTER 9

PERFORMANCE OF DRIP IRRIGATED TOMATO

RAJBIR SINGH, SATYENDRA KUMAR, D. D. NANGARE, and M. S. MEENA

CONTENTS

Printed and modified from Rajbir Singh, Satyendra Kumar, D. D. Nangare and M. S. Meena, 2009. *Drip irrigation and black polyethylene mulch influence on growth, yield and water-use efficiency of tomato. African Journal of Agricultural Research, 4(12), December.* Open Source Access, http://www.academic-journals.org/AJAR.

All prices in this chapter are based on 2003. The currency in this chapter is in Indian Rupees, with a conversion factor of Rs. 45.00 = 1.00 US$.

9.1 INTRODUCTION

Drip irrigation has proved its superiority over the conventional method of irrigation, especially in the cultivation of fruits and vegetables due to precise and direct application of water in root zone. A considerably saving in water, increased growth, development and yield of drip irrigated vegetables has been reported [2, 3, 8, 11]. The use of black polyethylene mulch in vegetable production has been reported to control the weed incidence, to reduce nutrient losses and to improve the hydrothermal regimes of soil [1, 4, 12]. However, limited information is available regarding the effects of drip irrigation alone and in conjunction with polyethylene mulch as compared with surface irrigation on growth and yield of tomato (*Lycopersicon esculentum* Miller). As tomato is the most important vegetable crop, such information is required for developing new strategies for intensive production of vegetables. Therefore, this chapter discusses the effects of different levels of drip irrigation with and without polyethylene mulch on growth, yield, water-use efficiency and economics of tomato.

9.2 MATERIALS AND METHODS

A field experiment was conducted during winter season of 2001–2002 and 2002–2003 at research farm of Central Institute of Post- Harvest Engineering and Technology, Abohar (latitude 30°09′ N, longitude 74°13′ E, 185.6 m above mean sea level), Punjab, India. The soil at the experimental plot was sandy loam with a pH of 8.4, poor in organic carbon and available nitrogen, medium in phosphorus and rich in potash content.

The following eight treatments were applied in a randomized black design and replicated thrice:

T1: drip irrigation with 1.00ET based on pan evaporation;
T2: drip irrigation with 0.80ET;
T3: drip irrigation with 0.60ET;
T4: surface irrigation;
T5: T1 + black polyethylene mulch;
T6: T2 + black polyethylene mulch;
T7: T3 + black polyethylene mulch; and
T8: T4 + black polyethylene mulch.

The volume of water for 1.00ET based on pan evaporation was computed using the following equation:

$$V = \text{Total } [(\text{Ep} \times \text{Kc} \times \text{Kp} \times \text{A} \times \text{N}) - (\text{Re} \times \text{A})] \quad (1)$$

where: V = volume of water required for 1.00ET = Ep; Ep = average monthly pan evaporation (mm/day); Kc = crop factor; Kp = pan factor; A = area of plot (m^2);

N = number of days in a month for which the volume of applied water needs to be calculated; Re = effective rainfall (mm); Total = signifies total of the crop season.

Table 9.1 shows the data on average pan evaporation, monthly effective rainfall, volume of water applied for 100% ET during the experimental period. The crop factor values for different crop stages were computed based on the existing relative humidity and wind velocity [5]. The pan factor value was 0.75 as suggested by USDA Class A pan. The area of plot was 9.0 m^2. A buffer zone spacing of 1.5 m was provided between the plots. In the treatment of surface irrigation (T4) and surface irrigation + black polyethylene mulch (T8), 14 irrigations each of 5 cm depth were applied.

TABLE 9.1 Pan Evaporation, Crop Factor, Effective Rainfall and Irrigation Depth At 1.00ET During the Experimental Period

Month	Average pan evaporation	Crop factor	Effective rainfall	Irrigation depth = 1.00ET
	mm/day	—	mm	mm
December (after 15 Dec)	2.37	0.90	-	25.6
January	1.37	0.90	0.6	28.7
February	2.56	1.02	10.7	54.3
March	3.34	1.02	11.6	78.6
April	6.28	1.05	9.3	147.6
May (upto 22 May)	7.69	0.85	6.8	151.2

Note: Pooled data of two years.

The 35 days old seedlings of tomato cv. Rupali were transplanted on the 15th of December for both years. This was done using row-to-row distance of 100 cm and plant-to-plant distance of 50 cm, respectively. All recommended cultural and plant protection operations were followed to raise the healthy crop according to package practices for tomato by Punjab Agricultural University. For mulching, black polyethylene film of 50 microns thickness was spread manually over the prepared field and tomato seedlings were transplanted by making holes of 5 cm diameter on the film. Lateral drip lines with emitter spacing of 50 cm with a discharge rate of 4 lph were placed in each row of plants both in unmulched and below the polyethylene mulch treatments. Volumetric method (the Eq. (1)) was used for calculating the uniformity coefficients (U_c) of drip irrigation system [11]. Data were recorded for plant height, leaf area index, fruit weight and fruit yield using standard methods. After the final harvest, the plants were cut at soil surface and the dry weight of top growth (stem and leaves) were determined after complete drying at 60 °C. The water use efficiency was computed by dividing tomato yield with total water applied (cm).

For economic analysis, total cost of production (fixed and operating costs of drip irrigation system) under different irrigation scheduling events with and without mulch was estimated [8]. The total cost of production was calculated by adding fixed cost, operating cost and production cost. The gross returns for different treatments were calculated taking into account the wholesale price of marketable tomatoes. The net returns were calculated considering gross returns and total cost of production. The benefit cost ratio (BCR) was estimated dividing gross return by total cost of production for each treatment.

9.3 RESULTS AND DISCUSSION

The uniformity coefficient (U_c) of drip irrigation system was found to be 92.5% during the experimentation. The high values of uniformity coefficients indicated excellent performance of drip irrigation system in supplying water uniformly throughout the lateral lines during the experiment. The data on growth parameters like plant height, leaf area index (LAI) and dry matter (Table 9.2) indicated that drip irrigated treatments, irrespective of mulch and unmulched treatments, produced significantly higher plant height, LAI and dry matter of the plant the corresponding values in surface irrigation.

TABLE 9.2 Effects of Eight Treatments on Growth Parameters and Fruit Weight of Tomato (Pooled Data of Two Years).

Treatments	Plant height	Leaf area index	Plant dry matter	Fruit weight
	cm	—	g	g
T1	80.4	3.14	34.1	76.4
T2	83.1	3.26	38.4	79.6
T3	74.6	2.86	27.5	69.8
T4	69.4	2.53	21.1	62.3
T5	83.4	3.71	41.3	81.7
T6	85.5	3.89	46.6	86.6
T7	79.4	3.38	34.3	75.9
T8	75.3	2.97	27.5	68.1
LSD (P = 0.05)	3.1	0.14	4.3	5.0

Drip irrigation without mulch with 1.00ET (T1), 0.80ET (T2) and 0.60ET (T3) treatments increased the dry weight of plant by 61.6 in T1, 82.0 in T2 and 30.3% in T3, respectively, compared to corresponding values in surface irrigation. The corresponding value for drip irrigation with black polyethylene mulch with these levels were 50.2 in T1, 69.5 in T2 and 24.7% in T3 higher than in surface irrigation plus mulch. Application of mulch in surface irrigation also increased the dry matter by 30% over unmulched surface irrigation. Plant height and leaf area index also followed similar trends of dry matter production (Table 9.2). Bhella [3], Bafna et al. [2] and Raina et al. [11] also reported significantly higher plant growth of tomato with drip irrigation compared to surface irrigation.

TABLE 9.3 Effect of Different Treatments on Fruit Yield (tones/ha) Water Use Efficiency and Benefit: Cost Ratio of Tomatoes (Pooled Data of Two Years).

Treatments	Yield	Water applied	Water use efficiency	Gross returns	Net returns
	t/ha	cm	t/(ha-cm)	Rs./ha	Rs./ha
T1	42.02	52.4	0.80	73535	27909
T2	45.57	43.1	1.06	79747	34431
T3	34.52	33.7	1.02	60410	15407
T4	29.43	70.0	0.42	51502	19146
T5	52.58	52.4	1.00	92015	41819
T6	57.87	43.1	1.34	101272	51386
T7	43.75	33.7	1.30	76562	26989
T8	36.06	70.0	0.51	63105	26183
LSD (P = 0.05)	2.85	-	-	-	-

Based on Indian Rupees (@ one US$ = 45.00 Indian rupees, 2003).

Irrespective of mulching, significantly higher fruit weight was observed with drip irrigation compared to surface irrigation (Table 9.2). Drip irrigation without mulch with 1.00ET (T1) and 0.80ET (T2) increased the fruit weight by 22.6 and 27.8%, respectively, compared to surface irrigation (Table 9.2). The corresponding increase with drip plus mulch (T5 and T6) was 19.9 and 27.1%, respectively compared to surface irrigation plus mulch (T8). Fruit weight was highest with 0.80ET irrigation level compared to other irrigation levels either with or without mulch. Surface irrigation recorded lowest fruit weight without mulch. Elkner and Kaniszewski [6] also observed significantly higher fruit weight of tomato under drip irrigation as compared to control treatment.

The data on fruit yield (t/ha) of tomato (Table 9.3) indicated that drip irrigation gave significantly higher yield compared to surface irrigation, irrespective of mulching. Drip irrigation without mulch with 1.00ET (T1), 0.80ET (T2) and 0.60ET (T3) increased the fruit yield by 42.8, 54.8 and 17.3%, respectively, compared to surface irrigation (T4). The corresponding value for drip irrigation plus plastic mulch with these levels (T5, T6 and T7) was 45.8, 60.5 and 21.3% higher than the yield in surface irrigation plus mulch (T8). Application of black plastic mulch in surface irrigation (T8) also increased fruit yield by 22.5% compared to unmulched surface irrigation (T4).

The increased yield under drip irrigation may be due to better water utilization [10], higher uptake of nutrients [2] and excellent soil-water relationship with higher oxygen concentration in the root zone [7]. Surface irrigation resulted in wastage of water due to deep percolation, leaching of available plant nutrients and poor aeration resulting in poor yield [11]. Our results are in accordance with the earlier findings of Bhella [3] who observed 70% higher tomato yield under drip irrigation compared to surface irrigation. Bafna et al. [2] and Raina et al. [11] also reported 40% increase in tomato yield with drip irrigation compared to surface irrigation.

A comparison of different levels of irrigation indicated maximum tomato yield with 0.80ET both in mulch (T6) and unmulched (T2) treatments (Table 9.3). Raina et al. [11] also observed highest tomato yield where irrigations through drip were applied at 0.80Epan. However, Locascio et al. [9] reported that drip irrigation requirement of tomato was about 50% of USDA Class A pan evaporation in fine sandy soil compared to 50 to 100% of pan evaporation in fine sandy loam soil.

Application of black polyethylene mulch increased the tomato yield in all levels of irrigation, however the response was comparatively higher in T5 and T6 treatments (Table 9.3). Higher yield under mulch treatments may be due to the weed control. In our study, there was complete elimination of weeds in black polyethylene mulch, whereas in unmulched plots weeding was done manually seven times during both years of experimentation. Chakaraborty and Sadhu [4] and Singh [12] also reported complete elimination of weeds with the use of black polyethylene. The higher fruit yield in polyethylene mulch may be ascribed to reduced nutrient losses due to weed control and improved hydrothermal regimes of soil [1, 3, 12].

The total irrigation depth varied from 33.7 to 70.0 cm in different treatments (Table 9.3). Drip irrigation, both with and without polyethylene mulch, registered higher water use efficiency (WUE) compared to WUE in surface irrigation. Highest WUE was 1.34 tons/ha-cm with drip irrigation at 0.80ET with mulch or 1.06 tons/ha-cm in plots without mulch compared to other levels of irrigation with or without mulch. Considering the average value for all levels of irrigation, drip irrigation without mulch gave WUE of 0.97 tons/ha-cm compared to 0.42 tons/ha-cm in surface irrigation. The corresponding value was 1.23 tons/ha-cm for drip plus mulch and 0.51 tons/ha-cm in surface irrigation plus mulch, respectively. Since, the losses due evaporation from soil surface was much lower in drip irrigation, WUE was higher compared to WUE in surface irrigation. A comparison of different levels of irrigation indicates maximum WUE with 0.80ET both in mulch (T6) and unmulched (T2) treatments (Table 9.3). These results are in agreement with the earlier findings of Bafna et al. [2] and Raina et al. [11] for WUE of drip irrigated tomato crop.

Table 9.3 indicates total cost of production, gross returns, net returns and BCR of tomato in all eight treatments. Drip irrigation with and without polyethylene mulch registered higher net returns and BCR compared to corresponding values in surface irrigation. Among different irrigation levels, drip irrigation at 0.80ET (T2) resulted in maximum returns (34431 Rs/ha) and higher BCR (1.76) in tomato. However, highest net returns (51386 Rs/ha) and BCR (2.03) was found with drip irrigation at 0.80ET with polyethylene mulch (T6). This was due to the fact that irrigation at 0.80ET with mulch resulted in optimum plant growth and development with improvement in water use efficiency.

9.4 CONCLUSIONS

The present study indicated that drip irrigation at 0.80ET with polyethylene mulch resulted in significantly highest yield, water use efficiency and maximum BCR in tomato. The drip system besides giving a saving of 39% water resulted in 55% higher fruit yield of tomato compared to surface irrigation. Therefore, drip irrigation system is a very effective and efficient method of irrigation for raising tomato crop especially on light texture sandy loam soil.

9.5 SUMMARY

During the winter season of 2001–2002 and 2002–2003, a two year field study on sandy loam soil was conducted to investigate the effects of drip irrigation and black polyethylene plastic mulch compared with the surface irrigation, on growth, yield, water-use efficiency and economics of tomato (*Lycopersicon esculentum* Miller). Drip irrigation application at 80% evapotranspiration (ET) crop based on pan evaporation gave significantly higher fruit yield (45.57 tons/ha) compared with the surface irrigation (29.43 tons/ha). Use of black polyethylene mulch plus drip irrigation further raised the fruit yield to 57.87 tons/ha. Plant height, LAI, dry matter production, fruit weight and yield were increased significantly with the use of drip irrigation alone and in conjunction with polyethylene mulch compared with surface irrigation alone or with mulch. WUE (tons/ha-cm) was 0.97 for drip irrigation alone, 1.23 for drip irrigation with polyethylene mulch and 0.42 for surface irrigation, respectively. Among different irrigation levels, drip irrigation at 0.80ET resulted in higher net returns (34431 Rs./ha) and benefit cost ratio (1.76). However, maximum net returns (51386 Rs./ha) and benefit cost ratio (2.03) was found with drip irrigation at 0.80ET coupled with polyethylene mulch compared with other treatments. Drip irrigation also resulted in a saving of 38% water and 55% higher fruit yield compared with surface irrigation.

KEYWORDS

- benefit cost ratio
- black polyethylene
- drip irrigation
- evapotranspiration, ET
- fruit yield
- India
- leaf area index LAI
- mulch
- pan evaporation
- sandy loam soil
- surface irrigation
- tomato
- water-use efficiency, WUE

REFERENCES

1. Ashworth, S., Harrison, H. (1983). Evolution of mulches for use in the home garden. *Hort. Sci., 18(2),* 180–182.
2. Bafna, A. M., Daftardar S. Y., Khade, K. K., Patel, P. V., Dhotre, R. S. (1993). Utilization of nitrogen and water by tomato under drip irrigation system. *J. Water Management, 1(1),* 1–5.
3. Bhella, H. S. (1988). Tomato response of trickle irrigation and black polyethylene mulch. *J. Am. Society for Hort. Sci., 113(4),* 543–546.
4. Chakarborty, R. C., Sadhu, M. K. (1994). Effect of mulch type and color on growth and yield of tomato. *Indian J. Agric. Sci., 64(9),* 608–612.
5. Doorenbos, J., Pruitt, W. O., Aboukhaled, A., Dastane, A. B. (1984). Guidelines for predicting crop water requirements. FAO Irrigation and Drainage Paper 24, FAO, Rome.
6. Elkner, K., Kaniszewski, S. (1995). Effect of drip irrigation and mulching on quality to tomato fruits. *Acta Hort., 379,* 175–80.
7. Gornat, B., Goldberg, D., Rimon, D., Asher, B. J. (1973). The physiological effect of water quality and method of application on tomato, cucumber and pepper. *J. Am. Society for Hort. Sci., 98(2),* 202–205.
8. Imtiyaz, M., Mgadla, N. P., Chepete, B., Mothobi, E. O. (2000). Yield and economic returns of vegetable crops under varying irrigation. *Irrigation Sci., 19,* 87–93.
9. Locascio, S. J., Olsen, S. M., Rhoads, F. M. (1989). Water quantity and time of N and K application of tickle irrigated tomatoes. *J. Am. Society for Hort. Sci., 114(2),* 265–268.
10. Manifrinato, H. A. (1974). Effect of drip irrigation on soil water plant relationships. Second International Drip Irrigation Congress, pages 446–451.
11. Raina, J. N., Thakur, B. C., Verma, M. L. (1999). Effect of drip irrigation and polyethylene mulch on yield, quality and water-use efficiency of tomato. *Indian J. Agric. Sci., 69,* 430–433.
12. Singh, R. (2005). Influence of mulching on growth and yield of tomato (*Lycopersicon esculentum*) in North India plains. *Vegetable Sci., 32(1),* 55–58.

CHAPTER 10

PERFORMANCE OF DRIP IRRIGATED NAVEL ORANGE IN NORTH DELTA, EGYPT

A. S. EL-HENAWY, M. R. KHALIFA, S. A. MASHALI, S. M. YOUSSEF, and A. RABIE

CONTENTS

In this chapter, area is in units of *feddans*. A *feddan* (Arabic: فدّان, *faddān*) is a unit of area. It is used in Egypt, Sudan and Syria. The feddan is not an SI unit and in Classical Arabic, the word means 'a yoke of oxen', implying the area of ground that could be tilled by oxen in a certain time. In Egypt the feddan is the only nonmetric unit which remained in use following the switch to the metric system. One fed. = 24 kirat = 60 m × 70 m = 4200 m^2 = 0.42 hectares.

10.1 INTRODUCTION

Surface irrigation method is the most common method of irrigating most of citrus orchards and field crops, particularly in Nile Delta lands of Egypt. National strategy of Egyptian Government includes saving in irrigation water by changing traditional irrigation systems to modern systems. Drip irrigation is one of the developed methods for irrigation.

Citrus production is the most important crop among the Egyptian growers. According to 2002 statistics, citrus acreage in Egypt represented approximately 39% of the total area of fruit trees. It is about 340,443 feddans from which 3047 feddans are cultivated in Kafr El-Sheikh Governorate. Surface irrigation is used in all areas [23]. On the other hand, rice straw poses an environmental problem in Egypt. Some farmers are burning rice straw in the fields directly, which causes phenomena called "Black Cloud" in the atmosphere of Cairo city. Rice straw can be used as soil mulch and to produce organic compost [21].

Nath and Sharma [26] found that mulching increased growth and yield of Assam lemon compared with bare soil. Khalifa et al. [19] mentioned that the highest values of fruit weight (220.4 g per fruit), total soluble solids percentage (TSS%), TSS/acid ratio and the lowest value of acidity (%) were found under border strip irrigation compared to basin, ring-shaped and basin furrow irrigation methods. However, vitamin C and peel thickness were not significantly affected by irrigation methods. Khalifa [20] reported that number of fruit drop per tree decreased on mulched soil (especially under soil surface covered with cutter weeds and soil surface covered with rice straw, 2 cm layer) compared with clean weeded (control).

Therefore, this chapter discusses the results to evaluate the effects of soil mulch and irrigation treatments on fruit growth and quality and yield of navel orange at Northern area of Nile Delta, Egypt.

10.2 MATERIALS AND METHODS

10.2.1 EXPERIMENTAL DESIGN

During two successive growing seasons of 2003 and 2004, this research was conducted at Citrus Orchard of Sakha Agricultural Research Station Farm, Kafr El-Sheikh Governorate, Egypt. Some chemical and physical properties of soil at the experimental site are shown in Table 10.1.

TABLE 10.1 Some Chemical and Physical Properties of Soil Samples At the Beginning of Experiment

Soil variable	Soil depth, cm			
	0–30	30–60	60–90	Mean
pH (1:2.5 soil suspension)	8.06	8.18	8.66	-
EC, dS/m(1:5 soil water extract)	0.723	0.485	0.356	0.521
Soluble cations and anions, meq /L				
Na^{+}	0.84	1.28	1.67	1.26
K^{+}	0.91	0.45	0.38	0.58
Ca^{++}	2.70	1.60	0.60	1.63
Mg^{++}	2.80	1.60	0.90	1.77
Cl^{-}	0.80	0.60	0.60	0.67
CO_3^{-}	0.00	0.00	0.00	0.00
HCO_3^{-}	1.40	1.30	1.50	1.40
SO_4^{-}	5.05	3.03	1.45	3.18
Total N, %	0.154	0.112	0.070	0.112
Available P, mg/kg soil	15.24	7.90	2.54	8.56
Available K, mg/kg soil	1154	800	624	859.3
Organic matter, %	1.31	0.99	0.76	1.02
Field Capacity, %	45.21	46.19	45.29	45.56
Wilting point, %	23.81	24.51	23.85	24.06
Available water, %	21.40	21.68	21.44	21.51
Bulk density, Mg / m^3	1.276	1.442	1.514	1.411
Particle size distribution, %				
Clay	66.57	67.43	63.37	65.79
Silt	27.94	27.29	31.14	28.79
Sand	5.49	5.28	5.49	5.42
Texture class	clay	clay	clay	clay

The citrus orchard was 10 years old, budded on sour orange rootstock and the tree spacing was 5×5 meters. The split plot design was used with four replicates in two seasons (a replication = one tree). The main plots were covered with 2 cm layer of rice straw as a whole and bare soil without any mulch. The subplots were subjected to surface irrigation (Basin method), drip irrigation 1.00 ET_C, drip irrigation 0.75 ET_C and drip irrigation 0.50 ET_C. The trees received eight irrigations during the study period of each season.

10.2.2 IRRIGATION WATER REQUIREMENTS OF NAVEL ORANGE

The climatic data were collected from Sakha Weather Station during 1993 to 2002. In this study, the data of the previous years was used to calculate the irrigation water requirements of navel orange in all irrigation treatments. Reference evapotranspiration (ET_o) with Penman-Montieth model was calculated using FAO-CROPWAT software computer program [27]. The crop evapotranspiration (ET_C) was estimated according to Doorenbos and Pruitt [10]. Gross irrigation water requirements (GIR) were calculated according to FAO [13] by using average of (ET_o) that was obtained with the previous meteorological data (1993–2002), 0.2 leaching fraction, 0.85 irrigation efficiency of drip irrigation and 0.65 reduction factor of navel orange trees. The GIR was estimated as 2918.1 for 1.00ETc, 2188.5 for 0.75ETc and 1459.0 m^3/feddan for 0.50ET_C, respectively.

Each tree was irrigated with three online emitters placed on polyethylene tubing of 16 mm OD. The emitters were placed at the middle of canopy tree cover. The drip irrigation interval (on/off) was timed every three days. The supply of water was from a storage tank (200 L) of about 0.5-meter height above the ground surface to give an adequate pressure. Each tank was able to irrigate two trees.

10.2.3 FERTILIZATION PRACTICES

All trees received regular fertilization dosages consisting of 960 of urea + 550 of super phosphate + 900 gm/tree of potassium sulfate, during the 1st season and alternated urea by 2170 gm/tree of ammonium sulfate during 2nd season. All fertilizers were broadcasted in surface irrigation and fertigated in drip irrigation treatments. All cultivation practices were performed according to recommendations by the Ministry of Agriculture and Land Reclamation [23].

10.2.4 MEASUREMENTS FOR PERFORMANCE OF NAVEL ORANGE

In October of each year, leaves of navel orange tree were taken, carefully washed with distilled water, then oven dried at 70°C for 48 h, grounded and wet digested by mixture of H_2SO_4 and H_2O_2 according to the standard method and were used for analysis [6]. Total nitrogen was determined by micro-Kjeldahi according to Cottenie et al. [9]. Total phosphorus and potassium were determined according to Carter [3]. Soil chemical analysis was conducted by using standard method according to Klute [22].

At harvest, fruit yield of navel orange tree was recorded for each treatment. Total yield was calculated (Kg/tree and tons/feddan). Number of fruits per tree was also recorded. The crop water use efficiency (WUE) was estimated by dividing the fruit yield (Kg/fed.) by actual evapotranspiration (expressed as irrigation depth) as described by Hukkeri et al. [17].

At harvest, 20 mature fruits were randomly collected from each tree in each treatment for the determination of TSS%, acidity, vitamin C, fresh weight, fruit volume, juice volume and peel thickness according to Association of official analytical chemists [1]. Average number of fruits falling was recorded for each treatment during the experimental period and calculated as % of mature fruits at the end of each season. Navel orange growth indicators included tree height starting from grafting point and diameter of trunk tree under grafting point [7]. Tree canopy was calculated according to Morse and Robertson [25].

The data were statistically analyzed according to Snedecor and Cochran [28].

10.3 RESULTS AND DISCUSSION

10.3.1 EFFECTS OF SOIL MULCHING AND IRRIGATION LEVELS ON PERFORMANCE OF DRIP IRRIGATED NAVEL ORANGE IN EGYPT

10.3.1.1 TREE VIGOR (M^3)

The variation in growth vigor of citrus tree is considered an important biological parameter, which is influenced by both the irrigation performance and the level of water application [5, 8]. Table 10.2 showed that the tree vigor was significantly affected by soil mulching and irrigation factor in 2003 and 2004 seasons, respectively. The highest values of tree vigor were found in drip irrigation 1.00 ET_C (also called full irrigation) and drip irrigation 0.75ET_C, treatments under bare soil and surface irrigation under soil mulching in the 2003 season. The highest value of tree vigor was 15.9 m^3 for drip irrigation 0.75ET_C and 15.8 m^3 for drip irrigation 1.00ET_C treatments under bare soil in season 2004, respectively. Tree vigor values were higher under drip irrigation than those under surface irrigation during the two seasons, except under soil mulching in 2003 season. These results are in agreement with those obtained by Castel et al. [4].

10.3.1.2 FRUIT YIELD

During both seasons, Table 10.2 indicates that the yield of Navel orange was higher under drip irrigation than those under surface irrigation. Also, yield under soil mulching with 2 cm layer of rice straw gave highest yield than that under non-mulched soil (bare soil). The highest yield was in drip irrigation 0.75 ET_C and drip irrigation 1.00 ET_C treatments under soil mulching and was 28.7 and 37.9 Kg/tree in seasons 2003 and 2004, respectively. The lowest yield was 21.5 and 26.1 Kg/tree in surface irrigation and drip irrigation 0.50 ET_C treatments under bare soil in seasons

2003 and 2004, respectively. The statistical analysis revealed that the mean yield (Kg/tree) under bare soil and soil mulching was nonsignificant, while irrigation factor was highly significant on mean yield of Navel orange. Interaction effect between mulching and irrigation factors on yield was not significant. These results are in agreement with Chung et al. [8], Khalifa [21] and El-Zawily [12].

TABLE 10.2 Effects of Soil Mulching and Irrigation Levels on Tree Vigor and Yield of Navel Orange During 2003 and 2004 Seasons

Mulch	Irrigation	Tree vigor (m^3)		No. of fruits falling / tree		Fruit yield (Kg/ tree)	
		2003	2004	2003	2004	2003	2004
Without	A	10.5^{a}	11.8^{b}	11.0^{a}	14.0^{a}	21.5^{a}	25.6^{b}
	B	14.2^{a}	15.8^{a}	9.0^{ab}	14.8^{a}	23.6^{a}	35.7^{a}
	C	14.2^{a}	15.9^{a}	5.8^{b}	13.3^{a}	24.3^{a}	33.1^{a}
	D	11.6^{a}	13.2^{ab}	11.3^{a}	11.5^{a}	23.0^{a}	26.1^{b}
	Mean	12.6	14.2	9.3	13.4	23.1	30.1
With	A	13.4^{a}	13.3^{a}	10.3^{ab}	13.3^{a}	22.1^{b}	28.3^{b}
	B	13.0^{a}	15.4^{a}	6.5^{bc}	14.0^{a}	24.1^{b}	37.9^{a}
	C	9.5^{a}	13.5^{a}	4.0^{c}	13.8^{a}	28.7^{a}	27.8^{b}
	D	10.3^{a}	14.0^{a}	10.8^{a}	12.3^{a}	23.0^{b}	26.8^{b}
	Mean	11.6	14.1	7.9	13.4	24.5	30.2
F-Test	M	*	Ns	Ns	Ns	Ns	Ns
	I	Ns	*	**	Ns	**	**
	M × I	Ns	Ns	Ns	Ns	Ns	Ns

A = surface irrigation, B = drip irrigation 1.00ET$_C$, C = drip irrigation 0.75ET$_C$, D = drip irrigation 0.50ET$_C$, With = soil mulching with 2 cm layer of rice straw, Without = bare soil with no mulch, Ns = nonsignificant, * = significant at 5% and ** = significant at 1%, M = soil mulching and I = irrigation.

10.3.1.3 NUMBER OF FRUITS FALLING PER TREE

In both seasons, Table 10.2 indicates that the number of fruits falling per tree of Navel orange was not significantly affected by irrigation, soil mulching and interaction among these in seasons 2003 and 2004, except irrigation factor in season 2003 was highly significant. The number of fruits falling was similar in surface irrigation and drip irrigation 0.50 ET_C treatments. The lowest number of fruits falling was obtained in drip irrigation 0.75 ET_C under soil mulching and drip irrigation 0.50 ET_C under bare soil treatments in seasons 2003 and 2004, respectively and was 4.0 and 11.5 fruit. These results are in agreement with those obtained by Khalifa [21], who found that soil mulching lowered fruit dropping comparing to bare soil.

10.3.1.4 TOTAL SOLUBLE SOLIDS (TSS, %)

Table 10.3 showed that the Total Soluble Solids (TSS) was not significantly affected by irrigation, soil mulching and interaction between them in two growing seasons 2003 and 2004. In season 2004, the values of TSS were lower under drip irrigation than that under surface irrigation while, in season 2003, the values of TSS did not indicate a clear trend. The highest TSS 12.1% in soil mulching in drip irrigation 0.75 ETc and 11.2% in 1.00 ETc, during 2003 and 2004 seasons, respectively.

TABLE 10.3 Some Chemical Characters (Quality) of Navel Orange Fruits During 2003 and 2004 Seasons As Affected by Soil Mulching and Irrigation Treatments

Mulch	Irrigation	TSS		Acidity (%)		Vitamin C (mg /100 mL juice)		
		2003	2004	2003	2004	2003		2004
Without	A	11.5^{a}	11.1^{a}	1.00^{a}	0.72^{a}	48.0	a	32.8^{c}
	B	11.7^{a}	10.8^{ab}	0.98^{a}	0.62^{a}	49.0	54.0^{a}	39.5^{ab}
	C	11.4^{a}	10.4^{ab}	0.98^{a}	0.64^{a}	51.0^{a}		42.4^{a}
	D	11.5^{a}	10.0^{b}	1.03^{a}	0.65^{a}	49.4^{a}		36.6^{bc}
	Mean	11.5	10.6	1.00	0.66	50.6		37.8
With	A	11.6^{a}	11.4^{a}	1.03^{ab}	0.67^{a}	45.4^{b}		36.1^{b}
	B	11.4^{a}	11.2^{a}	0.99^{ab}	0.74^{a}	50.8^{ab}		37.7^{b}
	C	12.1^{a}	10.7^{a}	0.81^{b}	0.67^{a}	54.7^{a}		43.5^{a}
	D	11.6^{a}	10.8^{a}	1.21^{a}	0.69^{a}	53.3^{a}		37.2^{b}
	Mean	11.7	11.0	1.01	0.69	51.1		38.6
F-Test	M	Ns	Ns	Ns	Ns	Ns		Ns
	I	Ns	Ns	Ns	Ns	*		**
	M × I	Ns	Ns	Ns	Ns	Ns		Ns

A = surface irrigation, B = drip irrigation 100% ET_C, C = drip irrigation 75% ET_C, D = drip irrigation 50% ET_C, With = soil mulching with 2 cm thickness of rice straw, Without = bare soil without any mulch, Ns = nonsignificant, * = significant at 5% and ** = significant at 1%, M= soil mulching and I= irrigation.

These results are in agreement with those obtained by Ghali and Nakhlla [16]. They reported that soil cover treatments gave a nearly stable percent of TSS. Also, Mohsen et al. [24] reported that TSS was increased with increasing soil moisture level in both flood and sprinkler irrigation.

10.3.1.5 ACIDITY

Table 10.3 showed that the fruit acidity (%) was not significantly affected by irrigation, soil mulching and interaction between them in two growing seasons 2003 and 2004. There are not marked variations between values of acidity in two growing season 2003 and 2004. The lowest values of acidity were obtained in drip irrigation 0.75 ET_C treatment with and without soil mulching and were 0.81 and 0.64% in season 2003 and 2004, respectively. Similar results have been reported by Fares and Alva [14], who noticed that acidity was increased, as a consequence of deficit irrigation (irrigating less than 1.00 ETc).

TABLE 10.4 Effects of Soil Mulching and Irrigation Treatments on Concentrations of N, P and K in Leaves of Navel Orange During 2003 and 2004 Seasons

Mulch	Irrigation	N %		P %		K %	
		2003	2004	2003	2004	2003	2004
Without	A	2.78^a	1.61^b	0.18^a	0.25^a	1.36^a	0.98^b
	B	1.41^b	1.73^{ab}	0.16^a	0.20^a	1.26^{ab}	1.20^{ab}
	C	1.54^b	2.12^a	0.16^a	0.23^a	0.81^b	1.34^a
	D	1.89^b	2.09^a	0.17^a	0.25^a	1.22^{ab}	1.32^a
	Mean	1.91	1.89	0.17	0.23	1.16	1.21
With	A	2.56^a	1.72^b	0.19^a	0.23^b	1.45^a	1.00^b
	B	1.77^a	1.79^b	0.16^{ab}	0.22^b	1.20^a	1.21^{ab}
	C	2.47^a	2.37^a	0.16^{ab}	0.34^a	0.98^a	1.18^{ab}
	D	1.87^a	2.07^{ab}	0.14^b	0.22^b	1.11^a	1.49^a
	Mean	2.17	1.99	0.16	0.25	1.19	1.22
F-Test	M	*	Ns	Ns	Ns	Ns	Ns
	I	**	**	Ns	Ns	*	**
	M × I	Ns	Ns	Ns	Ns	Ns	Ns

A = surface irrigation, B = drip irrigation 1.00 ET_C, C = drip irrigation 0.75 ET_C, D = drip irrigation 0.50 ET_C, With = soil mulching with 2 cm layer of rice straw, Without = bare soil without any mulch, Ns = nonsignificant, * = significant at 5% and ** = significant at 1%, M = soil mulching and I = irrigation.

10.3.1.6 VITAMIN C

Table 3 showed that vitamin C content (mg /100 mL of fruit juice) was not significantly affected by soil mulching and interaction between irrigation and soil mulching in two growing seasons 2003 and 2004, while irrigation factor was significant in 1st season and highly significant in 2nd season. The vitamin C content in fruit juice was higher under drip irrigation than that under surface irrigation. The highest values of vitamin C content (54.7 and 43.5) under the same conditions, were obtained in drip irrigation $0.75ET_C$. These results are in agreement with those reported by El-Zawily [12], who reported that increasing irrigation depth increased vitamin C content of Navel orange fruit juice.

10.3.2 CONCENTRATION OF NUTRIENTS IN LEAVES

10.3.2.1 NITROGEN CONTENT IN LEAVES

Table 4 showed that the irrigation factor had a highly significant effect on total N in Navel orange leaves in two growing season 2003 and 2004, while soil mulching had significant effect only in season 2003. Interaction effect was not significant in two growing seasons.

These values of total N were at minimum satisfactory level of nitrogen in orange leaves (2.5–2.8% according to Bennett [2]). The values of total N in Navel orange leaves were higher under soil mulching than that under bare soil except, in drip irrigation $0.50ET_C$ treatment in two growing seasons 2003 and 2004. In season 2004, the values of total N in Navel orange leaves were higher under drip irrigation than that under surface irrigation, but an opposite trend was found in season 2003.

These results are in agreement with those obtained by Yagev and Horesch [29], who showed that leaves of grapefruit under drip irrigation had a higher N content than that under sprinkler irrigation.

10.3.2.2 PHOSPHORUS CONTENT IN LEAVES

Table 10.4 showed that the irrigation, soil mulching and interaction between them had no significant effect on total P content of Navel orange leaves in two growing seasons 2003 and 2004. There is no variation between values of total P in two growing seasons, except in drip irrigation 0.75 ET_C treatment in season 2004 and was 0.34%, higher than the other one. These results agree with those obtained by Garcia et al. [15]. They showed that the amount of irrigation depth did not affect the leaf-P of lemon "Fino 49."

Data showed that the values of leaf-P content were in excess than the plant need: 0.10 to 0.17% of P according to Bennett [2]. Higher availability of soil-P (15.24 mg/kg) and added fertilizer can enhance the absorption of phosphate and its accumulation in orange leaves.

10.3.2.3 POTASSIUM CONTENT IN LEAVES

Table 10.4 showed that the irrigation factor had significant and high significant effect on total K of Navel orange leaves in season 2003 and 2004, respectively. The effect of soil mulching and interaction between irrigation and soil mulching were not significant in two growing seasons. In season 2004, the values of total K were higher under drip irrigation than that under surface irrigation. However, in season 2003, K-values had the opposite trend. The values of total K in Navel orange leaves were at optimum level according to Bennett [2], who reported that the optimum level of K in orange leaves ranged from 0.8 to 1.7%.

These results are in agreement with those obtained by Iobishvili and Mikautadze [18], who showed that irrigation increased K contents in both old and young leaves and higher application rate of irrigation water gave only slightly higher nutrient values.

10.3.3 WATER USE EFFICIENCY (WUE)

Table 10.5 indicates the effects of soil mulching and irrigation treatments on WUE of Navel orange fruits in two growing seasons. Data revealed that WUE values under soil mulching were higher than that under bare soil, except in 0.75 ETc drip irrigation treatment in 2nd season. Also, values of WUE were higher in 2nd season than that in 1st season, except in 0.75 ETc treatment in 2nd season. The highest value of WUE was 3.09 kg/m^3 under soil mulching in drip irrigation 0.50ET_C treatment in 2nd season. The lowest value of WUE was 0.86 kg/m^3 under bare soil in surface irrigation treatment in 1st season.

These results are in agreement with those obtained by Ghali and Nakhlla [16], who reported that soil mulch significantly increased fruit crop production and improved the values of water use efficiency under drip irrigation in sandy loam soil. Also, El-Araby [11] reported that WUE of mandarin under drip irrigation was almost 3.3 times higher than that under flood irrigation system.

TABLE 10.5 Effects of Soil Mulching and Irrigation Treatments on Water Use Efficiency of Navel Orange Fruits (Kg/m^3) in Two Growing Seasons 2003 and 2004

Irrigation	Soil mulching			
	Bare soil		Soil mulching with rice straw	
	2003	2004	2003	2004
Surface irrigation	0.86^d	1.07^d	1.04^d	1.33^d
Drip irrigation 100% ET_C	1.36^c	2.06^c	1.39^c	2.18^b
Drip irrigation 75% ET_C	1.87^b	2.54^b	2.20^b	2.13^c
Drip irrigation 50% ET_C	2.65^a	3.01^a	2.65^a	3.09^a

10.4 SUMMARY

The present investigation was conducted at orchard of Sakha Agriculture Research Station Farm, Kafr El-Sheikh Governorate during two successive seasons of 2003 and 2004. The effects of drip irrigation and soil mulching on growth, yield and yield quality of navel orange tree were compared to the corresponding parameters in surface irrigation. A spilt-plot design was used with four replicates in the two seasons. The main plots included soil mulching with 2 cm layer of rice straw and bare soil. The subplots were used for surface irrigation and drip irrigation at 100, 75 and 50% ET_C (deficit irrigation).

The results indicated that the tree vigor was significant affected by soil mulching and irrigation treatments in the both seasons. Tree vigor values were higher under drip irrigation than those under surface irrigation in the two seasons except under soil mulching in 1st season. Irrigation treatments had highly significant effect on yield of navel orange in the both seasons. The highest values of yield were in drip irrigation 75% ET_C and drip irrigation 100% ET_C treatments under soil mulching and were found to be 28.7 and 37.9 Kg/tree in seasons 2003 and 2004, respectively. The lowest number of fruit falling was obtained in drip irrigation 75% ET_C under soil mulching and drip irrigation 50% ET_C under bare soil treatments, in seasons 2003 and 2004, respectively. The highest values of vitamin C content (54.7 and 43.5 mg/100 mL juice) were recorded for treatment drip irrigation 75 and 50% ET_C in the 1st season, respectively. The irrigation had a highly significant effect on content of total N in Navel orange leaves in both growing seasons of 2003 and 2004 while soil mulching had significant effect only in season 2003. The irrigation had significant and highly effects on content of total K of navel orange leaves in season 2003 and 2004, respectively. Meanwhile, content of total P had no significant effect in the both growing seasons. The highest values of WUE were obtained under soil mulching in drip irrigation 50% ET_C followed by 75% ET_C treatments in 2nd and 1st seasons and were found to be 3.09 and 2.20 kg/m^3, respectively.

Based on results in this chapter, authors recommend using drip irrigation at 75% ETC in citrus orchards under soil mulching with rice straw which can increase WUE and improve yield and quality of navel orange fruits.

KEYWORDS

- acidity
- citrus
- crop evapotranspiration
- CROPWAT
- deficit irrigation

- **drip irrigation**
- **evapotranspiration**
- **FAO**
- **Feddans**
- **fertigation**
- **fertilization**
- **gross irrigation requirement**
- **mandarin**
- **mulching**
- **navel orange**
- **nitrogen**
- **North Delta**
- **Penman-Montieth**
- **phosphorous**
- **potassium**
- **reference evapotranspiration**
- **rice straw**
- **soil mulch**
- **surface irrigation**
- **tree height**
- **TSS**
- **urea**
- **water use efficiency**

REFERENCES

1. Association of Official Analytical Chemists, (1975). *Official Methods of Analysis*. 12th ed. Association of Official Analytical Chemists. Washington Press. *401,* 574–575 pages.
2. Bennett, William F. (1994). *Nutrient Deficiencies and Toxicities in Crop Plants*. 2nd edition, American Physical Society, APS Press, Washington D.C., Chapter 17, 165–170.
3. Carter, M. R. (1993). Soil Sampling and Methods of Analysis. *Canadian Soc. Soil Sci.,* Lewis Publishers.
4. Castel, J. R., Lidon, A. L., Ginetar, C., Ramos, C. (1995). Yield, growth and nitrate leaching in drip-irrigated citrus trees under different fertigation treatments. Proceedings of the 5th International Micro Irrigation Congress Orlando, Florida, April 2–6, 4, 461–968.

5. Chandaran, S. R., Salinas, K., Madeira, M. (2001). Weed Control Research At The Organic Farm. College of Agriculture, *Forestry and Consumer Sci.,* http://www.caf.wvu.edu/plsc/organic/frontpage/weed.html.
6. Chapman, H. D., Pratt, P. F. (1961). *Methods of Analysis for Soils, Plants and Waters*. University of California, USA.
7. Chartzoulakis, K., Michelakis, N., Stefanoudaki, E. (1999). Water use, growth, yield and fruit quality of "Bonanz" oranges under different soil water regimes. *Advances in Hort. Sci., 13(1),* 6–11.
8. Chung, S. O., Horton, R. (1987). Soil heat and water flow with a partial surface mulch. *Water Resour. Res., 23(12),* (2186).
9. Cottenie, A., Verloo, M., Velghe, G., Kiekens, L. (1982). *Biological and Analytical Aspects of Soil Pollution. Lab. of Analytical & Agron.* State University Ghent-Belgium.
10. Doorenbos, J., Pruitt, W. O. (1977). *Crop Water Requirements*. Irrigation and Drainage Paper No. 24, FAO. Rome, Italy.
11. El-Araby, A. A. (1997). Evaluation of drainage system in clay soils under flood and drip irrigation. M.Sc. Thesis, Faculty of Agriculture, Kafr El-Sheikh, Tanta University – Egypt.
12. El-Zawily, H. M. (2004). Influence of irrigation management, nitrogen and potassium fertilization on yield, fruit quality and storage life of "Washington Navel" orange fruits. M.Sc. Thesis, Fac. of Agric., Kafr El-Sheikh Tanta University.
13. FAO, (1980). *Localized Irrigation*. FAO Irrigation and Drainage Paper No. 36, Rome.
14. Fares, A., Alva, A. K. (1999). Water relations, fruit quality and yield of citrus under different irrigation and fertilization rates in a sandy soil. Dahlia Greidinger International Symposium: Nutrient Management Under Salinity and Water Stress. Technion-Israel Inst. of Technology, Haifa, 1–4 March, 411–421 pages.
15. Garcia-Sanchez, F., Carvajal, M., Porras, I., Botia, P., Martinez, V. (2003). Effects of salinity and rate of irrigation on yield, fruit quality and mineral composition of 'Fino 49' lemon. *Europ. J. Agron., 19*, 427–437.
16. Ghali, M. H., Nakhlla, F. G. (1996). Evaluation of perforated polyethylene mulch on loamy sand soil under drip irrigated orange trees, 2-Soil thermal regime and moisture, root distribution and tree productivity. *Ann. of Agric. Sci., Moshtohor, 34(3),* 1090–1116.
17. Hukkeri, S. B., Maji, C. C., Michael, A. M. (1977). *Water Requirement and Irrigation Management of Crops in India.* Water Technology Center, Indian Agric. Research Institute, New Delhi.
18. Iobishvili, N. P., Mikautadze, M. K. (1984). Effects of irrigation on the nutrient element content in mandarin leaves. *Sub-tropicheskie Kul'tury, 5,* 20–24.
19. Khalifa, M. R., Dawood, S. A., Zeerban, S. M. (2001). Yield, fruit quality and water consumptive use of Washington navel orange trees in relation to different flooding irrigation methods in Kafr El-Sheikh province. *J. Agricultural Research Tanta University, 27(2),* 358–370.
20. Khalifa, M. R. (1994a). Some soil properties and yield of Valencia orange trees as affected by different types of mulching. *J. Agricultural Research Tanta University, 20(2),* 361–371.
21. Khalifa, M. R., (1994b). Effect of different types of mulching on water consumptive use of Valencia orange trees. *J. Agricultural Research Tanta University, 20(3),* 591–602.
22. Klute, A. (1986). *Methods of Soil Analysis* – Part 1. 3rd edition. *Am. Soc. of Agron,* Madison – WI, USA.
23. Ministry of Agriculture and Land Reclamation, (2002). *Agricultural Statistics*. http://www.vercon.sci.eg
24. Mohsen, A. M., Gaafar, S. I., Meligi, M. A., Swellem, A. A. (1988). Effect of soil moisture levels and irrigation system on Valencia orange trees, I: Fruit set and yield. *Zagazig J. Agric. Res., 15(2).*

25. Morse, J. G., Robertson, C. A. (1987). Calculating canopy area of citrus trees and surface area o fruits. *J. Amer. Soc. Hort. Sci., 115(1),* 169–180.
26. Nath, J., Sharma, R. (1992). Effect of organic mulches on growth and yield of Assam lemon. *Hort. J., 5(1),* 19–23.
27. Smith, M. (1992). *CROPWAT8: A computer program for irrigation planning and management*. FAO Irrigation and Drainage Paper 46, Rome.
28. Snedecor, G. W., Cochran, W. G. (1972). Statistical Methods. 6th edition. Iowa State University, Ames-Iowa.
29. Yagev, E., Horesch, I. (1983). A comparison of drip and sprinkler irrigation in citrus orchards. *Proceedings of the International Society of Citriculture, 2,* 508–511.

CHAPTER 11

PHYTO-MONITORING TECHNIQUE FOR DRIP IRRIGATED CITRUS

P. PANIGRAHI and R. K. SHARMA

CONTENTS

11.1 INTRODUCTION

Throughout the world, water supply is a major constraint to crop production due to water demand for rapid industrialization and high population growth. The further scarcity of irrigation water for crop production should be addressed for sustaining the food supply through efficient water conservation and management practices even in high rainfall areas. Moreover, the crop yield per every drop of irrigation water should be enhanced while considering the best water use efficiency (WUE) associated with any crop.

In recent years, irrigated agriculture is shifting the paradigm of irrigation management from full to partial supply of water needs, which becomes a need in water scarce regions. Water scarcity in irrigation sector demands for the improvement in water use efficiency as a critical goal. One of the most promising techniques that can help to attain this objective is the use of deficit irrigation (DI) in crop production. In DI, the water loss can be controlled through reduction of evapotranspiration (ET) to improve WUE, while maintaining reproductive growth and development. Achieving higher WUE in any crop can be possible by enhancing yield and/or reducing the water losses due to deep percolation and evaporation from soil surface.

Past research has revealed the potential of DI technique as a way of reducing water consumption in tree crops and vines with little or no impact on yield and fruit quality [10, 17]. DI is mainly designed to restrict water supply to the crops to optimum level which produces absicic acid (ABA) in roots and it gets translocated to leaves that control transpiration by partial stomatal closure [5]. On the other hand, the wetted portion of root zone supplies water to shoots and leaves to maintain the turgidity to carry out optimum photosynthesis with reduced transpiration. Besides increased transpiration efficiency, another effect of DI is the limitation of vegetative growth of the trees [5]. Overall, under DI, optimum yield can be produced with higher WUE under limited water availability conditions.

Kinnow, an important citrus cultivar in India, is grown under irrigated condition in northern region of the country. The suboptimum productivity with poor fruit quality is the major pomological constraint, affecting the economics of Kinnow production in this region [2, 16]. Limited irrigation water availability following faulty irrigation method (surface irrigation) is one of the major reasons of low productivity of Kinnow. Micro irrigation or drip irrigation has been studied as an efficient irrigation method compared to surface irrigation in Kinnow [1]. Due to positive impact of drip irrigation on crop production with less water, the area under drip irrigation has increased substantially in last few years in the country. Scheduling irrigation with full crop water requirement including citrus is a common irrigation practice in different regions of the country. The irrigation water shortage in arid and semiarid areas of the country forces the orchard growers to impose deficit irrigation in crop production.

It has been recognized that the tree itself is best indicator of water stress [8]. Phyto-monitoring is one of the potential methods for irrigation scheduling in tree crops and forecasting yield using plant-based measurements and it can also benefit growers. In recent years, the role of phyto-monitoring techniques based on plant physiological parameters is emphasized in irrigation scheduling. The sensitivity of crops to water stress plays a major role in optimizing irrigation. Moreover, the remote sensing of the plants is another technique to quantify the water stress level. Therefore, there is an ample scope to use physiological responses and spectral signature of plants in irrigation scheduling. Further, the yield prediction under differential water stress condition is also limited in fruit crops. Keeping this in view, an experiment was conducted to study the effects of deficit irrigation on citrus production and to forecast citrus yield based on plant physiological parameters.

11.2 MATERIALS AND METHODS

The study was conducted with bearing 'Kinnow' mandarin (*Citrus reticulata* Blanco) plants budded on Jatti Khatti (*Citrus jambhiri* Lush) rootstock at Centre for Protective Cultivation Technology (CPCT) of IARI, New Delhi – India. The plant-to-plant spacing was 4 m, whereas row-to-row spacing was 5 m. The soil at the experimental site varied from sandy loam (top 40 cm soil depth) to sandy clay loam (40–100 cm soil depth) with bulk density of 1.47–1.61 g cm^{-3}. The irrigation water was free from salinity (EC, 1.15 dS m^{-1}), alkalinity (pH, 7.3) and sodicity (SAR, 4.4). The ground water contribution to plant water requirement is assumed to be negligible as water level in the nearby wells of the experimental plot was 15–18 m deep from ground surface.

The experimental site is located in the semiarid, subtropical climate with hot and dry summers. The hottest months of the year are May and June with mean daily temperature of 39 °C, whereas January is the coldest month with mean temperature of 14 °C. The mean annual rainfall at the site is 770 mm, out of which around 85% is concentrated mainly during June–September. Two irrigation regimes *viz.*, 50% and 75% of the crop evapotranspiration (ETc) were imposed and compared with irrigation at 100% of ETc (Full irrigation with no deficit). The treatments were:

DI_{50} Irrigation at 50% ETc
DI_{75} Irrigation at 75% ETc
FI Full irrigation (100% ETc)

The irrigation was continued from mid-January to June and mid-October to December in each year of the experiment. About 32 'Kinnow' plants were selected for this experiment and two treatments except FI were imposed following randomized complete block design, with four replicates per treatment and two plants per replication.

Irrigation water was applied on each alternate day using 6 on-line 8 lph pressure-compensating emitters per tree placed on two 12 mm diameter lateral pipes (3 emitters per lateral). The emitters were arranged at 1.0 m away from tree stem. The water quantity applied under FI was calculated based on 100% Class A pan evaporation rate for Kinnow mandarin in Delhi conditions, using the following formula:

$$ETc = Kp \times Kc \times Ep \tag{1}$$

where: ETc = the Crop-evapotranspiration (mm/day); Kp = the pan coefficient (0.8); Kc = the crop-coefficient (0.85 for mature Kinnow tree); and Ep = the 2-days cumulative pan evaporation (mm). The volume of water applied under FI was computed with the following formula:

$$V_{id} = [\pi\,(D^2/4) \times (ET_c - R_e)]/E_i \tag{2}$$

where: V_{id} is the irrigation volume (liter plant^{-1}) applied in each irrigation; D = the mean plant canopy diameter measured in N-S and E-W directions (m); ETc = the crop evapotranspiration (mm); R_e = the effective rainfall depth (mm); and E_i = the irrigation efficiency of drip system (90%).

The required amount of water to each irrigation treatment was regulated by adjusting the operating hours (irrigation duration) based on the actual discharge of the emitters from time to time. The flow of irrigation water in lateral pipes was controlled by lateral valves provided at the inlet end of lateral pipes. The recommended NPK-based fertilizers (354 g N, 160 g P_2O_5 and 345 g K_2O per tree) were fertigated at monthly intervals from January to June. Intercultural operation and the plant protection measures against insect pests and diseases were adopted uniformly for all trees in the experimental block, following the recommendations given for Kinnow mandarin in Delhi region of India.

Soils samples at 30 cm, 60 cm, 90 cm, 120 cm and 150 cm distance from tree trunk along and in between the drip emitters were collected from 0–20 cm, 20–40 cm, 40–60 cm, 60–80 cm and 80–100 cm soil depths during January of each year and analyzed for available macronutrients (N, P and K) and micronutrients (Fe, Mn, Cu and Zn) following the standard procedures. Four plant basins from each treatment were taken for soil sampling. For leaf nutrients (N, P, K, Fe, Mn, Cu and Zn) determination, 3- to 5-months old leaf samples (3rd and 4th leaf from tip of nonfruiting branches) were taken from each side of plant canopy, at a height of 1.5 m from ground surface during October and analyzed following the standard methods [4].

The mid-day leaf water potential was determined fortnightly taking two leaves per plant (sun-exposed) from the outer canopy using a pressure chamber (PMS instrument, Oregon, USA). For determination of stem water potential, two leaves per plant near to the trunk or a main scaffold branch was selected and covered by aluminum sheet and black polythene sheet to measure its potential at mid-day (12:00–13:00 clock hour). The leaves were enclosed in black polythene and aluminum sheet cover before 2 h of measurement for determination of both leaf and stem water potential. The water stress integral (S_ψ) for each treatment was calculated for midday leaf and stem water potential data, according to the equation defined by Myers [11]:

$$S\Psi = Absolute\,value\,of \sum_{i=0}^{i=1} \{(\Psi i, i+1) - c\}\, n \tag{3}$$

where: S_ψ is water stress integral (MPa day); $\psi_{i,\,i+1}$ is average midday leaf/stem water potential for any interval i and i+1 (MPa); c is maximum leaf/stem water potential measured during the study, and n is number of days in the interval.

Relative leaf water content (RLWC) and leaf water concentration (LWC) were determined for two leaves per plant (4 plants per treatment) following the standard procedures [3, 13]. The measurement of net photosynthesis rate (P_n), stomatal conductance (g_s) and transpiration rate (T_r) of leaves was performed fortnightly from 9:00 AM to 3:00 PM on a clear-sky day by portable photosynthesis meter (LI-COR-6400, Lincoln, Nebraska, USA). Four mature leaves per plant (3rd or 4th leaf from tip of shoot) from exterior canopy position (one leaf in each North, South, East and West direction) and two plants per treatment were taken for these measurements. Leaf water use efficiency (LWUE) was estimated as a ratio of P_n to T_r of leaves [14].

Reflectance of plant canopy was measured by a hand held ASD FieldSpec Spectroradiometer (Analytical Spectral Devices Inc., Boulder, CO, USA) from top of the plants at midday (12:00 to 13:00 clock hour) on cloudless days. From reflectance data, the spectral reflectance indices related to water deficit conditions were calculated as: water band index (WBI) = [(R_{900}) / (R_{970})]; normalized difference water index [NDWI= (R_{857} – R_{1241}) / (R_{857} + R_{1241}); moisture stress index [MSI = (R_{1599}) / (R_{819}]; normalized difference infrared index [NDII= (R_{819}-R_{1649}) / (R_{819}+R_{1649})], where: R and the subscript numbers indicate the light reflectance at the specific wavelength (nm).

The vegetative growth of plants (plant height, stem height, canopy diameter, stock girth diameter and scion girth diameter) was measured annually by using a metric tape. Plant canopy volume was estimated using the following formula [12]:

$$V_{pc} = 0.5238\, H\, (D)^2 \tag{4}$$

where: V_{pc} is the plant canopy volume (m^3); H is the plant canopy height (difference between plant height and stem height) in meter; and D is the mean plant canopy spread diameter (North-South and East-West) in meter.

The number of fruits harvested from each tree was counted and the total weight was recorded and the mean yield per tree under various treatments was estimated. Irrigation water use efficiency (IWUE) was worked out as the fruit yield per total tree water use and fruit yield per unit quantity of irrigation water applied, respectively.

The analysis of variance (ANOVA) for the data was done using SPSS statistical software and separation of means was obtained using Duncan multiple range test [9]. The SAS 9.2 statistical software was used for generating correlation matrix. Moreover, principal component analysis and multilinear regression were done with this software.

11.3 RESULTS AND DISCUSSION

11.3.1 AVAILABLE NUTRIENTS IN SOIL

The available N, P and K status in the soil under different irrigation strategies shows an increasing trend (Table 11.1). The increase in N, P and K was due to application of NPK-based fertilizers to the plants during irrigation seasons. The maximum increase in the available nutrients was observed in FI, whereas the minimum value was in DI_{50}. However, the annual increase in available nutrients under the treatments suggests for annual-soil nutrients based and yield-based fertilization strategies for Kinnow trees. Moreover, further studies on fertigation with deficit irrigation in Kinnow mandarin is suggested under drip irrigation.

The available micronutrients (Fe, Mn, Cu and Zn) in the soil showed a decreasing trend in all the irrigation strategies (Table 11.1). The maximum decrease in available micronutrients was observed with FI and minimum was with DI_{50}. The maximum reduction of available micronutrients in FI was probably caused by higher plant uptake of these nutrients under optimum soil water regime in this treatment. However, the consistent reduction of micronutrients in soil in both the years suggests the application of appropriate quantity of micronutrients-based fertilizers to mandarin trees for sustaining higher yield for long run.

TABLE 11.1 Changes in the Available N, P and K (mg.kg^{-1} Soil) in Soil (0–100 cm Soil Depth) and Micronutrients Under Different Irrigation Treatments in Kinnow Mandarin

Irrigation treatment	Macronutrients			Micronutrients			
	N	P	K	Fe	Mn	Cu	Zn
DI_{50}	+3.31^{a}	+0.66^{a}	+3.84^{a}	-0.80^{a}	-0.65^{a}	-0.11^{a}	-0.14^{a}
DI_{75}	+3.82^{b}	+0.78^{a}	+3.92^{c}	-0.89^{a}	-0.89^{a}	-0.22^{a}	-0.16^{a}
FI	+4.29^{c}	+0.93^{a}	+4.25^{e}	-1.19^{b}	-1.06^{b}	-0.22^{a}	-0.23^{b}

Data in the same column followed by different letter are significantly different at $P < 0.05$, based on Duncan's multiple range test.

TABLE 11.2 Total N, P and K in Leaves (%, Dry Weight Basis) of 'Kinnow' Mandarin As Affected by Various Irrigation Treatments

Treatments	Macronutrients			Micronutrients			
	N	P	K	Fe	Mn	Cu	Zn
DI_{50}	2.31^{a}	0.15^{a}	1.41^{a}	54.0^{a}	48.6^{a}	7.3^{a}	24.7^{a}
DI_{75}	2.46^{b}	0.19^{a}	1.54^{b}	58.4^{a}	57.8^{a}	7.4^{a}	25.6^{a}
FI	2.69^{c}	0.22^{a}	1.64^{c}	55.6^{a}	51.2^{a}	7.4^{a}	25.2^{a}

Data in the column followed by different letter are significantly different at $P < 0.05$, based on Duncan's multiple range test.

11.3.2 COMPOSITION OF LEAF NUTRIENTS

The leaf nutrient (N, P, K, Fe, Mn, Cu and Zn) analysis shows that all the nutrients except P and Cu were significantly affected by the irrigation treatments (Table 11.2). The highest concentration of the nutrients was registered with FI. The higher concentration of leaf-nutrients with fully irrigated trees was resulted by higher plant uptake with increased availability of such nutrients in root zone in FI. The concentrations of leaf nutrients were higher at 0.75 ETc irrigation regime than that at 0.50 ETc irrigation. However, the concentrations of P was not affected significantly ($P < 0.05$) by irrigation regimes. Among micronutrients, the magnitudes of all nutrients (Fe, Mn and Zn) were at par in DI_{50} and DI_{75}. The higher micronutrient concentration was observed with fully irrigated trees. However, the N, P and K concentrations in leaves in all the treatments were higher than the optimum quantity of N, P and K in leaves required for sustainable production of Kinnow mandarin in Northern India, whereas the concentration of Fe, Mn, Cu and Zn in leaves under all the treatments except FI were less than their optimum range [15].

11.3.3 LEAF AND STEM WATER POTENTIAL, RELATIVE LEAF WATER CONTENT AND LEAF WATER CONCENTRATION

The mid-day leaf water potential (Ψ_l) stem water potential (Ψ_s), leaf water stress integral ($S\Psi_l$) and stem water stress integral ($S\Psi_s$) of the mandarin trees were affected significantly by the irrigation treatments (Table 11.3). The mean Ψ_l and Ψ_s were higher in FI. The minimum values for Ψ_l and Ψ_s were observed in DI_{50}. The magnitudes of Ψ_s were observed to be higher than that of Ψ_l. The maximum values for $S\Psi_l$ and $S\Psi_s$ were observed in DI_{50}, whereas the minimum value was with FI.

The mean relative leaf water content (RLWC) and leaf water concentration (LWC) under different irrigation treatments were affected significantly under various irrigation treatments (Table 11.3). The highest value of RLWC and LWC were observed with fully irrigated plants (Fl), whereas the lowest values were observed with the plants under DI_{50}.

TABLE 11.3 Mean Seasonal Mid-Day Leaf Water Potential (Ψ_l), Stem Water Potential (Ψ_s), Leaf Water Stress Integral ($S\Psi_l$), Stem Water Potential Integral ($S\Psi_s$), Integrated Leaf Water Potential (Ψ_{intl}), Integrated Stem Water Potential (Ψ_{ints}), Relative Leaf Water Content (RLWC) and Leaf Water Concentration (LWC) of Kinnow Mandarin Under Deficit Irrigation Treatments

Treatments	Ψ_l (MPa)	Ψ_s (MPa)	$S\Psi_l$ (MPa day)	$S\Psi_s$ (MPa day)	RLWC (%)	LWC (%)
DI_{50}	–1.8[a]	–1.2[a]	52.6[a]	38.3[a]	79.3[a]	68.8[a]
DI_{75}	–1.6[c]	–1.0[c]	39.2[c]	29.2[c]	89.3[c]	72.7[c]
FI	–1.2[e]	–0.7[e]	24.5[e]	18.9[e]	92.7[e]	78.3[e]

Data in the column followed by different letter are significantly different at $P < 0.05$, based on Duncan's multiple range test.

11.3.4 LEAF PHYSIOLOGICAL PARAMETERS

The mean net photosynthesis rate (P_n), stomatal conductance (g_s), transpiration rate (T_r) and leaf water use efficiency (LWUE = $P_n \div T_r$) in three irrigation treatments were significantly affected (Table 11.4). The maximum P_n was registered with fully irrigated trees and the lowest value of P_n was recorded in DI_{50}. The g_s and T_r followed the same trend of P_n in different treatments. The highest values of g_s and T_r with FI attributed to higher soil water content in root zone of the trees in this treatment. Moreover, higher photosynthesis rate was probably caused by wider opening of stomata with higher stomatal conductance of fully irrigated trees. However, the LWUE was maximum, whereas the minimum LWUE was in DI_{75} treatment. Earlier similar results were observed, indicating higher Pn with lower Tr under DI with drip irrigation in citrus [18].

TABLE 11.4 Net Photosynthesis Rate, Stomatal Conductance, Transpiration Rate, Leaf Water Use Efficiency of 'Kinnow' Mandarin Under Different Irrigation Treatments

Treatments	Photosynthesis rate (μmol m^{-2} s^{-1})	Stomatal conductance (mmol m^{-2} s^{-1})	Transpiration rate (mmol m^{-2} s^{-1})	Leaf water use efficiency
DI_{50}	2.89[a]	21.07[a]	1.66[b]	1.74 [a]
DI_{75}	3.17c	24.80[d]	1.84[d]	1.72[a]
FI	3.88[d]	37.78[e]	2.08[e]	1.86[c]

Data in the column followed by different letter are significantly different at $P < 0.05$, based on Duncan's multiple range test.

TABLE 11.5 Mean Water Band Index (WBI), Normalized Difference Water Index (NDWI), Moisture Stress Index (MSI), Normalized Difference Infrared Index (NDII) and Simple Ratio Proposed (SR) of Kinnow Mandarin Under Different Irrigation Treatments

Treatments	WBI	NDWI	MSI	NDII	SR
DI_{50}	0.917	0.033	0.469	0.239	2.711
DI_{75}	0.966	0.035	0.472	0.243	2.802
FI	1.056	0.042	0.561	0.266	3.002

Data in the column followed by different letter are significantly different at $P < 0.05$, based on Duncan's multiple range test.

11.3.5 REFLECTANCE

The mean reflectance of mandarin plants under different irrigation treatments indicates that the maximum reflectance was observed in DI_{50} (9–39%), whereas the minimum reflectance (9–37%) was observed in FI. The lower reflectance in FI was due to better vegetative growth of trees with lower leaf water content in this treatment. The values for hyperspectral indices (WBI, water band index; NDWI, normalized difference water index; MSI, moisture stress index, NDII, normalized difference infrared index and SR, simple ratio) of Kinnow in different irrigation treatments are presented in Table 11.5. The minimum values of the indices were observed with DI. The higher irrigation regime resulted in higher values of indices.

11.3.6 TREE VEGETATIVE GROWTH

The tree vegetative growth parameters (tree height, PH; stem girth diameter, SD; canopy diameter, CD; and canopy volume, CV) were significantly affected by irrigation treatments during 2010 and 2011 (Table 11.6). The highest growth of the trees was observed with FI, followed by DI_{75}. The treatment DI_{50} produced the minimum growth of trees.

TABLE 11.6 Tree Growth of Kinnow Mandarin in Three Irrigation Regimes

Treatments	PH (cm)	SD (mm)	CD (cm)	CV (m^3)
DI_{50}	33.4^a	20.4^b	25.8^a	0.80^a
DI_{75}	36.2^b	22.5^d	31.3^d	0.83^b
FI	40.7^c	26.2^e	$48.^{7e}$	0.86^c

Data in the column followed by different letter are significantly different at $P < 0.05$, based on Duncan's multiple range test.

PH: tree height; SD: stem diameter; CD: canopy diameter; CV: canopy volume.

11.3.6 FRUIT YIELD AND IRRIGATION WATER PRODUCTIVITY

The number of fruits harvested per tree, average fruit weight and total fruit yield in various treatments are presented in Table 11.7. The maximum number of fruits was dropped in DI_{50}. The minimum fruit drop took place in FI. The fruit drop decreased with increase in irrigation regime in DI. The number of fruit harvested in different treatments followed the reverse trend of fruit drop. The increased number of fruits with FI can be a reason for smaller fruits in this treatment. The number of fruits per tree and mean fruit weight decreased with decreasing irrigation regime from 0.75 ETc to 0.50 ETc with DI.

The highest fruit yield was recorded in FI. The increased irrigation regime from 0.50 ETc to 0.75 ETc enhanced the fruit yield, resulting from less number of fruits with lower fruit weight in lower regime of irrigation. The similar results of lower fruit yield with DI have been reported by other researchers for citrus [7, 17]. The IWUE was maximum in DI_{50}. The higher IWUE resulted in DI_{50} was attributed to higher increase in fruit yield with comparatively less increase in irrigation water use under this treatment over other treatments.

TABLE 11.7 Fruit Drop (Number Fruit), Yield Harvested and Irrigation Water Use Efficiency (IWUE) of Kinnow Mandarin Under Different Irrigation Treatments

Treatments	No. fruits dropped/tree	No. fruits harvested/tree	Average fruit weight (g)	Fruit yield (t ha^{-1})	IWUE (t ha^{-1} mm^{-1})
DI_{50}	170	671[a]	152.7[a]	51.23[a]	0.108[c]
DI_{75}	135	718[c]	161.6[b]	58.01[b]	0.081[b]
FI	92	763[d]	162.3[b]	61.91[b]	0.065[a]

Data in the column followed by different letter are significantly different at $P < 0.05$, based on Duncan's multiple range test.

11.3.7 CORRELATION OF FRUIT YIELD WITH OTHER TREE-BASED PARAMETERS AND YIELD PREDICTION

Table 11.8 indicates the correlation matrix between fruit yield and other observations (SD, stem diameter; CV, canopy volume; leaf-N, leaf nitrogen content; leaf-K, leaf potassium content; Leaf- Fe, leaf iron content; Leaf-Zn, leaf zinc content; $S\Psi_l$, mid-day leaf water stress integral; $S\Psi_s$, mid-day stem water stress integral; RLWC, relative leaf water content; LWC, leaf water concentration; P_n, net leaf photosynthesis rate; T_r, leaf transpiration rate; g_s, leaf stomatal conductance; LWUE, leaf water use efficiency; WBI, water band index; NDWI, normalized difference water index; MSI, moisture stress index; NDII, normalized difference infrared index; SR, simple ratio).

TABLE 11.8 Correlation Matrix (Pearson's) For Plant-Based Observations Under DI Treatments

Parameters	Fruit yield	SD	CV	Leaf-N	Leaf-K	Leaf-Fe	Leaf-Zn	$S\Psi_l$	$S\Psi_s$	RLWC	LWC	Pn	Tr	gs	LWUE	WBI	NDWI	MSI	NDII
SD	0.25*																		
CV	0.33*	0.39*																	
Leaf-N	0.47^{+}	NS	0.29*																
Leaf-K	0.49^{+}	NS	0.41*	0.43*															
Leaf- Fe	NS	NS	NS	NS	NS														
Leaf-Zn	0.48*	NS	NS	NS	NS	0.41*													
$S\Psi_l$	0.59^{+}	0.21*	0.29*	0.33*	0.37*	NS	NS												
$S\Psi_s$	0.62^{+}	0.26*	0.32*	0.42*	0.39*	NS	NS	0.56^{+}											
RLWC	0.55^{+}	0.20*	0.17*	0.32	0.32*	NS	NS	0.44^{+}	0.49^{+}										
LWC	0.43^{+}	NS	NS	0.30	0.25*	NS	NS	0.49^{+}	0.39^{+}	0.54^{+}									
Pn	0.45^{+}	NS	0.23*	0.55^{+}	0.44*	0.48^{+}	0.36*	0.42^{+}	0.33^{+}	0.36^{+}	0.25^{+}								
Tr	0.41^{+}	NS	NS	0.59*	0.51^{+}	0.43*	0.29*	0.48^{+}	0.53^{+}	0.51^{+}	0.39^{+}	0.21^{+}							
gs	0.41^{+}	NS	NS	0.48*	0.55^{+}	0.45*	0.38*	0.49^{+}	0.46^{+}	0.35^{+}	0.36^{+}	0.39^{+}	0.51^{+}						
LWUE	0.48^{+}	NS	NS	0.47*	0.36*	0.42*	0.21*	0.43^{+}	0.39^{+}	0.39^{+}	0.28^{+}	0.19^{+}	0.49^{+}	0.37*					
WBI	0.47^{+}	0.29*	0.29*	0.49^{+}	0.47*	0.44*	NS	0.45^{+}	0.37^{+}	0.39^{+}	0.32^{+}	0.27*	0.35*	0.31*	0.30^{+}				
NDWI	0.43*	NS	NS	0.43*	NS	NS	NS	0.38*	0.28*	0.37*	0.20*	0.13*	0.29*	0.20*	0.21*	0.39^{+}			
MSI	0.49^{+}	0.22*	0.23*	0.41^{+}	0.40*	NS	NS	0.44*	0.21^{+}	0.32^{+}	0.27^{+}	0.12^{+}	0.23^{+}	0.25*	0.17^{+}	0.29*	0.24*		
NDII	0.49*	NS	NS	0.43*	0.36*	0.27*	NS	0.26*	0.22*	0.37*	0.29*	0.29*	0.27^{+}	0.19*	0.26*	0.15*	0.48*	0.41*	
SR	0.41^{+}	NS	NS	0.44^{+}	0.36*	NS	NS	0.57^{+}	0.32^{+}	0.33^{+}	0.38^{+}	0.27^{+}	0.30^{+}	0.24*	0.20*	0.34^{+}	0.49*	0.30*	0.37

The higher correlation between yield and $S\Psi_s$ indicates the use of stem water potential as a tool for irrigation scheduling. Previously in sweet orange, the maximum correlation of $S\Psi_s$ with fruit yield has been observed [7]. Moreover, a good correlation was observed in between Pn and leaf-N, P_n and leaf-Fe, $S\Psi_s$ and $S\Psi_l$, RLWC and $S\Psi_l$, T_r and $S\Psi_l$, g_s and $S\Psi_l$, RLWC and $S\Psi_s$, T_r and $S\Psi_s$, g_s and $S\Psi_s$, LWC and RLWC, g_s and T_r, MSI and WBI, SR and NDWI and SR and MSI, respectively. Similar pattern of correlation of fruit yield with T_r, g_s and WBI have also been observed in citrus under differential irrigation [6].

Principal component analysis (PCA) for 19 variables indicates that the first three PCs explained 89% and 84.3% variability of dataset in DI (Table 11.9). Therefore, the variables involved in these 3 PCs were considered for further analysis. The variables from PC1 ($S\Psi_s$, Leaf-N, Leaf-K, $S\Psi_l$, RLWC), PC2 (g_s, P_n) and PC3 (WBI, SR) were retained for interpretation, as their eigenvalues were > 1. A multiregression model developed between fruit yield and other selected plant variables ($S\Psi_s$, Leaf-N, Leaf-K, g_s and WBI) in DI was:

$$\text{Fruit yield} = [-0.836\,(\text{Leaf-N}) + 22.569\,(\text{Leaf-K}) - 0.115\,(S\Psi_s) + 0.123\,(g_s) + 11.675\,(\text{WBI}) - 14.675] \quad (5)$$

TABLE 11.9 Principal Components with Eigen Values and Variances in DI Treatments

PC	DI			
	Variables	**Eigen value**	**% variance**	**Cumulative % of variance**
1	$S\Psi_s$, Leaf-N, Leaf-K, $S\Psi_l$, RLWC	6.964	40.20	40.20
2	g_s, P_n	3.716	33.54	73.74
3	WBI, SR	2.449	15.28	89.02

The Eq. (5) is well validated to predict the fruit yield from the proposed tree-based variables with coefficient of determination (R^2) of 0.86, root mean square error (RMSE) value of 1.011% for DI (Fig. 11.1) and at $P > 0.05$.

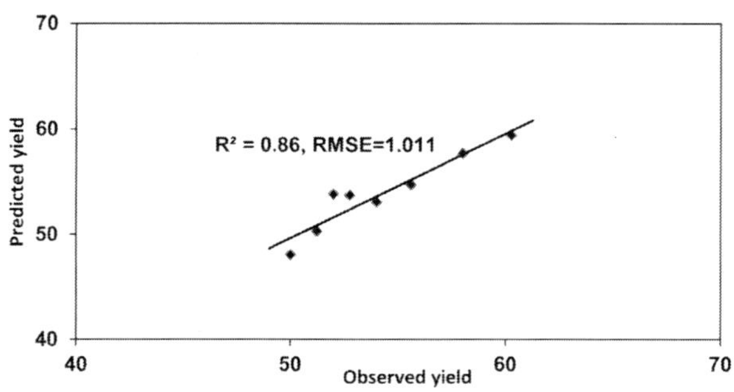

FIGURE 11.1 Relation between predicted yield and observed yield in DI.

11.4 CONCLUSIONS

The deficit irrigation is a potential water saving technique compared to full irrigation in drip-irrigated Kinnow mandarin. Both tree vegetative growth and yield parameters of Kinnow showed a need for higher soil moisture content that was evident from better growth and yield under full irrigation of the plants. However, the maximum values for leaf water use efficiency, irrigation water use efficiency and water use efficiency were obtained from irrigation at 50% crop-evapotranspiration in deficit irrigation, with some minor reduction in yield than that in full irrigation. Thus, the adoption of DI in drip irrigation scheduling at 75% crop water requirement is a viable option against traditional full irrigation for citrus cultivation in sandy loam soils in semiarid agro-climatic conditions, in India. The optimal NPK-fertigation strategy deficit irrigation was suggested for drip-irrigated Kinnow mandarin. Principal component regression model using tree-based variables is a potential technique to predict fruit yield in citrus.

11.5 SUMMARY

Increasingly serious shortages of water make it imperative to improve the irrigation efficiency in crop production in changing climate conditions. In recent years, the role of phyto-monitoring techniques based on plant physiological parameters are emphasized in irrigation scheduling. There is an ample scope to use physiological responses and spectral signature of the plants in irrigation scheduling. Drip irrigation is a potential water saving technique compared to traditional surface irrigation methods in citrus. DI is a recently proposed water saving technique in irrigated agriculture. The present study was planned with a hypothesis that drip irrigation scheduling with DI technique can save a substantial amount of water compared to full irrigation, without affecting the yield significantly.

The experiment was conducted for two years during 2010 and 2011, with drip-irrigated Kinnow mandarin at IARI, New Delhi. The crop response to DI scheduled at 50% and 75% of full irrigation (FI, 100% ETc) was recorded. DI at 75% ETc produced marginally lower fruit yield (8–9%), with lower vegetative growth of the trees compared to that in full irrigation. However, the irrigation water use efficiency in DI at 75% ETc was observed to be 81–83% higher, than in FI. The heavier fruits with better quality (higher TSS, ascorbic acid, total sugar and reducing sugar and lower acidity) were harvested in DI at 75% FI compared to FI. The tree water status (relative leaf water content, leaf water concentration, leaf water potential, stem water potential) was superior with fully irrigated trees. Likewise, in FI, the trees registered maximum rate of net-photosynthesis, stomatal conductance and transpiration in leaves. However, the trees in DI at 50% ETc exhibited the highest leaf water use efficiency (photosynthesis rate/transpiration rate). The leaf nutrient (N, P, K, Fe, Mn, Cu and Zn) analysis revealed that the concentration of all the nutrients was higher with fully irrigated trees, which was associated with higher availability of such nutrients in soil in this treatment. Yield prediction employing principal component-regression model include variables such as: leaf-N, leaf-K, stem water potential stress index, stomatal conductance and water band index as the predictors. This model gave satisfactory results.

KEYWORDS

- **deficit irrigation**
- **leaf nutrient**
- **leaf water potential**
- **leaf water stress integral**
- **leaf water use efficiency**
- **net photosynthesis**
- **photosynthesis**
- **stem water potential**
- **stem water stress integral**
- **stomatal conductance**
- **stress index**
- **surface irrigation**
- **transpiration rate**
- **water band index**
- **water potential**
- **water use efficiency**
- **yield parameters**
- **yield prediction**

REFERENCES

1. Anonymous, (1997). Annual Progress Report, 1997, Water Technology Centre, IARI, New Delhi.
2. Bhatt, A., Kachroo, J., Kachroo, D. (2011). Economic appraisal of kinnow production and its marketing under north-western Himalayan region of Jammu. *Agricultural Economic Research Review, 24,* 283–290.
3. Bowman, W. D. (1989). The relationship between leaf water status, gas exchange and spectral reflectance in cotton leaves. *Remote Sensing of Environment, 30,* 249–255.
4. Chahill, B. S., Dhatt, A. S., Singh, R. (1988). Standardization of foliar sampling techniques in kinnow mandarin. *Punjab Horticultural Journal, 20,* 118–129.

5. Dry, P. R., Loveys, B. R., During, H. (2000). Partial drying of the root zone of grapes. I. Transient changes in shoot growth and gas exchange. *Vitis*, *39*, 3–7.
6. Dzikiti, S., Verreynne, S. J., Stuckens, J., Strever, A., Verstraeten, W. W., Swennen, R., Coppin, P. (2010). Determining the water status of Satsuma mandarin trees [Citrus Unshiu Marcovitch] using spectral indices and by combining hyperspectral and physiological data. *Agricultural and Forest Meteorology*, *150*, 369–379.
7. García-Tejero, I., Jiménez-Bocanegra, J. A., Martínez, G., Romero, R., Durán-Zuazo, V. H., Muriel-Fernández, J. L. (2010). Positive impact of regulated deficit irrigation on yield and fruit quality in a commercial citrus orchard [Citrus sinesis (L.) Osbeck, cv. Salustiano]. *Agric. Water Manage.*, *97*, 614–622.
8. Goldhamer, D. A., Salinas, M., Crisosto, C., Day, K. R., Soler, M., Moriana, A. (2002). Effects of regulated irrigation and partial root zone drying on late harvest peach tree performance. *Acta Horticulturae*, *592*, 343–350.
9. Gomez, K. A., Gomez, A. A. (1984). *Statistical Procedures for Agricultural Research*. John Wiley and Sons: New York.
10. Kang, S., Hu, X., Goodwin, I., Jerie, P. (2002). Soil water distribution, water use and yield response to partial root zone drying under a shallow ground water table condition in a pear orchard. *Scientia Horticulture*, *92*, 277–291.
11. Myers, B. J. (1988). Water stress integral – a link between short-term stress and long-term growth. *Tree Physiology*, *4*, 315–323.
12. Obreza, T. A. (1991). Young Hamlin orange tree fertilizer response in south-west Florida. *Proc. Fla. State Hort. Soc.*, *103*, 12–16.
13. Peñuelas, J., Pinol, J., Ogaya, R., Filella, I. (1997). Estimation of plant water concentration by the reflectance Water Index WI (R900/R970). *International Journal of Remote Sensing*, 18, 2869–2875.
14. Ribeiro, R. V., Machado, E. C., Santos, M. G., Oliveira, R. F. (2009). Photosynthesis and water relations of well-watered orange plants as affected by winter and summer conditions. *Phtosynthetica*, *47(2)*, 215–229.
15. Srivastava, A. K. (2011). Diagnosis of nutrient starvation and its recommendation through microbial synergisms in citrus. In: *Manual of Winter School on Advances in Rootstock for Overcoming Biotic and Abiotic Stress in Fruit Crops*. November 17 to December 7, IARI – New Delhi. Pages 74–79.
16. Thakur, D.S., Saibabu, M. V. S., Sharma, K. D. (1986). Economics of Kinnow cultivation in Himachal Pradesh. *Him. J. Agric. Res.*, *12(1)*, 47–51.
17. Treeby, M. T., Henriod, R. E., Bevington, K. B., Milne, D. J., Storey, R. (2007). Irrigation management and rootstock effects on navel orange (Citrus Sinensis L. Osbeck) fruit quality. *Agric. Water Manage.*, *91*, 24–32.
18. Vu, J. V., Yelenosky, G. (1988). Water Deficit and associated changes in some photosynthetic parameters in leaves of Valencia orange (*Citrus sinensis* L. Osbeck). *Plant Physiol.*, *88*, 375–378.

CHAPTER 12

IRRIGATION SCHEDULING FOR *CITRUS RETICULATA* BLANCO

P. PANIGRAHI and A. K. SRIVASTAVA

CONTENTS

12.1 INTRODUCTION

Deficit irrigation (DI) is an irrigation strategy to reduce water use over full water requirement of the crop. The correct application of DI requires thorough understanding of the yield response to water and of the economic impact on crop production [19]. In regions of limited water resources, DI can be more profitable for a farmer to maximize crop water productivity instead of maximizing the harvest per unit land [6]. The saved water can be used for other purposes or to irrigate extra units of land. In other words, DI aims at stabilizing yields and at obtaining maximum crop water productivity rather than maximum yields [8, 10].

Citrus is a high water requiring evergreen perennial crop, grown in tropics. Once the tree attains a desirable canopy, it is not advisable to supply water for its vigorous growth [5]. Moreover, the higher vegetative growth in mature trees reduces the productivity and quality of citrus fruits. Once the crop develops a wide spread root-zone, it draws some quantity of water from soil, beyond irrigation. If water is supplied in full to fill the total evapotranspiration of the trees, the water uptake from nonirrigated rhizosphere reduces. Thus, applying irrigation water throughout the whole season as per the tree water requirement may cause low water productivity in mature citrus trees. However, the plants undergo severe stress when soil-water is very low and the water uptake by the roots fails to compensate the optimal water requirement of the trees. Hence the accuracy in water application, creating a desirable stress is important for citrus production in water scarce areas.

Nagpur mandarin (*Citrus reticulata* Blanco), a loose skin citrus cultivar, is commercially grown in around of 0.185 million hectares area of central India as an irrigated crop [16]. The irrigation water shortage is one of the major *abiotic* constraints for higher and quality production of citrus in this region. Under such water scarce conditions, it becomes necessary to find new irrigation strategies to reduce the water consumption and make more efficient use of the available water resources, focusing on maximizing water savings and improving its final productivity. One of the potential water saving strategies is DI. The aim of application of different DI strategies should boost water productivity without sacrificing yield. In recent years, several contributions have documented the advantages of using DI strategies to improve the water use efficiency (WUE) and fruit quality in different citrus species in various citrus growing regions of the world [7, 10, 12]. It has been reported that daily irrigation at 80% of open pan (Class A) evaporation rate through drip system is optimum to supply full water requirement of bearing Nagpur mandarin trees in central India [15]. However, the information regarding the effects of DI on water use, fruit yield and quality under drip irrigation is not reported in any citrus cultivar in India.

Therefore, keeping this in view, this study was undertaken to evaluate various DI levels through drip irrigation taking 'Nagpur' mandarin as a test crop grown in hot sub-humid tropical climate of central India.

12.2 MATERIAL AND METHODS

The field experiment was conducted at experimental farm of National Research Centre for Citrus, Nagpur (21° 08′45″ N, 79° 02′ 15″ E and 340 m above mean sea level) during 2006–2008 with 16 year-old Nagpur mandarin (*Citrus reticulata* Blanco) trees budded on rough lemon (*Citrus Jambhiri* Lush) root stock at a spacing of 6 *x* 6 m. The experimental soil was clay loam (31.65% sand, 23.6% silt and 44.8% clay) with field capacity and permanent wilting point of 29.3% (v/v) and 18.5% (v/v), respectively, with bulk density of 1.18 $g.cm^{-3}$. The mean daily USWB Class-A pan evaporation varied from 2.0 mm in month of December to 12.0 mm in May at the experimental site. The treatments imposed to irrigate the trees were:

T1 drip irrigation at 30% of full irrigation (FI).
T2 drip irrigation at 50% of full irrigation (FI).
T3 drip irrigation at 70% of full irrigation (FI).
T4 drip irrigation at 100% full irrigation (control).

The plots were drip irrigated through four numbers of 8 lph pressure compensating on-line drippers per tree, placed at 1.0 m away from tree girth. FI was estimated as daily irrigation supply at 80% of class-A pan evaporation rate (Ep) [15]. The experiment was in randomized block design (RBD) with five replications and three adjacent trees in a row per replication. Irrigation quantity for different drip irrigation treatments was calculated using the formula:

$$V= [S \times Kp \times K_C \times (Ep - ER)]/r \quad (1)$$

where: V is the irrigation volume (liters/day/tree), S is the tree canopy area (m^2), Kp is the pan factor (0.7), Kc is the crop factor (0.6) as suggested by Allen et al. [1], Ep is the daily Class A pan evaporation (mm), ER is the cumulative effective rainfall for corresponding two days (mm) and r is the water application efficiency of drip irrigation system (»90%). The orchard floor was kept cleaned and all the experimental trees were grown under uniform cultural and management practices.

The soil moisture content was monitored twice a week at 30, 45 and 60 cm soil depths by neutron moisture meter (Troxler model-4300, USA). Indexed leaf samples (2nd–4th leaf from tip of branches) surrounding the trees at a height of 1.5–1.8 m from ground surface were collected at the end of irrigation seasons as suggested by Srivastava et al. [17] and were subjected to analysis of various nutrients (N, P, K, Fe, Mn, Cu and Zn). The samples were thoroughly washed and ground using a Willey grinding machine to obtain homogenous samples and subsequently digested in tri-acid mixture of 2 parts $HClO_4$ + 5 parts HNO_3 + 1 part H_2SO_4 [3]. Analyzes made in acid extracts of leaves consisted of: N by auto-nitrogen analyzer (Model Perkin Elmer-2410), P using vanadomolybdo-phosphoric acid method, K by flame photometry and micronutrients (Fe, Mn, Cu and Zn) by atomic absorption spectrophotometer (Model GBC-908).

The vegetative growth parameters (tree height, stem height, canopy spread and stem stock, stem scion and girth diameter) were measured for all trees and the pooled annual incremental magnitudes of these parameters were compared. The canopy volume was calculated based on the formulae:

$$\text{The canopy volume} = 0.\ 5233\ H\ W^2 \tag{2}$$

where: H = (tree height – stem height) and W = the canopy width [11]. The weight of total fruits from each tree for all treatments was recorded and the total yield was estimated considering 278 trees per hectare. The water productivity was calculated as the ratio of total fruit yield ($t.ha^{-1}$) to total irrigation water used per hectare ($m^3.ha^{-1}$) in different treatments. Five fruits per tree were taken randomly for determination of fruit quality parameters (juice percent, acidity and total soluble solids). Juice was extracted manually by juice extractor and the percent content was estimated on weight basis with respect to fruit weight. The total soluble solids (TSS) were determined by digital refractometer (Atago model-PAL 1, Japan) and acidity was measured by volumetric titration with standardized sodium hydroxide, using phenolphthalein as an internal indicator [13]. All the observed data were subjected to analysis of variance (ANOVA). The Least Significant Difference (LSD) at 5% probability level was obtained according to the method described by Gomez and Gomez [9].

12.3 RESULTS AND DISCUSSION

12.3.1 IRRIGATION WATER

The monthly irrigation water application ($liters.day^{-1}.tree^{-1}$) in four drip irrigation regimes was highest in May (48–160) and lowest in December (8–26), due to increasing rate of pan evaporation from December to May during study years (Table 12.1). Earlier studies by Autak et al. [2] and Shirgure et al. [15] recorded the same trend of water requirement of Nagpur mandarin from December to June under Central Indian conditions. On the whole, the total quantity of water application ($m^3\ ha^{-1}\ yr^{-1}$) was 1704 for T1, 2840 for T2, 3976 for T3 and 5679 for T4 irrigation regimes, respectively.

TABLE 12.1 Mean Daily Irrigation Water Applied ($liters.day^{-1}.tree^{-1}$) For Four Irrigation Treatments During Each Month of the Study Period

	Months								TWA*
Treatment	**Nov**	**Dec**	**Jan**	**Feb**	**Mar**	**Apr**	**May**	**Jun**	**$m^3.ha^{-1}.yr^{-1}$**
+DI at 30% FI	14	8	13	20	29	38	48	40	1704
DI at 50% FI	23	13	21	33	48	64	80	60	2840
DI at 70% FI	32	18	29	46	67	90	112	84	3976
FI	45	26	42	65	96	128	160	120	5679

TWA = Total yearly water application per hectare.

12.3.2 SOIL MOISTURE VARIATION

The mean monthly soil moisture at 30, 45 and 60 cm soil depths during irrigation periods indicated that the FI showed significantly higher soil water content (25.5–28.5%, v/v) compared to DI treatments at 30 cm depth (Fig. 12.1). The soil water content at 30 cm depth increased invariably in all the treatments during January-February due to some un-seasonal rains (10–15 mm) in these months. The soil water fluctuation between two measurements in a week in FI was wider than DI treatments. It was due to higher evapotranspiration (ET) rate of the trees under increased soil water availability in FI compared to DI treatments, reported by Cohen [4]. However, the soil water fluctuations under different irrigation regimes were affected negligibly at 45 cm and 60 cm soil depths, suggesting the confinement of effective root zone of the plants within top 30 cm soil profile. The fluctuation of soil water content at 0.30 cm depth in FI was relatively higher during April to June than November to March, indicating the higher tree water consumption under higher quantum of irrigation water supply in FI during April to June than November to March.

12.3.3 LEAF NUTRIENT COMPOSITION

The imposed irrigation treatments showed a differential response on leaf nutrient status of mandarin trees (Table 12.2). The trend of leaf nutrient composition registered in all irrigation treatments was similar with little variation in magnitude during two years of study. The mean data of leaf nutrient composition indicated that the N (2.17%) and K (1.62%) in FI were significantly higher than the DI treatments (1.80–2.10% of N, 1.41–1.54% of K). This is due to higher nutrient uptake by trees under increased availability of nutrients in tree rhizosphere caused by superior soil water content in FI than DI. However, phosphorous content in the leaves did not show any significant variation, probably due to its lower solubility and slow movement in soil-water continuum.

In our studies, DI at 30% FI registered a suboptimal leaf N (1.8%) and K (1.41%), as per the standard foliar diagnostic chart developed by Srivastava et al. [17] for optimum Nagpur mandarin productivity in central India. The irrigation had no significant effect on the fluctuation in leaf micronutrient content, except Fe statistically ($P < 0.05$). The highest leaf Fe was registered in FI (140.6 ppm), followed by DI at 70% FI. Higher uptake of Fe in FI is attributed to increased solubility of iron (Fe^{2+}) in the tree rhizosphere under increased water availability conditions in this treatment [14]. However, the leaf-Fe content (135.8 ppm) under DI at 70%FI was at optimum level for Nagpur mandarin production, as reported by Srivastava et al. [18].

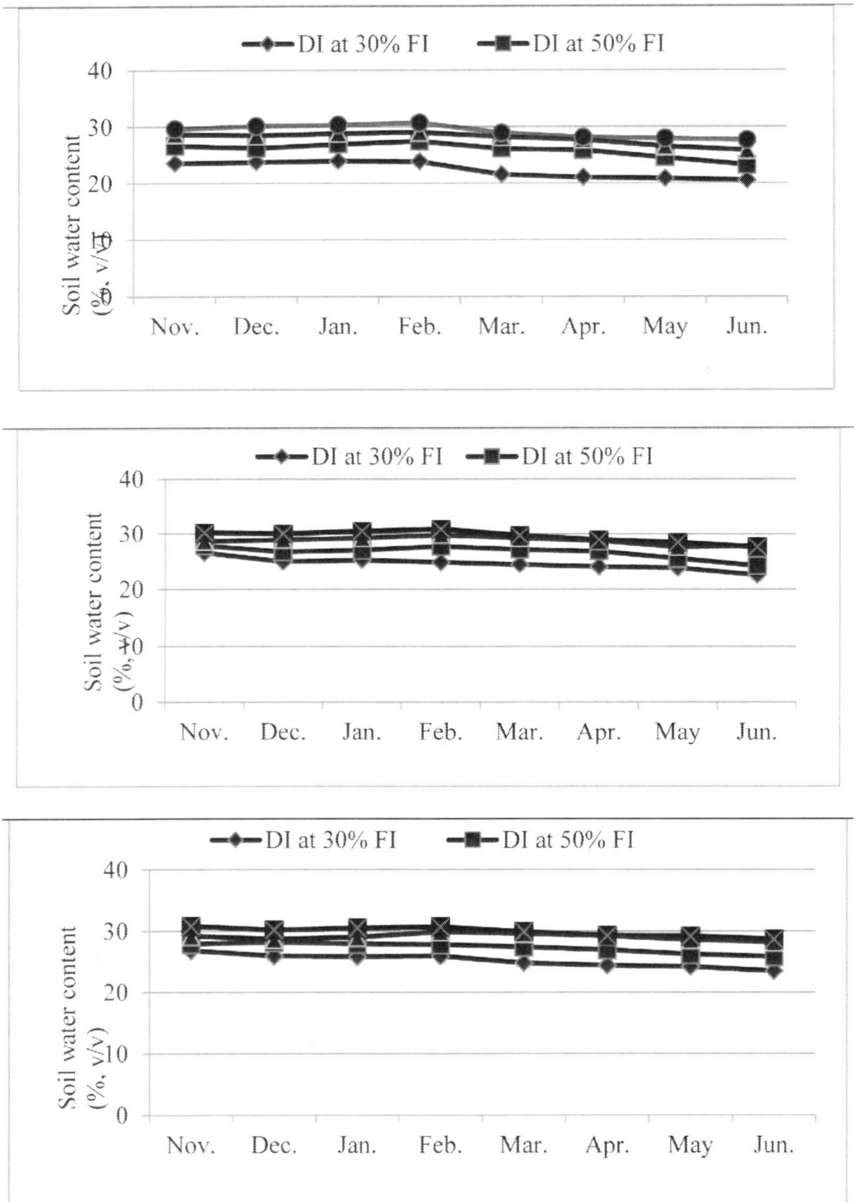

FIGURE 12.1 Soil water variation at 30 cm (top, Fig. 12.1a), 45 cm (center, Fig. 12.1b) and 60 cm (bottom, Fig. 12.1c) soils depths during various months in different irrigation treatments. (a) Soil water variation at 30 cm depth. (b) Soil water variation at 45 cm depth. (c) Soil water variation at 60 cm depth

TABLE 12.2 Leaf Nutrient Composition of Nagpur Mandarin For Different Irrigation Treatments

Treatment	Macro-nutrients (%)			Micro-nutrients (ppm)			
	N	P	K	Fe	Mn	Cu	Zn
+DI at 30% FI	1.80	0.08	1.41	115.1	27.3	6.7	16.7
DI at 50%FI	1.91	0.07	1.51	127.2	33.2	7.0	16.8
DI at 70%FI	2.10	0.09	1.54	135.8	42.7	9.3	19.7
FI, Control	2.17	0.10	1.62	140.6	45.2	9.7	17.3
$CD_{0.05}$	0.10	NS	0.08	5.6	NS	NS	NS

TABLE 12.3 Annual Increment in Tree Growth Parameters of Nagpur Mandarin For Different Irrigation Treatments

Treatment	Tree height (m)	Stock girth (cm)	Scion girth (cm)	Canopy volume (m^3)
+DI at 30% FI	0.21	2.4	2.0	7.23
DI at 50%FI	0.32	2.6	2.3	8.89
DI at 70%FI	0.37	3.1	2.8	9.42
FI (Control)	0.41	3.3	2.9	9.53
$CD_{0.05}$	0.08	NS	NS	0.6

12.3.4 TREE GROWTH, FRUIT YIELD AND WATER PRODUCTIVITY

The measurement of annual incremental growth characteristics of the tree (tree height, canopy volume, stock girth, scion girth) showed that only tree height and canopy volume were significantly influenced by irrigation regimes (Table 12.3). The maximum increase in tree height (0.41 m) and canopy volume (9.53 m^3) was observed in FI followed by DI at 70% FI. This may be due to better metabolic activities of the tree under favorable soil moisture in the root zone in these treatments. The minimum vegetative growth was observed in DI at 30% FI. These results agree with findings of Garcıa-Tejero et al. [7] in 'Salustiano' orange in Spain.

The fruit yield for all irrigation treatments revealed that FI produced a marginally higher (5.6%) yield than DI at 70% FI (Table 12.4). More number of fruits (515 per tree) with less fruit weight (98.5 g) were observed in FI compared with DI at 70% FI (468 number of fruits per tree; 102.6 g per fruit). The similar results were also reported by Perez-Perez et al. [12] in 'lane late' orange and Garcıa-Tejero et al. [7] in 'Salustiano' orange. However, DI at 70% FI resulted in a substantial increase in water productivity (0.0034 tons.m^{-3}) than FI (0.0025 tons.m^{-3}). The higher water productivity in DI at 70%FI was attributed to higher increase in fruit yield with comparatively less increase in irrigation water supply than FI. Fruit quality (juice content, TSS, acidity percentage) assessment for all irrigation treatments showed that DI at 70% FI produced the fruits having higher TSS and lower acidity, followed by FI. Earlier studies comparing FI with DI also demonstrated comparatively better fruit quality (higher TSS and lower acidity) of citrus fruits in optimal DI the full irrigation [7].

TABLE 12.4 Fruit Yield, Water Productivity and Fruit Quality of Nagpur Mandarin in Four Irrigation Treatments

Treatment	Yield parameters			Water used (m^3.ha^{-1})	Water productivity (t.m^{-3})	Quality parameters		
	No. of fruits/ tree	Average fruit weight (g)	Total yield (t.ha^{-1})			Juice (%)	Acidity (%)	TSS (°Brix)
DI at 30% FI	210	67.2	3.92	1704	0.0023	38.6	0.88	9.5
DI at 50% FI	329	96.2	8.80	2840	0.0031	39.9	0.84	10.1
DI at 70% FI	468	102.6	13.35	3976	0.0034	40.2	0.85	10.2
FI, Control	515	98.5	14.10	5679	0.0025	40.4	0.86	9.7
$CD_{0.05}$	6.5	2.6	0.80	-	-	0.3	0.04	0.06

12.4 SUMMARY

The deficit irrigation is found to be an effective water saving technique in drip-irrigated Nagpur mandarin. This study demonstrated that irrigation water quantity of 18–112 L day^{-1} tree^{-1} applied through drip system during December to June is optimum for 16 to 17 year-old mandarin trees in central India. The significant variation of soil water content at 0–30 cm soil profile suggested that the soil water depletion measured at 30 cm depth may be used in drip irrigation scheduling for mature Nagpur mandarin. The higher (36%) water productivity and improved fruit quality in optimum deficit irrigation regime (70% of full irrigation) than full irrigation warrants the adoption of deficit irrigation in Nagpur mandarin orchards of central India. This will help in bringing more area under irrigation, resulting in large increase in production of citrus with prolonged orchard longevity.

A study was conducted to assess the response of deficit irrigation (DI) in 16-year-old drip-irrigated Nagpur mandarin (*Citrus reticulata*) trees budded on rough lemon (*Citrus jambhiri* L.) root stock at Nagpur, Maharashtra, India. DI imposed were at

30, 50 and 70% of full irrigation (FI) and compared with FI (control). The annual increase in vegetative growth parameters (tree height and canopy volume) of trees under FI was found to be superior over DI treatments. The soil moisture status measured at 30 cm, 45 cm and 60 cm depths showed that the mean monthly soil moisture content was significantly affected at 30 cm depth, with highest magnitude (27.6–30.7%, v/v) under FI. Leaf nutrient (N, P, K, Fe, Mn, Cu and Zn) analysis indicates that FI registered the maximum leaf N, K and Fe, followed by 70% FI. The highest fruit yield (14.10 tons.ha^{-1}) recorded in FI was at par with DI at 70% FI (13.35 tons.ha^{-1}). More number and smaller size of fruits were recorded in FI compared to DI at 70% FI. However, DI at 70% FI produced 36% higher water productivity with superior quality fruits (more fruit weight, higher TSS and lower acidity) than FI.

KEYWORDS

- **deficit irrigation, DI**
- **drip irrigation**
- **fruit quality**
- **fruit acidity**
- **irrigation scheduling**
- **Nagpur mandarin**
- **soil moisture**
- **TSS**
- **vegetative growth**
- **Vertisols**
- **water productivity**

REFERENCES

1. Allen, R. G., Pereira, L. S., Raes, D., Smith, M., (1998). *Crop Evapotranspiration. Guidelines For Computing Crop Water Requirements.* FAO Irrigation and Drainage Paper 56 by FAO, Rome, Italy.
2. Autkar, V. N., Kolte, S. O., Bagade, T. R., (1989). Distribution of active rooting zones in Nagpur mandarin and estimates of water requirement for Vertisoles of Maharashtra. *Ann. Pl. Physiol., 2(2),* 219–222.
3. Chapman, H. D., Prat, P. F., (1961). Methods of analysis for soil, plants and waters. In: *Plant analysis.* University of California, Division of Agricultural Sciences, Berkeley, CA, USA. Pages 56–64.
4. Cohen, M., (2001). Drip irrigation in lemon orchards. *International Water and Irrigation, 21(1),* 18–21.

5. Davies, F. S., Albrigo, L. G., (1994). *Citrus*. CAB International, Wallingford, U.K.
6. English, M., (1990). Deficit irrigation I: Analytical framework. *Journal of Irrigation and Drainage Engineering* (ASCE), *116(1),* 399–412.
7. García-Tejero, I., Jiménez-Bocanegra, J. A., Martı'nez, G., Romero, R., Dura'n-Zuazo, V. H., Muriel-Ferna'ndez, J. L., (2010). Positive impact of regulated deficit irrigation on yield and fruit quality in a commercial citrus orchard [Citrus sinensis (L.) Osbeck, cv. Salustiano]. *Agric. Water Manage.*, *97*, 614–622.
8. Geerts, S., Raes, D., (2009). Deficit irrigation as an on-farm strategy to maximize crop water productivity in dry areas. *Agric. Water Manage., 96*, 1275–(1284).
9. Gomez, K. A., Gomez, A. A., (1984). *Statistical Procedures for Agricultural Research.* John Wiley and Sons, Inc., New York.
10. Gonzalez-Altozano, P., Castel, J. R., (1999). Effects of regulated deficit irrigation on 'Clementina de Nules' citrus trees, I. Yield and fruit quality effect. *J. Am. Soc. Hort. Sci., 74*, 706–713.
11. Obreza, T. A., (1991). Young Hamlin orange tree fertilizer response in south-west Florida. *Proceedings of Florida State Horticultural Society*, *103*, 12–16.
12. Perez-Perez, J., Romero, P., Navarro, J., Botia, P., (2008). Response of sweet orange cv. 'lane late' to deficit irrigation strategy in two rootstocks, II: Flowering, fruit growth, yield and fruit quality. *Irrigation Science, 26(6),* 519–529.
13. Ranganna, R., (2001). *Handbook of Analysis and Quality Control for Fruit and Vegetable Products*. 2nd edition, Tata Mcgraw Hill, pages 860–861.
14. Schaffer, B., (1992). Flood tolerance of 'Tahiti' lime rootstocks in South Florida soil. *Proc. Fla. State Hort. Soc., 104*, 31–32.
15. Shirgure, P. S., Srivastava, A. K., Singh, S., (2001). Effect of pan evaporation based irrigation scheduling on yield and quality of drip irrigated Nagpur mandarin. *Indian Journal of Agricultural Sciences, 71(4),* 264–266.
16. Singh, S., Srivastava, A. K., (2004). Citrus industry of India and overseas. In: *Advances in Citriculture* edited by Singh, S., Shivankar, V. J., Srivastava, A. K., Singh, I. P. Jagmander Book Agency, New Delhi.
17. Srivastava, A.K., Singh, S., Huchche, A.D., Ram, L., (2001). Yield based leaf and soil test interpretations for Nagpur mandarin in central India. *Commun. Soil Science Plant Annal., 32(3/4),* 585–99.
18. Srivastava, A. K., Kohli, R. R., Dass, H. C., Huchche, A. D., Ram, L., Singh, S., (1999). Evaluation of the nutritional status of Nagpur mandarin (*Citrus reticulata* Blanco) by foliar sampling. *Tropical Agriculture, 76(2),* 93–103.
19. Zhang, H., Oweis, T., (1999). Water-yield relations and optimal irrigation scheduling of wheat in the Mediterranean region. *Agric. Water Manage.*, *38*, 195–211.

CHAPTER 13

RESPONSE OF SUGAR BEET TO THREE IRRIGATION REGIMES

HANY S. GHARIB and AHMED S. EL-HENAWY

CONTENTS

Printed from Hany S. Gharib and A.S. El-Henawy, 2011. Response of sugar beet (*Beta vulgaris*, L.) to irrigation regime, nitrogen rate and micronutrients application. Alexendria Science Exchange J., Open Access article, *32(2)*, 140–156.

In this chapter, area is in units of *feddans*. A *feddan* (Arabic: فدّان, *faddān*) is a unit of area. It is used in Egypt, Sudan and Syria. The feddan is not an SI unit and in Classical Arabic, the word means 'a yoke of oxen', implying the area of ground that could be tilled by oxen in a certain time. In Egypt the feddan is the only nonmetric unit which remained in use following the switch to the metric system. One fed. = 24 kirat = 60 m × 70 m = 4200 m^2 = 0.42 hectares.

13.1 INTRODUCTION

Increasing agricultural production by effective use of available water resources or more production per drop is a major challenge during the coming decade. The deficit irrigation is an option that may increase water use efficiency (WCUE). Mahmoodi et al. [22] found that the optimum soil water content is 70% of field capacity for 78.5 tons/ha of root yield of sugar beet. The minimum root yield (52.5 t/ha) was observed at 90% of field capacity. Irrigation at 30, 50 and 70% of field capacity had same effects on sugar content while sugar content was decreased at 90% field capacity. At available soil water content of 70% of field capacity, maximum root yield and better quality were observed. Fabeiro et al. [16] reported that moderate water consumption rate (6898 m^3ha^{-1}) was achieved with yields up to 117.64 $t.ha^{-1}$). Excessive irrigation does not increase yield and maximum WCUE was 7.2 $kg.m^{-3}$ for nearly 500 mm of water use [21]. Water deficit decreased root yield of sugar beet, but increased the sugar content, amount of potassium and α-amino N amount, and total irrigation depth increased sugar content in sugar beet [2].

Nitrogen plays an important role in sugar beet production. It affects the root yield and sucrose content and the constituents of sugar yield [7]. Nitrogen deficiency can reduce root and sucrose yields but will increase sucrose content and juice purity [33]. High levels of nitrogen stimulated vegetative growth and consequently increased fresh root weight, but reduced the quality of roots [7, 23, 27]. Fresh root and sugar yields, and nonsugar impurities (K, Na and α-amino N) were positively related to increased rate of N, and sucrose content was reduced by increasing rate of N [23, 32].

Sugar beet response to use of various micronutrients has been the focus of several research studies. Generally, the importance of specific micronutrient for sugar beet production is often related to soil characteristics. The Egyptian soil is deficient in micronutrients due to intensive cropping, low percentage of soil organic matter and soil alkalinity that may decrease availability of cation trace elements such as Mn, Zn and Fe [10]. El-Fouly et al. [11] reported that spraying with micronutrients Fe, Mn, Zn and B significantly increased sugar beet root yield and sugar content. Shaban and Negm [29] found that foliar spraying with the combination of Zn and B increased significantly root, shoot and sugar yields compared to the control. Moustafa and Omran [25] found that foliar spray with B increased root diameter, fresh and dry weight of roots and tops, root and sugar yields, sucrose % and impurity of K.

This chapter discusses effects of irrigation regimes, nitrogen and micronutrients applications on growth, yield quality and water relations of sugar beet.

13.2 MATERIALS AND METHODS

Two field experiments were conducted in a clay soil at Water Management Research Station at El-Karada – Kafrelsheikh of Egypt, during 2007/2008 and 2008/2009 seasons. The sugar beet cultivar "Farida" was used followed by cotton, in both seasons. A split-split plot design with four replications was used. The plot size was 44.1 m^2 (6.3 × 7 m^2). Each plot included seven ridges at 90 cm apart and 7 m long. To avoid the effect of lateral movement of irrigation water, the main plots were isolated by levees of 1.5 m wide.

Seeds of multigerm sugar beet cultivar "Farida" were sown in hills 20 cm apart on both sides of the ridge at the rate of 3–4 seeds per hill, on September 15, 2007 and September 20, 2008. The main plots were assigned to three irrigation regimes and the subplots to two nitrogen rates and the sub-subplots to application of micronutrients. The three irrigation regimes were applied at 40, 55 and 70% available soil moisture depletion (ASMD).

The representative soil samples were taken from each plot at a 0–30 cm depth from the soil surface. Samples were air-dried then grounded to pass through a two mm sieve and were well mixed. The soil analysis was done following the methods of Black et al. [4]. Results of chemical analysis in both seasons are shown in Table 13.1. The soil bulk density, field capacity and wilting point were also determined and are given in Table 13.2. Temperature, relative humidity and rainfall at El-Karada station from sowing to harvest are presented in Table 13.3.

TABLE 13.1 Soil Chemical Analysis at 0–30 cm Soil Depth, in 2007/8 and 2008/9 Seasons

Season	pH (1:2.5)	EC (ds/m)	$CaCO_3$ (%)	OM (%)	Available nutrients, ppm			DTPA extract, ppm		
					N	P	K	Zn	Mn	Fe
2007/8	7.9	2.63	3.33	1.25	22.84	13.07	345	0.71	3.29	4.92
2008/9	8.2	2.54	3.09	1.51	21.10	16.46	319	0.63	4.60	6.34

Note: pH was determined in soil suspension 1:2.5; EC was determined in soil paste extract.

TABLE 13.2 Field Capacity, Wilting Point and Bulk Density of Soil, in 2007–2008 and 2008–2009 Seasons

Soil depth (cm)	Field capacity %		Wilting point %		Bulk density (g/cm^3)	
	2007/8	2008/9	2007/8	2008/9	2007/8	2008/9
0–20	44.21	43.76	24.06	23.82	1.10	1.05
20–40	39.68	39.03	21.25	21.60	1.22	1.15
40–60	35.83	36.42	19.51	19.83	1.33	1.30
Mean	39.91	39.74	21.61	21.75	1.22	1.17

TABLE 13.3 Mean Monthly of Air Temperature, Relative Humidity and Rainfall, During 2007/8 and 2008/9 Seasons

	Temperature (°C)		Relative humidity (%)		Total rainfall (mm)	
Month	**2007/8**	**2008/9**	**2007/8**	**2008/9**	**2007/8**	**2008/9**
November	17	17.0	65.1	67.5	28	-
December	20.6	14.5	64.85	65.3	-	6
January	9.7	13.4	66.0	65.0	12	35
February	11.7	15.2	68.2	70.0	13	45
March	15.4	14.5	65.0	62.1	-	-
April	18.1	19.0	58.0	62.5	-	-
May	19.5	20.7	56.5	58.8	-	-

The experimental field was fertilized with 31 kg of P_2O_5/feddan in the form of superphosphate fertilizer (15.5% P_2O_5) and 24 kg of K_2O/feddan in the form of potassium sulfate (48% K_2O) during soil preparation. The two nitrogen rates were 75 and 90 kg N/feddan. Solution of micronutrients mixture were applied through seed soaking (SS), foliar spraying (FS) and SS+FS as well as control (untreated). Solution of micronutrients consisted of 2 g per liter from each of $ZnSO_4$ (26% Zn), $MnSO_4$ (24% Mn), $FeSO_4$ (20% Fe) and boric acid. Seeds were soaked in solution of micronutrients for 24 h and then dried at room temperature for 24 h. Foliar spraying with solution of micronutrients mixture was done twice, at 80 and 100 days after sowing.

All plots were irrigated immediately after sowing. Light irrigation was given after 8 days after sowing to ensure high seed emergence. Thirty-five days after sowing, the plants were thinned to one plant per hill. The nitrogen fertilizer in the form of urea (46% N) was applied as split into two equal doses, half before the second irrigation after thinning and the remaining half 15 days later before the third irrigation. Irrigation treatments started after the third irrigation. Other cultural practices were done as usual.

Desired depth of irrigation was determined by drying the soil samples for 24 h at 110 °C and soil moisture was expressed on an oven dry weight basis in percent. Soil samples were obtained from 0–60 cm soil depth at an interval of 20 cm depth, before and after each irrigation event to calculate water consumptive use (WCU) of sugar beet from sowing to harvest according to Israelsen and Hansen [19]:

$$WCU = \frac{\theta_2 - \theta_1}{100} \times Bd \times D \times 4200 \qquad (1)$$

where: WCU = amount of water consumptive use (m^3/feddan) of sugar beet; θ_2 = soil moisture content % after irrigation; Θ_1 = soil moisture content % before the next irrigation; Bd = bulk density (g/cm^3); D = depth of soil layer (m); and 4200 is a conversion constant.

Water use efficiency (WCUE) was calculated according to Doorenbos and Pruitt [6] as follows:

$$WUE = \frac{\text{Yield (kg/ feddan)}}{\text{water consumpitive use } (m^3 / \text{feddan})} \quad (2)$$

The number of germinated hills was counted at 30 days after sowing for two ridges in each plot before seedling thinning and the percentage of emerged seeds was calculated.

In each plot, 2 ridges were used for plant growth sampling and 5 ridges for determining root and top yields at harvest. Five guarded plants were randomly taken from each plot at 136, 151 and 165 days after sowing (DAS) to determine leaf area and dry weight of root and top dry weight per plant. The different plant samples were oven dried at 70 °C. For leaf area measurements, the disk method was used. The leaf area index (LAI), crop growth rate (CGR) and net assimilation rate (NAR) were computed as follows [35]:

$$LAI = \text{leaf area per plant/ surface area occupied by one plant} \quad (3)$$

$$CGR = [W_2 - W_1] / [t_2\text{-}t_1] \quad (4)$$

$$RGR = [(\log_e W_2 - \log_e W_1)] / [t_2\text{-}t_1] \quad (5)$$

$$NAR = [(W_2 - W_1)(\log_e A_2 - \log_e A_1)] / [(A_2\text{-}A_1)(t_2\text{-}t_1)] \quad (6)$$

where: W_1 and W_2 refer to dry weight at t_1 and t_2 time (weeks), respectively, A_1 and A_2 refer to leaf area at t_1 and t_2 (weeks), respectively.

At harvest (190 days after sowing), the area of 18.9 m^2 of selected ridges for yield data were harvested to obtained root and top yields. Ten guarded plants were taken at random and were screened for root and top yields/plant, root diameter and root length.

Sugar and other chemical contents in roots were determined by Delta Company of Sugar with an automatic sugar polarimeter, described by McGinnus [24]. Corrected sugar content (white sugar) of sugar beet was calculated by linking nonsugars K, Na and α-amino-N (expressed as milliequivalents/100 g of beet) as described by Harvey and Dutton [18]:

$$Z_B = \{Pol - [0.343(K+Na) + 0.094\, N_{BI} + 0.29]\} \quad (7)$$

where: Z_B = corrected sugar content (% beet) and N_{BI} = α-amino-N determined by the "blue number" method. Juice purity percentage (QZ) was calculated by the Delta Company:

$$QZ = \frac{Z_B}{Pol} \tag{8}$$

The observed data were subjected to analysis of variance according to Gomez and Gomez [17]. Means of each parameter were compared by Duncan's Multiple Range Test [8]. All statistical analysis was performed using analysis of variance technique by means of "MSTATC" computer software package.

13.3 RESULTS AND DISCUSSION

13.3.1 EFFECT OF SEED SOAKING ON SEED GERMINATION

The Fig. 13.1 shows effects of seed soaking on the percentage of germinated hills at 30 DAS, in 2007–2008 and 2008–2009 seasons. Soaking seeds in micronutrient solution significantly enhanced seed emergence compared to untreated seeds (dry seed) at 30 DAS in two seasons. Seed soaking resulted in a substantial increase in hill germination by 4.1 and 5.8% than untreated seeds in the first and second seasons, respectively. This may be due to leaching of inhibitor substance from seeds by soaking in micronutrients solution, hence improving seed emergence, plant vigor and growth attributes. These results are in harmony with those of El-Hindi et al. [12], who found that soaking sugar beet seeds in water for 24 h increased emergence percentages. Sorour et al. [31] reported that mechanical or manual plantings with soaked seeds increased number of germinated hills per m^2 than planting with dry seed.

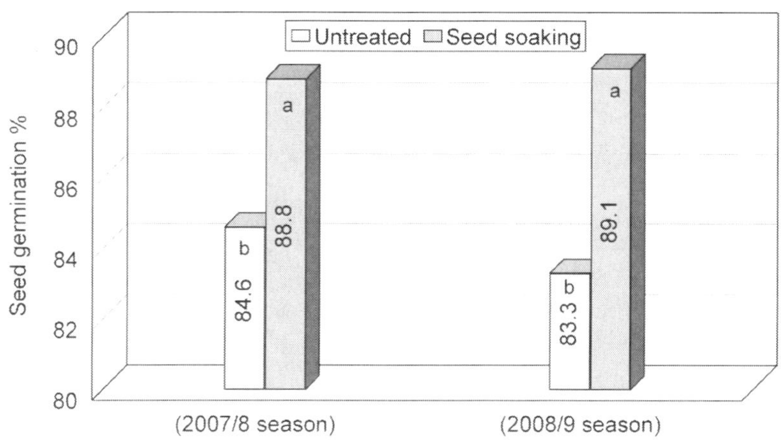

FIGURE 13.1 The percentage of germinated seeds at 30 days after sowing as affected by seed soaking in solution of micronutrients mixture, in 2007/8 and 2008/9 seasons.

13.3.2 EFFECTS OF IRRIGATION REGIMES AND FERTILIZATION ON GROWTH PARAMETERS

Means of dry seed weight, LAI, CGR and NAR of sugar beet were affected by irrigation regime, nitrogen rates and micronutrients during 2007/2008 and 2008/2009 seasons (Tables 13.4 and 13.5).

13.3.2.1 EFFECTS OF IRRIGATION REGIME

The abundance of available soil moisture in the root zone resulted in a substantial increase in dry matter accumulation (g/plant) and LAI at 137, 151 and 165 DAS and CGR at two periods of 137–151 and 151–165 DAS, in both seasons. Scheduling irrigation at 40% ASMD produced largest dry weight, LAI and CGR, compared to the lowest values of these parameters at 70% ASMD produced. Certainly the sufficient soil moisture content at 40% ASMD favored cell division and elongation and thus the expansion of leaves which in turn resulted in more photosynthates available for dry matter accumulation per unite area (CGR). These results confirm the findings of El-Zayat [14] and Sorour [30]. On the contrary, root/top ratio and NAR was significantly influenced by irrigation regime in favor of plants irrigated at 55 and 70% ASMD compared to those irrigated at 40% ASMD. Abundance of available soil moisture content pushed the plants towards the top growth that in turn may have decreased the efficiency of assimilation of translocation from tops to roots and in turn decreasing root/top ratio [30]. Such reduction in NAR obtained from high soil moisture level may be attributed to very large leaf area which led to increase mutual-shading and transpiration and in turn caused a reduction in rate of assimilation per unit of leaf area (NAR). El-Zayat [14] and Sorour [30] reported similar results.

TABLE 13.4 Effects of Irrigation Regimes, Nitrogen Rate, Micronutrients and Their Interactions on Dry Matter Accumulation and Root/Top Ratio of Sugar Beet, in 2007/8 and 2008/9 Seasons

Treatment	2007–2008 season				2008–2009 season			
	Dry weight (g/plant)			Root/Top ratio	Dry weight (g/plant)			Root/Top ratio
	Days after sowing				Days after sowing			
	136	151	165	165	136	151	165	165
ASMD at Irrigation, (I)	**	*	*	*	**	**	**	*
40%	192 a	233 a	305 a	1.82 b	211 a	244 a	306 a	1.98 b
55%	173 b	211 b	286 b	2.66 a	201 a	241 a	304 b	2.41 a
70%	147 c	176 c	235 c	2.38 a	166 c	200 c	255 c	2.29 a

TABLE 13.4 *(Continued)*

Treatment	2007–2008 season				2008–2009 season			
	Dry weight (g/plant)			Root/Top ratio	Dry weight (g/plant)			Root/Top ratio
	Days after sowing				Days after sowing			
	136	151	165	165	136	151	165	165
Kg of N/fed., (N)	*	*	*	*	**	*	**	*
75	161 b	195 b	260 b	2.42 a	176 b	209 b	262 b	2.51 a
90	181 a	218 a	290 a	2.15 b	209 a	248 a	315 a	1.94 b
Micronutrient, (M)	**	**	**	NS	**	*	*	NS
Control (C)	153 c	186 c	246 c	2.4	168 c	200 c	261 c	2.32
Seed soaking (SS)	169 b	204 b	274 b	2.34	197 b	237 ab	294 a	2.25
Foliar spraying (FS)	166 b	200 b	268 b	2.18	195 b	227 b	289 a	2.16
SS + FS	194 a	236 a	313 a	2.21	211 a	251 a	310 a	2.18
Interactions among I, N and M								
I X N	NS	*	NS	*	NS	*	NS	*
I X M	NS	NS	*	NS	NS	NS	NS	NS
N X M	NS	NS	NS	NS	NS	NS	NS	NS
I X N X M	NS	NS	NS	NS	NS	NS	NS	NS

*, ** and NS indicate $P< 0.05$, $P< 0.01$ and not significant, respectively. Means of each value designated by the same latter are not significantly different at $P<5\%$ level using Duncan's MRT.

TABLE 13.5 Effects of Irrigation Regimes, Nitrogen Rate, Micronutrients and Their Interactions on Leaf Area Index (LAI), Crop Growth Rate (CGR) and Net Assimilation Rate (NAR) of Sugar Beet, in 2007/8 and 2008/9 Seasons

Factor	LAI			CGR (g/m^2/week)		NAR (g/m^2/week)	
	Days after sowing, DAS						
	136	151	165	136–151	151–165	136–151	151–165
2007–2008 season							
ASMD at Irrigation (I)	**	**	**	*	**	*	*
40%	2.71^a	4.07^a	4.66^a	224^a	413^a	66^b	92^b
55%	2.34^b	3.15^b	3.5^b	213^a	401^a	79^a	125^a
70%	1.64^c	2.46^c	2.74^c	162^b	331^b	80^a	127^a
Kg N/fed. (N)	*	*	*	**	*	NS	NS

TABLE 13.5 *(Continued)*

Factor	LAI			CGR (g/m^2/week)		NAR (g/m^2/week)	
	Days after sowing, DAS						
	136	151	165	136–151	151–165	136–151	151–165
75	2.13^{b}	3.02^{b}	3.41^{b}	191^{b}	361^{b}	77	116
90	2.32^{a}	3.43^{a}	3.85^{a}	208^{a}	402^{a}	74	114
Micronutrient (M)	**	*	*	*	**	NS	*
Control (C)	1.82^{c}	2.95^{b}	3.14^{c}	182^{c}	336^{c}	76	112^{b}
Seed soaking (SS)	2.36^{ab}	3.27^{b}	3.63^{b}	195^{b}	385^{b}	73	116^{ab}
Foliar spraying (FS)	2.19^{b}	2.96^{b}	3.55^{b}	189^{b}	380^{b}	76	121^{a}
SS + FS	2.54^{a}	3.73^{a}	4.22^{a}	233^{a}	425^{a}	76	111^{b}
Interactions among I, N and M							
I X N	NS	*	NS	*	NS	*	NS
I X M	NS	NS	*	NS	NS	NS	NS
N X M	NS	*	NS	NS	NS	NS	NS
I X N X M	NS	NS	NS	NS	NS	NS	NS
2008–2009 season							
ASMD at Irrigation (I)	**	**	*	**	**	*	*
40%	2.2^{a}	3.35^{a}	3.51^{a}	213^{a}	349^{a}	66^{b}	95^{b}
55%	2.0^{a}	3.30^{a}	3.49^{a}	202^{a}	340^{a}	85^{a}	105^{a}
70%	1.7^{b}	2.93^{b}	3.24^{b}	183^{b}	307^{b}	86^{a}	107^{a}
Kg N/fed. (N)	*	*	**	**	**	NS	*
75	1.85^{b}	2.97^{b}	3.12^{b}	184^{b}	294^{b}	78	115^{a}
90	2.09^{a}	3.42^{a}	3.71^{a}	215^{a}	372^{a}	80	89^{b}
Micronutrient (M)	NS	*	*	**	*	NS	**
Control (C)	1.85	2.90^{c}	3.08^{c}	169^{c}	317^{c}	76	108^{a}
Seed soaking (SS)	2.02	3.36^{a}	3.55^{ab}	217^{a}	341^{a}	84	93^{b}
Foliar spraying (FS)	1.98	3.15^{b}	3.37^{b}	191^{b}	329^{b}	73	113^{a}
SS + FS	2.04	3.38^{a}	3.65^{a}	221^{a}	344^{a}	83	95^{b}
Interaction among I, N and M							
I X N	NS	NS	*	NS	NS	NS	NS
I X M	NS	*	NS	NS	NS	NS	NS
N X M	NS	NS	NS	NS	NS	NS	NS
I X N X M	NS	NS	NS	NS	NS	NS	NS

*, ** and NS indicate $P < 0.05$, $P < 0.01$ and not significant, respectively. Means of each parameter designated by the same letter are not significantly different at 5% level using Duncan's MRT.

13.3.2.2 EFFECTS OF NITROGEN RATE

Dry matter accumulation (g/plant), LAI and CGR at all sampling dates in both seasons were with 90 kg N/feddan significantly superior to those with 75 kg N/feddan in. This reflects the important role of nitrogen in building up the photosynthetic matter of plants and consequently accumulation of more dry matter per plant or per unite of ground area. The inverse was true in root/top ratio at 159 DAS in both seasons. Such decrease in root/top ratio may be due to the fact that top growth was favored more than root growth, with more nitrogen. No significant differences in NAR were observed between two nitrogen rates for all growth periods, except during 151–165 DAS in the second season, where increasing nitrogen rate significantly decreased NAR. This may be due to overlap between the large leaves of adjacent plants at higher nitrogen rate which decreased the dry matter accumulation efficiency per unit leaf area as a result of competition for light and in turn reduced the NAR. Also, the data reflects the negative correlation between LAI and NAR, whereas the rate of 75 kg N/feddan produced higher NAR and lower LAI at the mentioned period. Attia and Abd-Motagally [3], El-Zayat [14] and Selim et al. [28] found that increasing nitrogen rate increased root and top dry weight and vegetative growth of sugar beet.

13.3.2.3 EFFECTS OF MICRONUTRIENTS

Application of micronutrients resulted in a significant increase in dry matter accumulation (g/plant), LAI and CGR compared with the control (untreated treatment) at all sampling dates, except LAI at 136 DAS in the second season. Application of micronutrients through seed soaking and foliar spraying (SS+FS) recorded the highest values of these traits. This trend may be attributed to the role of micronutrients as a cofactor in the enzymatic reaction of the anabolic pathways in plant growth [1]. There was no significant difference in root/top ratio due to micronutrients, in two seasons (Table 13.4). However, NAR was significantly influenced by micronutrients in the second period in both seasons (Table 13.5). The relative ranking of micronutrients treatments for NAR was inconsistent in two seasons. Foliar spraying (FS) produced higher NAR in this period in both seasons. The lowest NAR was obtained from beet plants of SS+FM treatment in both seasons. This may be attributed to the increase in mutual shading and/or the dilution effect caused by the large leaf area formed at SS+FS treatment, which in turn decreased NAR. This reflects the negative correlation between LAI and NAR. These results agree with those reported by Ebrahim [9]. Sorour et al. [31] reported that seed soaking increased dry matter, LAI and CGR.

13.3.2.4 EFFECTS OF INTERACTIONS

The interaction between irrigation regimes and nitrogen rates had a significant effect on dry weight per plant at 151 DAS and root/top ratio at 165 DAS, in the two seasons: LAI at 151 DAS in the first season and at 165 DAS in the second season as well as CGR and NAR during 136–151 DAS in the first season. The interaction between irrigation regimes and micronutrients had a significant effect on dry weight per plant and LAI at 165 DAS in the first season as well as LAI at 151 DAS in the second season. However, the other interactions did not reach the level of significance for these cases.

13.3.3 ROOT AND TOP YIELDS AND THEIR COMPONENTS

Table 13.6 indicates the effects of irrigation regime, nitrogen rate and micronutrients on the means of root length, root diameter, root weight, root yield and top yield, in 2007/8 and 2008/9 seasons.

TABLE 13.6 Effects of Irrigation Regime, Nitrogen Rate, Micronutrient and Their Interactions on Root Yield, Top Yield and Root Dimensions of Sugar Beet, in 2007/8 and 2008/9 Seasons

Treatment	Root length (cm)	Root diameter (cm)	Top yield		Root yield	
			Kg/plant	t/fed.	Kg/ plant	t/fed.
2007–2008 season						
ASMD at Irrigation (I)	**	**	**	**	**	**
40%	28.1^{b}	10.8^{a}	0.499^{a}	14.254^{a}	0.995^{a}	28.498^{a}
55%	29.7^{a}	11.0^{a}	0.385^{b}	11.523^{b}	1.014^{a}	30.344^{a}
70%	30.8^{a}	9.3^{b}	0.350^{b}	9.915^{b}	0.754^{b}	21.431^{b}
Kg N/fed. (N)	**	NS	*	*	**	**
75	28.9^{b}	10.3	0.370^{b}	10.770^{b}	0.879^{b}	25.674^{b}
90	30.2^{a}	10.4	0.453^{a}	13.025^{a}	0.963^{a}	27.842^{a}
Micronutrient (M)	**	*	*	*	**	**
Control (C)	28.0^{c}	9.6^{b}	0.315^{b}	10.269^{b}	0.819^{c}	24.326^{c}
Seed soaking (SS)	30.0^{ab}	10.8^{a}	0.430^{a}	12.042^{a}	0.948^{ab}	27.785^{ab}
Foliar spraying (FS)	28.9^{bc}	10.2^{ab}	0.421^{a}	12.224^{a}	0.915^{b}	26.322^{b}
SS + FS	31.3^{a}	10.8^{a}	0.479^{a}	13.054^{a}	1.001^{a}	28.598^{a}
Interaction among I, N and M						
I X N	NS	NS	*	*	*	*
I X M	NS	NS	NS	NS	*	*
N X M	NS	NS	NS	NS	*	**
I X N X M	NS	NS	NS	NS	**	**

TABLE 13.6 *(Continued)*

Treatment	Root length (cm)	Root diameter (cm)	Top yield		Root yield	
			Kg/plant	t/fed.	Kg/ plant	t/fed.
2008–2009 season						
ASMD at Irrigation (I)	**	**	*	**	**	*
40%	28.2[c]	10.3[a]	0.472[a]	13.969[a]	0.963[a]	28.582[a]
55%	29.9[b]	10.2[a]	0.449[a]	11.910[b]	1.117[a]	29.675[a]
70%	32.1[a]	9.5[b]	0.307[b]	9.835[c]	0.731[b]	20.538[b]
Kg N/fed. (N)	**	NS	**	**	*	*
75	28.6[b]	9.9	0.359[b]	10.860[b]	0.859[b]	24.958[b]
90	31.6[a]	10.2	0.460[a]	12.949[a]	1.016[a]	27.572[a]
Micronutrient (M)	**	NS	*	*	**	**
Control	27.5[c]	9.7	0.342[b]	10.359[b]	0.801[c]	23.593[c]
Seed soaking (SS)	30.9[ab]	10.1	0.426[a]	11.966[a]	0.994[ab]	26.898[ab]
Foliar spraying (FS)	29.5[b]	9.8	0.406[a]	12.148[a]	0.911[b]	26.136[b]
SS + FS	32.4[a]	10.4	0.464[a]	13.145[a]	1.042[a]	28.433[a]
Interaction among I, N and M						
I X N	NS	NS	*	*	*	*
I X M	NS	NS	NS	NS	*	*
N X M	NS	NS	NS	NS	**	**
I X N X M	NS	NS	NS	NS	*	*

*, ** and NS indicate $P < 0.05$, $P < 0.01$ and not significant, respectively. Means of each value designated by the same letter are not significantly different at 5% level using Duncan's MRT.

13.3.3.1 EFFECTS OF IRRIGATION REGIME

Root yield and its attributes were significantly different among irrigation regimes, in two seasons. Plants irrigated at 70% ASMD produced longer roots than those irrigated at 40% ASMD. Results show that water stress enhanced the deep rooting. These results are in agreement with those reported by Sorour [30], Emara [15], El-Zayat [14] and Vamerali et al. [34]. However, root diameter, root weight and root yield per feddan with irrigation at 40 or 55% ASMD were almost the same and were significantly superior than those with irrigation at 70% ASMD. Top yield per plant or per feddan was significantly increased by increasing available soil moisture.

Such increase in root yield with irrigation at 40 or 55% ASMD can be attributed to improved beet growth, in terms of thicker roots, higher crop growth rate and heavier root weight. Also, the abundance of soil moisture increased top yield by increasing dry matter accumulation and leaf area. These results are in agreement with those reported by Sorour [30], Emara [15], El-Zayat [14] and Kenter et al. [20]. Mahmoodi et al. [22] reported that the optimum soil water content for root yield is 70% of field capacity, while the minimum root yield was observed at 90% of field capacity. El-Sarag [13] found that increasing irrigation intervals from 5 to 11 days sharply reduced top fresh weight, while irrigation every 8 days was superior in root yield.

13.3.3.2 EFFECTS OF NITROGEN RATE

Nitrogen application exerted a significant effect on root yield, root length, root weight and top yield in favor of 90 kg N/feddan compared with 75 kg N/feddan, in the two seasons. Thus, the highest nitrogen rate increased root yield by increasing LAI, dry matter accumulation, CGR, root length and root weight. However, there was no evidence for significant difference in root diameter due to nitrogen rate. The positive effect of nitrogen on root yield is supported by studies by Nemeat Alla and El-Geddawy [26], Tsialtas and Maslaris [32], Attia and Abd-Motagally [3], El-Sarag [13], Vamerali et al. [33], Marinkovic et al. [23] and Selim et al. [28].

13.3.3.3 EFFECTS OF MICRONUTRIENTS

Micronutrients significantly affected root yield and all yield attributes in both seasons, except root diameter in the second season. The early supply of micronutrients by seed soaking enabled beet plants to have deep (in both seasons) and thick (in the first season) roots compared to the control. Application of micronutrients resulted in a significant increase in top and root yields compared with control (untreated) in both seasons. The beets of seed soaking and foliar micronutrients (SS+FS) produced the greatest root and top yields in the two seasons. The root and top yields of SS were statistically at par with those of SS+FS. This indicates that seed soaking method was more effective in these cases than the foliar application. This may be due to the considerable increase in early growth, which was reflected in higher root yield and its components, i.e. root length, diameter and weight. Nemeat Alla and El-Geddawy [26] found that foliar spraying twice with micronutrients mixture significantly increased root length, root diameter, top yield and root yield. Shaban and Negm [29] reported that combination of Zn and B increased significantly root and shoot yields over the control. Sorour et al. [31] reported that seed soaking increased root yield.

13.3.3.4 EFFECTS OF INTERACTION

Root yield was significantly affected by all interactions, in both seasons. The interaction, irrigation regimes × nitrogen rate, had a significant effect on top yield, in the two seasons. Means of root yield/feddan as influenced by first and the second order interactions are presented in Table 13.7.

TABLE 13.7 Effects of Interactions Among Irrigation Regime, Nitrogen Rate and Micronutrient on Root Yield (t/fed.) of Sugar Beet, in 2007/8 and 20089/9 Seasons

Irrigation at ASMD (I)	Micro-element (M)	2007–2008 season			2008–2009 season		
		Kg N/fed., (N)		Mean	Kg N/fed., (N)		Mean
		75	90		75	90	
		I × N-Mean		**I-Mean**	**I × N-Mean**		**I-Mean**
40%		27.37^{b}	29.62^{ab}	28.498^{a}	27.24^{c}	29.92^{ab}	28.58^{a}
55%		28.84^{b}	31.84^{a}	30.344^{a}	28.19^{bc}	31.16^{a}	29.68^{a}
70%		20.80^{c}	22.06^{c}	21.431^{b}	19.44^{d}	21.63^{d}	20.54^{b}
		N × M-mean		M-mean	N × M-mean		M-mean
	C*	23.55^{d}	25.11^{cd}	24.33^{c}	22.56^{e}	24.63^{de}	23.59^{c}
	SS**	26.62^{bc}	28.95^{ab}	27.79^{ab}	25.47^{cd}	28.32^{ab}	26.90^{ab}
	FS***	25.23^{cd}	27.42^{bc}	26.32^{b}	24.81^{de}	27.46^{bc}	26.14^{b}
	SS+FS	27.3^{bc}	29.90^{a}	28.60^{a}	26.99^{bcd}	29.88^{a}	28.43^{a}
		I × N × M-Mean		I × M-Mean	I × N × M-Mean		I × M-Mean
40%	C	25.95^{e-h}	27.52^{c-g}	26.74^{c}	25.04^{efg}	26.58^{c-f}	25.81^{d}
	SS	28.63^{c-f}	30.33^{a-e}	29.48^{bc}	27.95^{b-f}	31.20^{abc}	29.58^{abc}
	FS	26.87^{d-g}	29.10^{b-e}	27.98^{c}	26.89^{c-f}	29.74^{a-d}	28.31^{bcd}
	SS+FS	28.05^{c-f}	31.54^{a-d}	29.79^{bc}	29.09^{b-e}	32.16^{ab}	30.63^{ab}
55%	C	26.12^{e-h}	28.7^{c-f}	27.41^{c}	25.29^{d-g}	27.97^{b-f}	26.63^{cd}
	SS	29.67^{a-e}	33.51^{ab}	31.59^{ab}	28.86^{b-e}	31.95^{ab}	30.41^{ab}
	FS	27.88^{c-f}	31.24^{a-d}	29.56^{bc}	27.99^{b-f}	30.88^{abc}	29.44^{abc}
	SS+FS	31.71^{abc}	33.94^{a}	32.82^{a}	30.60^{abc}	33.84^{a}	32.22^{a}
70%	C	18.57^{j}	19.10^{j}	18.83^{e}	17.35^{i}	19.32^{hi}	18.33^{f}
	SS	21.57^{hij}	23.01^{g-j}	22.29^{d}	19.60^{hi}	21.82^{ghi}	20.71^{ef}
	FS	20.93^{ij}	21.92^{hij}	21.42^{de}	19.56^{hi}	21.76^{ghi}	20.66^{ef}
	SS+FS	22.15^{hij}	24.21^{f-i}	23.18^{d}	21.26^{ghi}	23.64^{fgh}	22.45^{e}

*Control, ** Seed soaking, *** Foliar spraying.

Irrigation Regime* × *Nitrogen Rate Interaction (I×N): The highest root yield per feddan was obtained in plots with at 55% ASMD and fertilized with 90 kg N/feddan, while the lowest root yield per feddan was obtained in plots at 70% ASMD and fertilized with 75 kg N/feddan, in the two seasons.

Irrigation Regime* × *Micronutrients Interaction (I×M): The combination of irrigation at 55% ASMD and SS+FS or SS recorded the highest root yield without significant differences between them. However, the low irrigation regime × untreated (C) recorded the lowest yield in both seasons.

Nitrogen Rate* × *Micronutrients Interaction (N×M): It is clear that beets with 90 kg N/feddan along with SS+FS or SS produced the highest root yield, while those with 75 kg N/feddan and without micronutrients produced the lowest yield, in both seasons.

Irrigation Regime* × *Nitrogen Rate* × *Micronutrients Interaction (I×N×M): The combination of medium irrigation regime × high N rate × SS+FS produced the maximum root yield in both seasons. Application of SS or FS separately along with medium irrigation regime and high N rate was statistically at par with the mention combination in root yield. The combination of low irrigation regime × low N rate × without micronutrients produced the lowest root yield in both seasons.

13.3.4 SUGAR YIELD AND ROOT QUALITY

The soluble nonsugars, potassium, sodium and α-amino nitrogen in the roots are regarded as impurities because they interfere with the sugar extraction. Table 13.8 shows effects of irrigation regime, nitrogen rate and micronutrients means of these impurities, gross sugar %, extractable white sugar %, sugar loss %, juice purity % and white sugar yield per feddan, in 2007/8 and 2008/9 seasons.

TABLE 13.8 Effects of Irrigation Regime, Nitrogen Rate and Micronutrients Means of These Impurities, Gross Sugar %, Extractable White Sugar %, Sugar Loss %, Juice Purity % and White Sugar Yield Per Feddan, in 2007/8 and 2008/9 Seasons

Treatment	Gross sugar (%)	K+Na (meq/100 g)	α-N	White sugar (%)	Sugar loss	Juice purity	Sugar yield (t/fed.)
2007–2008 season							
ASMD at Irrigation (I)	**	**	**	*	**	*	*
40%	17.53[b]	7.68[b]	3.97[b]	14.23[b]	3.30[b]	81.18[a]	4.053[a]
55%	18.34[a]	8.30[a]	4.16[a]	14.81[a]	3.53[a]	80.76[ab]	4.500[a]
70%	18.59[a]	8.61[a]	4.19[a]	14.95[a]	3.64[a]	80.43[b]	3.208[b]
Kg N/fed. (N)							
75	18.22[a]	8.15[b]	4.07[b]	14.75[a]	3.47[b]	80.98[a]	3.786[b]

TABLE 13.8 *(Continued)*

Treatment	Gross sugar (%)	K+Na	α-N	White sugar	Sugar loss	Juice purity	Sugar yield (t/fed.)
		(meq/100 g)		(%)			
90	18.08[b]	8.24[a]	4.15[a]	14.57[b]	3.51[a]	80.60[b]	4.055[a]
Micronutrient (M)							
Control	17.84[b]	8.22[b]	4.08[b]	14.35[c]	3.49[ab]	80.43[b]	3.481[c]
Seed soaking (SS)	18.33[a]	8.04[c]	4.03[b]	14.90[a]	3.43[b]	81.32[a]	4.134[a]
Foliar spraying (FS)	18.15[a]	8.35[a]	4.21[a]	14.60[b]	3.55[a]	80.46[b]	3.841[b]
SS + FS	18.28[a]	8.19[bc]	4.11[b]	14.80[ab]	3.48[ab]	80.95[a]	4.226[a]
Interactions among I, N and M							
I X N	NS	NS	*	*	NS	*	*
I X M	NS	NS	*	*	NS	*	**
N X M	NS	NS	*	NS	NS	NS	*
I X N X M	NS	NS	NS	NS	NS	NS	**
2008–2009 season							
ASMD at Irrigation (I)	**	**	*	**	**	*	*
40%	17.40[c]	6.65[c]	3.29[b]	14.53[c]	2.88[b]	83.46[a]	4.150[b]
55%	18.85[b]	7.33[b]	3.60[a]	15.71[a]	3.14[a]	83.33[a]	4.661[a]
70%	19.22[a]	7.68[a]	3.65[a]	15.96[a]	3.27[a]	83.00[b]	3.278[c]
Kg N/fed. (N)							
75	18.63[a]	7.15[b]	3.44[b]	15.57[a]	3.07[b]	83.55[a]	3.876[b]
90	18.35[b]	7.29[a]	3.58[a]	15.22[b]	3.13[a]	82.97[b]	4.184[a]
Micronutrient (M)							
Control (C)	18.22[b]	7.21[b]	3.43[c]	15.15[c]	3.09[b]	83.10[bc]	3.567[c]
Seed soaking (SS)	18.61[a]	7.03[c]	3.48[bc]	15.58[a]	3.03[b]	83.72[a]	4.172[ab]
Foliar spraying (FS)	18.55[a]	7.40[a]	3.59[a]	15.39[b]	3.17[a]	82.93[c]	4.001[b]
SS + FS	18.59[a]	7.23[b]	3.56[ab]	15.46[ab]	3.10[ab]	83.29[ab]	4.378[a]
Interactions among I, N and M							
I X N	NS	NS	*	*	NS	*	*
I X M	NS	NS	NS	*	NS	NS	*
N X M	NS	NS	*	NS	NS	NS	NS
I X N X M	NS	NS	NS	NS	NS	NS	NS

*, ** and NS indicate $P < 0.05$, $P < 0.01$ and not significant, respectively. Means of each value designated by the same letter are not significantly different at 5% level using Duncan's MRT.

13.3.4.1 EFFECTS OF IRRIGATION REGIMES

Irrigation regimes had a significant effect on sugar yield and quality of root juice, in two seasons. Data show that water stress significantly increased total sugar content, impurities (K+ Na and α-amino-N), white sugar % and sugar loss %. Irrigation at 70% ASMD recorded the highest values of these traits. Irrigation at 55% ASMD was statistically at par with irrigation at 70% ASMD in white sugar %, in both seasons. It may be due to that some impurities in roots of stressed plants resulted from osmotic adjustment in sugar beet in response to soil drying [5]. On the contrary, juice purity % and white sugar yield/feddan were significantly decreased by the water stress. Although, water stress increased the concentration of gross sugar in roots, it decreased juice purity %. This might be due to increasing impurities in the roots of stressed plants, which cause problems during juice purification and crystallization and in turn can decrease purity. Sorour [30] found that irrigation improved the quality of sugar beet by reducing the K, Na and N contents. Such increase in white sugar yield (obtained from the plants irrigated at 55% ASMD) may be attributed to the increases in root yield and white sugar extraction %. These results agree with those obtained by Sorour [30] El-Zayat [14], Vamerali et al. [33] and El-Sarag [13]. Mahmoodi et al. [22] reported that irrigation at 30, 50 and 70% of field capacity had same effects on sugar content while sugar content decreased at 90% of field capacity. These researchers indicated that irrigation at 70% of field capacity produced the maximum root quality.

13.3.4.2 EFFECTS OF NITROGEN RATE

All traits were significantly different between the two nitrogen rates in both seasons, except sugar loss % in the second season. The concentration of K + Na and α-amino-N in roots and sugar loss % were significantly increased by increasing the nitrogen rate. Total sugar content, extraction of white sugar and juice purity were decreased as nitrogen rate increased, in both seasons. White sugar yield was significantly increased by increasing nitrogen rate from 70 to 90 kg N/feddan. This may be due to increase in root yield. These findings are in agreement with those by Nemeat Alla and El-Geddawy [26], Tsialtas and Maslaris [32], Attia and Abd-Motagally [3], Marinkovic et al. [23] and Selim et al. [28]. Marinkovic et al. [23] stated that the sugar content was significantly decreased, while the content of α-amino-N and sodium were significantly increased with increasing N dose.

13.3.4.3 EFFECTS OF MICRONUTRIENTS

Application of micronutrients as SS, FS and SS+FS resulted in significant increase in total sugar content compared with the control, in both seasons. However, the soluble nonsugars, potassium + sodium and α-amino nitrogen in the roots were significantly increased by foliar spraying with micronutrients (FS) compared with the control treatment. Seed soaking in micronutrients solution (SS) increased extractable white sugar % and juice purity % by improving sugar beet quality due to increase of gross sugar % and reduction of K+ Na and N contents and sugar loss%. The treatments SS or/and FS out-yielded the control treatment in white sugar yield. The maximum white sugar yield was obtained in SS treatments. This may be due to the considerable increase in root yield and white sugar extraction percentage. These results are in accordance with those reported by Ebrahim [9]. Shaban and Negm [29] reported that combination of Zn and B increased significantly sugar yield than the control. Sorour et al. [31] reported that seed soaking increased dry matter, LAI and CGR.

13.3.4.4 EFFECTS OF INTERACTIONS

The Table 13.9 shows the interactions between irrigation regimes × nitrogen rates for the concentration of amino-nitrogen and white sugar % in both seasons, irrigation regimes × micronutrients for white sugar % in both seasons and amino-nitrogen in the first season, nitrogen rates × micronutrients for amino-nitrogen in both seasons. The entire first and second order interactions had a significant effect on white sugar yield in both seasons.

TABLE 13.9 Effects of Irrigation Regimes, Nitrogen Rate and Micronutrient Sugar Yield (t/fed.) of Sugar Beet, in 2007/8 and 2008/9 Seasons

Irrigation at ASMD (I)	**Micro-nutri-ents (M)**	**2007–2008 season**			**2008–2009 season**		
		Kg N/fed. (N)		**Mean**	**Kg N/fed. (N)**		**Mean**
		75	**90**		**75**	**90**	
		I × N-Mean		**I-Mean**	**I × N-Mean**		**I-Mean**
40%		3.952^{b}	4.155^{b}	4.053^{a}	4.022^{c}	4.278^{b}c	4.150^{b}
55%		4.289^{b}	4.712^{a}	4.500^{a}	4.484^{b}	4.838^{a}	4.661^{a}
70%		3.117^{c}	3.299^{c}	3.208^{b}	3.121^{e}	3.435^{d}	3.278^{c}
		N × M-mean		M-mean	N × M-mean		M-mean
	C*	3.420^{c}	3.543^{c}	3.481^{c}	3.457^{e}	3.676^{de}	3.567^{c}
	SS**	3.976^{ab}	4.291^{a}	4.134^{a}	4.001^{bcd}	4.343^{ab}	4.172^{ab}
	FS***	3.687^{bc}	3.995^{ab}	3.841^{b}	3.848^{cd}	4.155^{bc}	4.001^{b}
	SS+FS	4.06^{ab}	4.392^{a}	4.226^{a}	4.196^{abc}	4.561^{a}	4.378^{a}
		I × N × M-Mean		I × M-Mean	I × N × M-Mean		I × M-Mean
40%	C	3.712$^{c–h}$	3.806$^{c–h}$	3.759^{def}	3.652$^{f–j}$	3.772$^{e–i}$	3.712^{ef}
	SS	4.178$^{b–e}$	4.297^{bcd}	4.238^{cd}	4.176$^{c–g}$	4.479$^{b–e}$	4.327^{cd}

TABLE 13.9 *(Continued)*

Irrigation at ASMD (I)	Micro-nutri-ents (M)	2007–2008 season			2008–2009 season		
		Kg N/fed. (N)		Mean	Kg N/fed. (N)		Mean
		75	90		75	90	
		I × N-Mean		I-Mean	I × N-Mean		I-Mean
	FS	$3.864^{c–g}$	$4.088^{b–f}$	3.976^{cde}	$3.967^{d–h}$	$4.240^{c–f}$	4.104^{de}
	SS+FS	$4.052^{b–f}$	$4.430^{a–d}$	4.241^{cd}	$4.294^{c–f}$	$4.620^{a–d}$	4.457^{bcd}
55%	C	$3.785^{c–h}$	$4.086^{b–f}$	3.936^{cde}	$3.973^{d–h}$	$4.257^{c–f}$	4.115^{de}
	SS	4.469^{abc}	5.061^{a}	4.765^{ab}	$4.648^{a–d}$	5.054^{ab}	4.851^{ab}
	FS	$4.122^{b–e}$	4.634^{ab}	4.378^{bc}	$4.445^{b–e}$	4.758^{abc}	4.602^{bc}
	SS+FS	4.78^{ab}	5.065^{a}	4.923^{a}	4.868^{abc}	5.283^{a}	5.076^{a}
70%	C	2.764^{i}	2.736^{i}	2.750^{h}	2.744^{k}	3.000^{jk}	2.872^{g}
	SS	3.282^{ghi}	$3.516^{e–h}$	3.399^{fg}	3.18^{ijk}	$3.497^{g–j}$	3.339^{f}
	FS	3.073^{hi}	3.265^{ghi}	3.169^{gh}	3.132^{ijk}	3.465^{hij}	3.299^{fg}
	SS+FS	$3.349^{f–i}$	$3.679^{d–h}$	3.514^{efg}	3.427^{hij}	$3.779^{e–i}$	3.603^{f}

*Control, ** Seed soaking, *** Foliar spraying.

Irrigation regime* × *nitrogen rate interaction (I×N): The highest white sugar yield per feddan was obtained with irrigation at 55% ASMD and fertilized with 90 kg N/feddan, while the lowest value was obtained with irrigation at 70% ASMD and fertilized by 75 kg N/feddan in the two seasons.

Irrigation regime* × *micronutrients interaction (I×M): The higher white sugar yield was achieved with irrigation at 55% ASMD along with SS+FS or SS in both seasons. However, the lowest one was achieved with irrigation at 70% ASMD without micronutrients.

Nitrogen rate* × *micronutrients interaction (N×M): Data show clearly that treatment 90 kg N/feddan along with SS+FS or SS produced the highest value of white sugar yield, while 75 kg N/feddan without micronutrients produced the lowest value in both seasons.

Irrigation regime* × *nitrogen rate* × *micronutrients interaction (I×N×M): The combination of medium irrigation regime × high N rate × SS+FS produced the maximum white sugar yield in both seasons. Application of SS or FS independently along with medium irrigation regime and high N rate was statistically at par with the mentioned combination in white sugar yield. The combination of low irrigation regime × low N rate × without micronutrients produced the lowest white sugar yield in both seasons.

13.3.5 WATER RELATIONS FOR SUGAR BEET CULTIVATION

Tables 13.10–13.12 indicate water consumptive use (WCU) by sugar beet crop from sowing to harvest, water consumptive use efficiency for root (WCUER), and water consumptive use efficiency for white sugar (WCUES).

13.3.5.1. EFFECTS OF IRRIGATION REGIMES

As soil moisture regime increased, WCU was increased due to more improved growth and perhaps luxury consumptive of water [13, 14, 30]. Water use efficiency for root or white sugar production was increased with irrigation at 55% ASMD, then it decreased. This may be attributed to increase of root and white sugar yields at 55% ASMD. Sorour [31] stated that water use efficiency for root or white sugar production were increased by increasing depletion of available soil moisture up to 60%, then it decreased. El-Sarag [13] found that increasing irrigation intervals from 5 to 11 days sharply reduced consumptive use, while, irrigation every 8 days was superior in water use efficiency, on sandy soil.

TABLE 13.10 Effects of Irrigation Regimes, Nitrogen Rate, Micronutrient And Their Interactions On Seasonal Consumptive Use (m^3/feddan) of Sugar Beet, in 2007/8 and 2008/9 Seasons

Irrigation at ASMD (I)	Micro-nutrients (M)	2007–2008 season			2008–2009 season		
		Kg N/fed. (N)		Mean	Kg N/fed. (N)		Mean
		75	90		75	90	
		I × N-Mean		I-Mean	I × N-Mean		I-Mean
40%		2277	2378	2328	2304	2418	2361
55%		2113	2202	2157	2160	2218	2189
70%		1545	1622	1584	1559	1614	1587
		N × M-mean		M-mean	N × M-mean		M-mean
	C*	1959	2048	2003	1985	2060	2022
	SS**	1977	2066	2021	2006	2081	2044
	FS***	1972	2061	2017	2001	2076	2038
	SS+FS	2005	2095	2050	2040	2117	2078
		I × N × M-Mean		I × M-Mean	I × N × M-Mean		I × M-Mean
40%	C	2250	2350	2300	2286	2398	2342
	SS	2277	2378	2328	2299	2413	2356

TABLE 13.10 *(Continued)*

Irrigation at ASMD (I)	Micro-nutrients (M)	2007–2008 season			2008–2009 season		
		Kg N/fed. (N)		Mean	Kg N/fed. (N)		Mean
		75	90		75	90	
		I × N-Mean		I-Mean	I × N-Mean		I-Mean
	FS	2265	2366	2316	2297	2411	2354
	SS+FS	2316	2419	2367	2334	2449	2392
55%	C	2096	2185	2141	2131	2188	2160
	SS	2115	2204	2159	2166	2223	2194
	FS	2113	2202	2157	2151	2209	2180
	SS+FS	2127	2217	2172	2193	2252	2223
70%	C	1531	1608	1570	1538	1593	1565
	SS	1539	1615	1577	1553	1608	1580
	FS	1539	1616	1578	1554	1609	1581
	SS+FS	1572	1651	1611	1592	1648	1620
N-Mean		1978	2068		2008	2083	

*Control, ** Seed soaking, *** Foliar spraying.

TABLE 13.11 Effects of Irrigation Regimes, Nitrogen Rate, Micronutrient And Their Interactions On Water Use Efficiency For Root Yield (Kg root/m^3 water) of Sugar Beet, in 2007/8 and 2008/9 Seasons

Irrigation (I) ASMD	Micro-nutrients (M)	2007–2008 season			2007–2008 season		
		Kg N/fed. (N)		Mean	Kg N/fed. (N)		Mean
		75	90		75	90	
		I × N-Mean		I-Mean	I × N-Mean		I-Mean
40%		12.02	12.45	12.24	11.82	12.37	12.10
55%		13.65	14.46	14.05	13.04	14.04	13.54
70%		13.46	13.59	13.52	12.46	13.39	12.93
		N × M-mean		M-mean	N × M-mean		M-mean
	C*	12.04	12.24	12.14	11.37	12.00	11.68
	SS**	13.54	14.07	13.80	12.70	13.62	13.16
	FS***	12.88	13.35	13.12	12.44	13.28	12.86
	SS+FS	13.70	14.34	14.02	13.26	14.17	13.71
		I × N × M-Mean		I × M - Mean	I × N × M-Mean		I × M - Mean

TABLE 13.11 *(Continued)*

Irrigation (I) ASMD	Micro-nutrients (M)	2007–2008 season			2007–2008 season		
		Kg N/fed. (N)		Mean	Kg N/fed. (N)		Mean
		75	90		75	90	
		I × N-Mean		I-Mean	I × N-Mean		I-Mean
40%	C	11.54	11.71	11.62	10.95	11.08	11.02
	SS	12.57	12.75	12.66	12.16	12.93	12.54
	FS	11.86	12.30	12.08	11.71	12.33	12.02
	SS+FS	12.11	13.04	12.57	12.46	13.13	12.80
55%	C	12.46	13.13	12.80	11.87	12.78	12.33
	SS	14.03	15.20	14.62	13.33	14.37	13.85
	FS	13.20	14.19	13.69	13.01	13.98	13.49
	SS+FS	14.91	15.31	15.11	13.95	15.03	14.49
70%	C	12.12	11.88	12.00	11.28	12.13	11.71
	SS	14.01	14.25	14.13	12.63	13.57	13.10
	FS	13.60	13.56	13.58	12.59	13.53	13.06
	SS+FS	14.09	14.67	14.38	13.35	14.34	13.85
N-Mean		13.04	13.50		12.44	13.27	

*Control, ** Seed soaking, *** Foliar spraying

TABLE 13.12 Effects of Irrigation Regimes, Nitrogen Rate, Micronutrient And Their Interactions On Water Use Efficiency For White Sugar Yield (Kg of white sugar/m^3 of water) of Sugar Beet, in 2007/8 and 2008/9 Seasons

Irrigation (I) ASMD	Micro-nutrients (M)	2007–2008 season			2007–2008 season		
		Kg N/fed. (N)		Mean	Kg N/fed. (N)		Mean
		75	90		75	90	
		I × N-Mean		I-Mean	I × N-Mean		I-Mean
40%		1.74	1.75	1.74	1.75	1.77	1.76
55%		2.03	2.14	2.08	2.07	2.18	2.13
70%		2.02	2.03	2.02	2.00	2.13	2.06
		N × M-mean		M-mean	N × M-mean		M-mean
	C*	1.75	1.73	1.74	1.75	1.80	1.77
	SS**	2.03	2.09	2.06	2.00	2.10	2.05
	FS***	1.88	1.95	1.92	1.94	2.02	1.98
	SS+FS	2.04	2.12	2.08	2.07	2.17	2.12

TABLE 13.12 *(Continued)*

		I × N × M-Mean		I×M-Mean	I × N × M-Mean		I×M-Mean
40%	C	1.65	1.62	1.63	1.60	1.57	1.59
	SS	1.84	1.81	1.82	1.82	1.86	1.84
	FS	1.71	1.73	1.72	1.73	1.76	1.74
	SS+FS	1.75	1.83	1.79	1.84	1.89	1.86
55%	C	1.81	1.87	1.84	1.86	1.95	1.90
	SS	2.11	2.30	2.20	2.15	2.27	2.21
	FS	1.95	2.10	2.03	2.07	2.15	2.11
	SS+FS	2.25	2.28	2.27	2.22	2.35	2.28
70%	C	1.80	1.70	1.75	1.78	1.88	1.83
	SS	2.13	2.18	2.15	2.05	2.17	2.11
	FS	2.00	2.02	2.01	2.02	2.15	2.08
	SS+FS	2.13	2.23	2.18	2.15	2.29	2.22
N-Mean		1.93	1.97		1.94	2.02	

*Control, **Seed soaking, ***Foliar spraying.

13.3.5.2 EFFECTS OF NITROGEN RATES

Water consumptive use was increased as nitrogen rate increased from 75 to 90 kg N/feddan. This may be attributed to considerable increase in leaf area index at high nitrogen rate, which resulted in a greater transpiration and in turn higher WUC. Increasing nitrogen rate slightly increased WUC for root and white sugar. Similar results were obtained by El-Zayat [14].

13.3.5.3 EFFECTS OF MICRONUTRIENTS

Application of micronutrients slightly increased WCU in both seasons. However, it substantially increased WCUER and WCUES. The treatment SS+FS recorded the best values of WCUER and WCUES. This may be due to increase of root yield and white sugar yield at application of micronutrients through SS+FS and in turn water use efficiency in both seasons.

13.3.5.4 EFFECTS OF INTERACTIONS

Means of WCU, WCUER and WCUES as influenced by the first and the second order interactions are presented in Table 13.10–13.12.

Irrigation regime × nitrogen rate interaction (I×N): WCU was increased by increasing irrigation water regimes and N rate. However, water stress resulted in increased WCUER and WCUES at any N rate. The best WUCER and WCUES were obtained with irrigation at 55% ASMD along with application of 90 kg N/feddan.

Irrigation regime × micronutrients interaction (I×M): Application of micronutrients resulted in a slight increase in WCU with any irrigation rate. Abundance soil moisture increased WCU in any micronutrients treatment. However, WCUER and WCUES were improved by application of micronutrients at the same irrigation regime. The higher WCUER and WCUES were achieved by irrigation at 55% ASMD along with SS+FS or SS in both seasons.

Nitrogen rate × micronutrients interaction (N×M): Application of micronutrients had slight effect on WCU in the same nitrogen rate. However, WCU was increased by increasing nitrogen rate in any of micronutrients treatments. Data show clearly that plants with 90 kg N/feddan along with SS+FS recorded the highest values of WCUER and WCUES.

Irrigation regime × nitrogen rate × micronutrients interaction (I×N×M): Table 13.10 shows that abundance of soil moisture in root zone substantially increased WCU for any combination of nitrogen rate and micronutrients in both seasons. The highest values of actual WCU (2419 and 2449 m^3/feddan) were obtained from the combination of high irrigation regime × high N rate × SS+FS, while the lowest WCU values (1531 and 1538 m^3/feddan) were obtained with low irrigation regime × low N rate × without micronutrients in the two seasons, respectively. The increase of actual WUC at the combination of high irrigation regime × high N rate × SS+FS can be attributed to increase in evaporation at high available moisture, because supplying plants with sufficient moisture led to an increase in green cover and hence increase in transpiration. Although, medium irrigation regime was equivalent to high irrigation regime in root and sugar yields at the combination of high N rate × SS+FS, yet medium regime was lower in water consumptive use. It saved 202 and 197 m^3 of WUC than the values for high irrigation regime in the two seasons, respectively.

Tables 13.11 and 13.12 showed that WCUER and WCUES were increased by increasing depletion of available soil water from 40 to 55% and then it decreased at any combination of nitrogen rate and micronutrients in both seasons. The combination of medium irrigation regime × high N rate × SS+FS recorded the highest values of WCUER 15.31 in 2007–2008 and 15.03 kg root/m^3 water use in 2008–2009 and of WCUES 2.28 in 2007–2008 and 2.35 kg white sugar /m^3 water use in second season, respectively. This may be due to increase in root and white sugar yields. However, the combination of high irrigation regime × low N rate × without micronutrients recorded the lowest values of WCUER 11.54 and 10.95 kg root/m^3 water use, in the first and second seasons, respectively. The combination of high irrigation regime × high N rate × without micronutrients recorded the lowest values of WCUES 1.62 and 1.57 kg whit sugar/m^3 water use in the first and second seasons, respectively.

Application of SS did not differ than SS+FS in WCU, WCUER, WCUES, root yield and sugar yield at medium irrigation regime and high N rate.

It can be concluded from this study that the irrigation at 55% ASMD along with 90 kg N/feddan and SS+FS or SS was the recommended treatment for optimum root and extractable white sugar yield per unit area with less WUC at Kafrelsheikh Governorate

13.4 SUMMARY

The sugar beet (*Beta vulgaris*, L. cv. Farida) was grown on a clay soil at Water Management Research Station at El-Karada, Kafrelsheikh, Egypt, in 20007/2008 and 2008/2009 seasons, to determine the effects of three irrigation regimes (at 40, 55, 70% ASMD), two nitrogen rates (75 and 90 kg N/feddan) and four micronutrients treatments (by SS, FS, (SS+FS) and control) on growth, yields, quality and water relations. Solution of micronutrients contained 2 g/L of each of Zn SO4 (26% Zn), Mn SO4 (24% Mn), Fe SO4 (20% Fe) and boric acid.

Abundance of the available soil moisture significantly increased dry weight/plant, LAI, root diameter, top yield, CGR and WCU. The inverse was true for root/top ratio, root length and concentration of gross sugar in roots. Increasing soil moisture level improved the juice purity by decreasing impurities (K, Na and α-amino-N) in roots. The plants irrigated at 55% ASMD produced the highest NAR and WCUER and WCUES production compared to those irrigated at 40 or 70% ASMD.

Increasing nitrogen rate from 75 to 90 kg N/feddan significantly increased dry weight, LAI, CGR, root length, root weight, top yield, root yield, concentration of α-amino-N% and Na + K in roots, sugar loss %, sugar yield and WCU. The inverse was true in root/top ratio, gross sugar%, white sugar % and juice purity %. Nitrogen rate slightly affected WCUER and WCUES.

Application of micronutrients by seed soaking and foliar spraying (SS+FS) produced the greatest dry weight, root/top ratio, LAI, CGR, root length, root diameter, root weight, top yield, root yield, gross sugar %, white sugar % and juice purity %, sugar yield, WCUER and WCUES. The SS was at par with SS+FS in most of these traits. FS increased concentration of α-amino-N% and Na + K in roots and the most of mentioned traits compared with those in the control.

All interactions had a significant effect on root and white sugar yields/feddan. The maximum root and white sugar yields and the best WCUER and WCUES were achieved from plants with irrigation at 55% ASMD and 90 kg N/feddan along with SS+FS.

It can be concluded that the irrigation at 55% ASMD along with 90 kg N/feddan and SS+FS or SS is the recommended treatment for optimum root and extractable white sugar yield per unit area with less water consumptive use at Kafrel Shiekh Governorate.

KEYWORDS

- **available soil moisture deficit**
- **boric acid**
- **CGR**
- **deficit irrigation**
- **dry weight**
- **Egypt**
- **furrow irrigation**
- **irrigation regime**
- **available soil moisture depletion**
- **extractable white sugar**
- **feddan**
- **fertilization**
- **foliar spraying**
- **gross sugar**
- **irrigation**
- **juice purity**
- **LAI**
- **micronutrient**
- **Nitrogen**
- **Phosphorous**
- **Potassium**
- **root diameter**
- **root length**
- **root weight**
- **root yield**
- **root/top ratio**
- **sandy soil**
- **seed soaking**
- **sugar beet**
- **sugar yield**
- **top yield**
- **water consumptive use**
- **water relation**
- **water use efficiency**
- **white sugar %**

REFERENCES

1. Alloway, B. J. (2004). Zinc in soils and crop nutrition.1st Ed., International Zinc Association Communications, IZA Publications, Brussels. Belgium.
2. Almani, M. P., Abd-Mishani, C.; Smadi, B. Y. (1997). Drought resistance in sugar beet genotypes. *Iranian J. Agric. Res., 28*, 15–25.
3. Attia, K. K., Abdel-Motagally, M. F. (2009). Response of sugar beet plants to nitrogen and potassium fertilization under sandy soil conditions. International Conference on: "World Perspectives for Sugar Crops as Food and Energy Suppliers," 1–4 March 2009, Luxor, Egypt. P.17/1–15.
4. Black, C. A., Evans, D. D., Ensminger, L. E., White, L. E., Clark, F. E., Dinauer, R. C. (1965). Methods of Soil Analysis. Ser. Agron. No.9 American Soc. Agron., Inc., Publisher, Madison, Wisconsin, USA.
5. Brown, K. F., McGowan, M., Armstrogn, M. J. (1987). Response of the components of sugar beet leaf water potential to a drying soil profile. *J. Agric. Sci., 109*, 437–444.
6. Doorenbos, J., Pruitt, W. O. (1979). Yield response to water. FAO Irrigation and Drainage Paper No. 33, Rome.
7. Draycott, A. P. (1993). Nutrition. In: The Sugar Beet Crop, D. A. Cooke, R. K. Scott, (eds.). pp. 239–278. Chapman and Hall, London, UK.
8. Duncan, B. D. (1955). Multiple Range and Multiple F-Test. *Biometrics, 11*, 1–42.
9. Ebrahim, M. H., Sorour, S. G. R., Omar, A. M. (1988). Varietal response of sugarbeet to foliar application of some micronutrients. Proc. 3rd. Egyptian Conf. Agron. Kafr El-Sheikh. pp. 324–334.
10. El-Fouly, M. M. (1983). Micronutrients in arid and semi arid areas: Level in soils and plants and the need for fertilizers with reference to Egypt. Proc.17th Colloquium of the International Potash Institute (IPI), Bern, Switzerland, pp. 163–173.
11. El-Fouly, M. M., Rezk, A. L., Omer, M. A., Shalaby, E. E., El-Nasharty, A. B. (2005). Effect of potassium and foliar application of micronutrients on the yield and quality of sugar beet grown in calcareous soil. Egyptian *J. Agric. Res., 2(2),* 775–784.
12. El-Hindi, M. H., Sultan, M. S., Attia, A. N., Selem, E. H. (1990). Effect of seed soaking in water, gibbrilic acid and growth regulators on emergence, yield and quality of sugar beet. *Proc. 4th Conf. Agron., Cairo., 11,* 405–413.
13. El-Sarag, E. I. (2009). Maximizing sugar beet yield, quality and water use efficiency using some agricultural practices under North Sinai conditions. *Bull. Faculty of Agriculture Cairo University, 60(2),* 155–167.
14. El-Zayat, M. M.T. (2000). Effect of irrigation regime and fertilization on sugar beet. Ph.D. Thesis, Faculty of Agriculture, Tanta University, Egypt.
15. Emara, T. K.S. E. (1996). Studies on the effect of spme cultural practices on water parameters of sugar beet (*beta vulgaris*, L) in North Delta region. Ph.D. Thesis, Faculty of Agriculture, Mansoura University, Egypt.
16. Fabeiro, C., Olalla, F. M. S., Lopez, R., Domınguez, A. (2003). Production and quality of the sugar beet (*Beta vulgaris* L.) cultivated under controlled deficit irrigation conditions in a semiarid climate. *Agric. Water Manag., 62,* 215–227.
17. Gomez, K. A., Gomez, A. A. (1984). Statistical procedures for agricultural research. An International Rice Research Institute Book, John Willy and Sons. Inc., New York.
18. Harvey, C. W., Dutton, J. V. (1993). Root quality and processing. pp. 571–617. In Cooke, D. A., Scott, R. K. The sugar beet crop. Chapman & Hall, 2–6 Boundary Row, London SE1 8HN, UK.
19. Israelsen, O. W., Hansen, V. E. (1962). Irrigation principles and practices. 3rd Ed. John Wiley and Sons Inc. New York.

20. Kenter, C., Hoffmann, C. M., Marlander, B. (2006). Effects of weather variables on sugar beet yield development (*Beta vulgaris* L.). *Europ. J. Agron., 24(2),* 62–69.
21. Koksal, E. S., Gungor, Y., Yildirim, Y. E. (2011). Spectral reflectance characteristics of sugar beet under different levels of irrigation water and relationships between growth parameters and spectral indexes. *Irrig. Drain., 60,* 187–195.
22. Mahmoodi, R., Maralian, H., Aghabarati, A. (2008). Effects of limited irrigation on root yield and quality of sugar beet (*Beta vulgaris* L.). *African J. Biotechnol., 7(24),* 4475–4478.
23. Marinkovic, B., Crnobarac, J., Jacimovic, G., Rajic, M., Latkovic, D., Acin, V. (2010). Sugar yield and technological quality of sugar beet at different levels of nitrogen fertilization. *Res. J. Agric. Sci., 42(1),* 162–167.
24. McGinnus, R. A. (1971). Sugar beet technology. 2nd Ed. Sugar Beet Development Foundation, For Collins, Colo., USA.
25. Moustafa, Z. R., Omran, S. E. H. (2006). Effect of foliar spray with boron or magnesium in combination with nitrogen fertilization on sugar beet plants. Egypt. *J. Soil Sci., 46(2),* 115–129.
26. Nemeat Alla, E. A. E., El-Geddawy, I. H. M. (2001). Response of sugar beet to foliar spraying time with micronutrients under different levels of nitrogen and phosphorus fertilization. *J. Agricultural Research Tanta University, 27(4),* 670–681.
27. Oliveira, M. D., Carranca, C. F., Oliveira, M. M., Gusmao, M. R. (1993). Diagnosing nutritional status of sugarbeet by soil and petiole analysis. In: M. A. C. Fragoso, M. L. van Beusichem, eds. Optimization of Plant Nutrition, pp. 147—151. Kluwer Academic Publishers, The Netherlands.
28. Selim, E. M., El-Sirafy, Z. M., Taha, A. A. (2010). Effect of Irrigation Methods and N-applications on the Utilization of Nitrogen by Sugar Beet Grown under Arid Condition. *Australian J. Basic and Applied Sci., 4(7),* 2114–2124.
29. Shaban, KH. A., Negm, M. A. (2008). Effect of Zn and B foliar application to sugar beet crown on a calcareous soil on root, sugar yields and nutrient contents. Egypt. *J. Soil Sci. 48(1),* 17–29.
30. Sorour, S. G. R. (1995). Response of sugar beet to irrigation regime and plant population. Egypt. *J. Appl. Sci., 20(8),* 648–668.
31. Sorour, S. G. R., Allam, A. Y., Drawish, Y. Y., Abd-Elsalam, M. K. M. (2009). Effect of land leveling, planting methods and time of nitrogen application on growth, yield and quality of sugar beet. International Conference on: "World Perspectives for Sugar Crops as Food and Energy Suppliers," 1–4 March, Luxor, Egypt. P.9/1–22.
32. Tsialtas, J. T., Maslaris, N. (2005). Effect of N fertilization rate on sugar yield and nonsugar impurities of sugar beets (*beta vulgaris*) grown under mediterranean conditions. *J. Agron. Crop Sci., 191,* 330–339.
33. Vamerali, T., Ganis, A., Bona, S., Mosca, G. (1999). An approach to minirhizotron root image analysis. *Plant Soil, 217(1–2),* 183–193.
34. Vamerali, T., Guariseb, M., Ganisb, A., Moscab, G. (2009). Effects of water and nitrogen management on fibrous root distribution and turnover in sugar beet. *Europ. J. Agron., 31,* 69–76.
35. Watson, D. J. (1952). The physiological basis of variation in yield. *Adv. Agron. 4,* 101–145.

CHAPTER 14

SOYBEAN IRRIGATION REQUIREMENTS UNDER VARYING IRRIGATION REGIMES

M. R. KHALIFA, IMAN M. SOLTAN, and A. S. EL-HENAWY

CONTENTS

In this chapter, area is in units of *feddans*. A *feddan* (Arabic: فدّان, *faddān*) is a unit of area. It is used in Egypt, Sudan and Syria. The feddan is not an SI unit and in Classical Arabic, the word means 'a yoke of oxen', implying the area of ground that could be tilled by oxen in a certain time. In Egypt the feddan is the only nonmetric unit which remained in use following the switch to the metric system. One fed. = 24 kirat = 60 m × 70 m = 4200 m^2 = 0.42 hectares.

14.1 INTRODUCTION

Because of limited water resources in Egypt, all irrigation users must work towards effective rationalization of irrigation at farm level. The simple procedure is to have knowledge of how much water should be applied. Irrigation can significantly increase soybean seed yield, [10] and can increase profits [17] under soil moisture deficit conditions. However, Cox and Jolliff [2] found that soybean plants are unable to withstand prolonged droughts. They also found that evapotranspiration of soybean plants was 17 and 68% less in deficit-irrigated and nonirrigated plants than that for the well-irrigated plants, respectively.

Soybean (*Glycirre max L.*) is one of the most important protein and edible oil crops, throughout the world. Foroud et al. [7], Eck et al., [5] and Speck et al., [19] have shown that soybean is amenable to limited irrigation supply. Stegman et al., [20] indicated that short-term water stress in soybean in the lower canopy was able to increase the number of pods in the upper nods, where there is a resumption of normal irrigation. The oil production poses a challenging problem, because of lack 90% of edible oil of national consumption. Therefore, greater attention must be paid on edible crops such as soybean, to increase the oil production [17].

Besides limitation of water resources in Egypt, there is also a challenge to Egyptian agriculture because of increasing prices of mineral fertilizer in addition to the negative effects of pollution of soil/ water and air due to leaching and oxidation of these fertilizers. We can overcome the pollution problem by using biofertilizers instead of mineral ones. Douka et al. [3] reported that inoculation of soybean could save more than 84 kg-N/fed. Many published reports have revealed that the relationship between soybean cultivars and rhizobium strain is one of the most important factors influencing biological N2-fixation [8]. Rhizobium is a genus of Gram-negative soil bacteria that fix nitrogen. Rhizobium forms an endosymbiotic nitrogen fixing association with roots of legumes and Parasponia. The bacteria colonize plant cells within root nodules, here the bacteria convert atmospheric nitrogen to ammonia and then provide organic nitrogenous compounds, such as, glutamine or ureides to the plant. The plant provides the bacteria with organic compounds made by photosynthesis.

This chapter presents the research results for soybean cultivation in North Delta region in Egypt. Effects of irrigation regimes based on available soil moisture depletion (ASMD) and use of mineral fertilizers and biofertilizers in soybean cultivation on the growth parameters, yield and its components and irrigation water application (WA), water consumptive use (WCU), crop water use efficiency (CWUE) and water productivity (WP).

14.2 MATERIALS AND METHODS

14.2.1 SITE LOCATION

During summer of 2008 and 2009, field experiments were conducted at El-Karada Station, Kafr El-Sheikh Governorate of Egypt. Kafr El-Sheikh is located at 31° 07′ latitude, 30° 52′ longitude and at an elevation of 6 m above sea level. The physical and chemical characteristics of soil at the experimental site were determined according to the standard methods outlined by Klute [14] and Page et al. [16]; and these values are tabulated in Tables 14.1 and 14.2. The experimental design was a split-plot with three replicates. The main plots were used three irrigation treatments based on 25, 50 and 75% of available soil moisture depletion. The subplots were used for fertilizer treatments.

TABLE 14.1 Physical Properties of Soil

Soil depth	Particle size distribution			Texture	Bulk density	Soil water constants		
	Clay	Silt	Sand			FC	PWP	AW
cm	%	%	%	—	Mg/m^3	%	%	%
0–15	53.4	25.40	21.20	clay	1.12	45.28	24.60	20.68
15–30	49.95	26.80	23.25	clay	1.18	44.12	23.73	20.39
30–45	40.40	34.30	25.30	clay	1.26	39.77	21.52	18.25
45–60	37.80	34.80	27.40	clay	1.33	39.42	21.43	17.99
Mean	45.40	30.33	24.29	clay	1.22	42.15	22.82	19.33

TABLE 14.2 Soil Chemical Properties at the Experimental Site

Soluble anions				Soluble cations				EC	pH	Soil depth
SO_4^-	Cl^-	HCO_3^-	CO_3^-	K^+	Na^+	Mg^{++}	Ca^{++}			
meq/L								dS/m	–	cm
6.56	8.52	2.50	-	0.10	12.10	2.10	3.28	1.34	7.56	0–15
7.05	8.84	2.80	-	0.10	12.63	2.60	3.36	1.47	7.62	15–30
4.95	9.12	2.70	-	0.04	12.81	1.20	2.72	1.42	7.61	30–45
3.07	9.02	2.75	-	0.04	11.43	1.92	1.45	1.21	7.58	45–60
5.41	8.88	2.69	-	0.07	12.24	1.96	2.70	1.36	—	Mean

14.2.2 FERTILIZATION TREATMENTS

The fertilization treatments in the subplots consisted of: biological fertilizer inoculation with phosphorien-containing phosphate-dissolving bacteria (PDB), biological fertilizer inoculation with rhizobial, mixture of inoculated seeds with phosphorien and rhizobial and conventional NPK.

Phosphorien, as a PDB (Bacillus-magaterium var. phosphaticum) at rate of 500 g/fed, was used as an inoculant, which converts the insoluble tricalcium to the soluble monocalcium phosphate. Phosphorien as a commercial compound is produced by Organization for Agricultural Equalization, Ministry of Agriculture, Egypt. Inoculation with Rhizobia, strain of *Rhizobial Japonicum* namely 110 and 1577 Okadin inoculation were obtained from Soil Microbiology Department of Sakha Agricultural Research Station. The strain was grown separately in a liquid media 79. The resulting culture was used as seed inoculants at rate of 399 g/fed.

14.2.3 PACKAGE PRACTICES

Soybean seeds (*Glycine maxl.* var. Giza 21) were obtained from Field Crops Research Institute, Agricultural Research Center, Department of Legumes, Sakha Agricultural Research Station. The seeds were planted on June 15th in 2008) and June 17th in 2009. The crop was harvested on October 15th in both seasons, respectively. Each subplot consisted of 6 rows, 60 cm apart and 5 meters long, with an area of 18 m^2. All practices were followed as recommended for soybean production.

14.2.4 IRRIGATION TREATMENTS

Soil moisture content was gravimetrically determined for soil samples at 15–60 cm soil depth. Soil samples were taken periodically, until it reached the desired level of allowable moisture. The amount of water applied for each irrigation regime was determined on the basis of raising the soil moisture content to its field capacity plus 10% due to leaching requirement. Irrigation water was pumped from the main canal near the experimental field into a settling basin with a baffle wall, to maintain a constant head over the crest of affixed rectangular weir. Irrigation discharge was calculated with a weir as follows:

$$Q = 1.84\ LH^{1.5} \tag{1}$$

where: Q = rate of discharge, m^3/min.; L = length edge of weir, cm; and H = height column of water above edge of weir, cm. Irrigation water was controlled by a steel gate for each experiment plot, as well as, those fixed at the side of each feeder canal.

To compute the actual water consumed by the growing soybean plants, soil moisture percentage was determined gravimetrically before and 48 h after each irrigation watering, as well as, harvesting. Soil samples were taken from the effective root zone of 60 cm with 15 cm for each successive soil layer. Water consumptive use for each soil depth was calculated using the following equation of Hansen et al. [9]:

$$CU = \sum \frac{\theta_2 - \theta_1}{100} \times Dbi \times Di \tag{2}$$

where: CU = actual water consumptive use of the growing plants (cm) in the effective root zone (60 cm); θ_2 = Soil moisture content for each layer in percent at 48 h after irrigation; θ_1 = Soil moisture content in percent before the next irrigation; Dbi = Bulk density of the specific soil layer (Mgm^{-3}); and Di = Depth of each soil layer, 15 cm. Water consumed per feddan was calculated. Crop water use efficiency (CWUE) was calculated according to Jensen [11].

Soybean plots were subjected to three irrigation treatments based on 25, 50 and 75% of available soil moisture depletion.

14.2.5 PARAMETERS FOR SOYBEAN PERFORMANCE

At harvest, 10 soybean plants were collected from each subplot to determine the yield and yield components: plant height (cm), pods weight/plant (g), seeds weight/plant (g), 1000-seeds weight (g), number of branches/plant, number of pods/plant and stem weight/plant (g). Seed yield was obtained from central area of each plot (two rows) and calculated as kg/fed. Data collected were subjected to statistical analysis according to Snedecor and Cochram [18]. The differences between the means were compared by Duncan's multiple range tests.

14.3 RESULTS AND DISCUSSION

14.3.1 GROWTH PARAMETERS

Table 14.3 indicates that most of vegetative characteristics of soybean plant were significantly affected by three irrigation regimes in both growing seasons, except plant height and number of branches per plant (in 1st season) were not significantly affected by irrigation regimes. Highest mean values of plant height, number of branches per plant, number of pods per plant and stem weight per plant (g) were recorded three irrigation levels based on 25% ASMD, 50% ASMD and 75% ASMD for both growing seasons.

TABLE 14.3 Effects of Irrigation Regime, Biofertilizers, Mineral Fertilizers On Growth Parameters of Soybean During Two Growing Seasons, 2008 and 2009

Treatment	Plant height cm		No. of branches per plant		No. of pods per plant		Stem weight g per plant	
	2008	2009	2008	2009	2008	2009	2008	2009
Irrigation (I) treatments based on ASMD								
25% ASMD	54.5^{b}	51.3^{b}	5.1^{a}	3.9^{b}	48.0^{b}	56.6^{b}	17.9^{b}	14.8^{b}
50% ASMD	61.7^{a}	63.4^{a}	5.0^{a}	4.6^{a}	83.6^{a}	77.9^{a}	24.4^{a}	25.2^{a}
75% ASMD	59.1^{ab}	55.7^{b}	4.8^{a}	4.3^{ab}	57.4^{b}	49.9^{b}	21.6^{ab}	15.8^{b}
F-test	Ns	**	Ns	*	*	*	*	**
Fertilizers (F) and biofertilizers								
Phos.	51.0^{b}	52.2^{c}	4.2^{b}	4.2^{b}	40.4^{c}	56.8^{b}	13.4^{c}	17.0^{a}
Rhiz.	59.0^{a}	54.5^{bc}	4.9^{b}	4.1^{b}	60.6^{b}	52.7^{b}	17.8^{bc}	17.1^{a}
Pho.+Rhiz.	63.4^{a}	61.9^{a}	4.4^{b}	3.7^{b}	51.3^{bc}	49.8^{b}	22.1^{b}	18.3^{a}
NPK	60.3^{a}	58.6^{ab}	6.5^{a}	5.2^{a}	99.7^{a}	86.6^{a}	31.9^{a}	19.4^{a}
F-test	**	**	**	**	**	**	**	Ns
Interaction (I x F)								
F-test	**	*	Ns	**	Ns	Ns	Ns	Ns

* = significant at $p < 0.05$; ** = significant at = $p < 0.01$ and ns = not significant. Means are not significantly different at 5% level, if each value designed by the same letter: using Duncan's MRT [4].

The Table 14.3 also indicates that mean values of the aforementioned vegetative parameters of soybean plant were highly significant, and were affected by biofertilizers in both growing seasons, except stem weight per plant (in 2nd season). The highest mean values of plant height (63.4 and 61.9 cm) were recorded under treated plants with phosphoriene + rhizobial in both seasons, respectively. However, the highest values of number of branches per plant (6.5 and 5.20), number of pods per plant (99.7 and 86.6) and stem weight per plant (31.9 and 19.4 g) were observed for treatments with recommended dose of NPK in both seasons, respectively. Data showed that no significant differences were recorded due to inoculation with phosphorien, rhizobian and their mixture on number of branches per plant (1st and 2nd seasons), number of pods per plant (2nd season) and stem weight per plant (2nd season).

On the other hand, the interaction effect between irrigation regimes and biofertilizers was highly significant, except the differences were not significant in case of: number of branches per plant and number of pods per plant (1st season) and stem weight per plant (1st and 2nd seasons). The results are in agreement with those reported by Karam et al., [13] and Moursi et al., [15].

14.3.2. SOYBEAN SEED YIELD AND ITS COMPONENTS

Table 4 indicates that soybean yield and its components were significantly affected by all irrigation and bio-fertilizers treatments in both growing seasons. Regarding the effect of irrigation regimes, data indicated that pods weight per plant, seed weight per plant, weight per 1000 seeds and seed yield (kg/fed.) were significantly affected by all irrigation regimes in both seasons. Irrigation at 50% ASMD produced the highest values of pods weight per plant (40.3 and 38.6), seed weight per plant (19.8 and 17.4 g), weight per 1000 seeds (15.4 and 16.6 g) and seed yield (1986.3 and 1734.7 kg/fed) in both growing seasons, respectively. On the other hand, the lowest values of the aforementioned parameters were recorded under irrigation at 75% ASMD in both seasons. Increasing soybean seed yield under irrigation at 50% ASMD may be due to improving the rate of aeration, which will increase soil organic matter, and hence increasing availability of nutrients, therefore forming strong plants with good vegetative growth. These results are in agreement with those obtained by Balasubramanian and Chari [1] and Moursi et al. [15].

With respect to biofertilizers and mineral fertilization, the results in Table 14.4 indicated that weight of pods per plant, seed weight/plant, weight per 1000 seeds and seed yield (kg/fed) were highly significantly affected by these treatments in both growing seasons. The highest mean values for soybean pods weight per plant (43.5 and 38.5 g), seed weight per plant (20.8 and 19.7 g), 1000 seed weight (16.1 and 14.1) and seed yield (2075.9 and 1917.2 kg/fed.) were recorded under application of recommended NPK in both seasons, respectively followed by rhizobial inoculation.

TABLE 14.4 Effects of Irrigation Regimes and Biofertilizers, Mineral Fertilizers On Yield and Its Components of Soybean, During the Two Growing Seasons

Treatment	Pods weight g/plant		Seed weight g/plant		Weight of 1000 seeds g		Seed yield kg/fed	
	2008	2009	2008	2009	2008	2009	2008	2009
Irrigation (I) treatments based on ASMD								
25% ASMD	26.3^b	24.9^b	13.3^b	12.5^b	12.6^b	12.3^b	1324.8^b	1250.6^b
50% ASMD	40.3^a	38.6^a	19.8^a	17.4^a	15.4^a	16.6^a	1986.3^a	1734.7^a
75% ASMD	25.4^b	18.7^c	9.6^c	7.2^c	11.4^c	9.4^b	958.1^c	720.8^c
F-test	*	**	**	**	**	**	**	**
Fertilizers (F) and biofertilizers								
Phos.	20.1^c	21.8^b	8.7^c	9.1^b	11.5^c	12.4^b	869.9^c	906.2^b
Rhiz.	34.3^b	26.0^b	15.7^b	11.4^b	11.9^{bc}	11.8^b	1572.6^b	1140.0^b
Pho.+Rhiz.	24.8^c	23.5^b	11.7^c	9.8^b	131^b	12.2^b	1173.9^b	978.0^b
NPK	43.5^a	38.5^a	20.8^a	19.7^a	16.1^a	14.6^a	2075.9^a	1917.2^a
F-test	**	**	**	**	**	**	**	**
Interaction (I x F)								
F-test	*	Ns	*	Ns	**	**	*	Ns

* = significant at $p < 0.05$; ** = significant at = $p < 0.01$ and ns = not significant. Means are not significantly different at 5% level, if each value designed by the same letter: using Duncan's MRT [4].

It is evident that inoculation of soybean plant with rhizobial strain gave the highest values of the aforementioned parameters compared to with phosphoriene or their mixture in both seasons. These results are in harmony with those obtained by Moursi et al., [15]. They reported that application of microbial inoculants (rhizobian) produced the highest seed yield and its components. Our data showed that the interaction between irrigation regimes and fertilization was significantly different in both growing seasons, except that pods weight per plant, seed weight per plant and seed yield (2nd season) were not significantly affected.

14.3.3 IRRIGATION PARAMETERS FOR SOYBEAN CROP

Under conditions of this study, irrigation water applied for soybean as a summer crop is the only component for water applied since no rainfall was observed during the summer growing season, in Egypt. Table 14.5 showed that the average volume of water applied was 3450.5, 3005.4 and 2516.30 m^3/fed (1st season) and 3452.3, 3008.3 and 2514.7 m^3/fed (2nd season) for irrigating soybean plants at 25%, 50% and 75% ASMD, respectively. It is obvious that irrigation at 25%ASMD had the highest value of water applied in both seasons, while irrigation at 75%ASMD was accompanied with least value of water applied. On the other hand, irrigation at 50%ASMD was in between. These results are in harmony with those obtained by Jiamin et al. [12], Moursi et al. [15] and Zhen et al. [21].

TABLE 14.5 Effects of Different Irrigation Regimes on Irrigation Water Application (WA), Water Consumptive Use (WCU), Crop Water Use Efficiency (CWUE) and Water Productivity (WP) For Soybean, During the Two Growing Seasons

Irrigation regimes	WA		WCU		CWUE		WP	
	2008	2009	2008	2009	2008	2009	2008	2009
	m^3/fed		m^3/fed		kg/m^3		kg/m^3	
25%ASMD	3450.50	3452.2	2271.6	2274.2	0.61	0.55	0.40	0.36
50%ASMD	3005.40	3008.3	2073.1	2075.2	0.96	0.84	0.66	0.58
75%ASMD	2516.30	2514.7	1858.2	1860.1	0.50	0.39	0.38	0.29

Seasonal rates of water consumption by soybean plants in all irrigation regimes and fertilizer treatments during the two growing seasons are presented in Table 14.5. Data showed that irrigating soybean plants at 25% ASMD gave the highest value of seasonal water consumptive use and was (2271.6 and 2274.2 m^3/fed), followed by irrigation at 50% ASMD (2073.1 and 2075.2 m^3/fed); and minimum value of seasonal water consumption was (1858.2 and 1860.1 m^3/fed) at 75% ASMD in both seasons, respectively. These results demonstrate that water consumptive use was increased as soil moisture was maintained high by frequent irrigations. The cause may be that higher frequent irrigations (25% and 50% ASMD) provided a chance for more consumption of water, which ultimately resulted in an increasing transpiration and evaporation from the soil surface. These results are in agreement with those reported by Karam et al. [13].

Data illustrated that in Table 14.5 show that irrigating soybean plants at 50% ASMD achieved the highest value of crop water use efficiency (0.96 and 0.84 Kg/ m^3 of water consumed), followed by irrigation at 25% ASMD (0.61 and 0.55 Kg / m^3) and at 75% ASMD (0.57 and 0.39 Kg/m^3) in both seasons, respectively. These findings are in harmony with the scientific approach that the plant roots can extract more soil water from greater depth under moderate stress (50% ASMD) compared to those irrigated at relatively wet level (25% ASMD). This implies that the stored water in soil at a moderate irrigation can be more available for roots, as well as, can be used more efficiently. Also, these findings may be attributed to the differences among seed soybean yield, as well as, differences between water consumptive uses.

Water use efficiency (WAUE) is an indicator to find out the yield per unit water applied (WA). Table 5 indicates that highest values of WAUE (0.66 and 0.58 kg of seed per m^3 of WA) were recorded under irrigating at 50% ASMD, in both seasons, respectively. Meanwhile, the lowest values of WP were 0.38 and 0.29 Kg per m^3 of WA in both seasons, respectively. These findings may be attributed to the differences among seed soybean yield, as well as, differences between applied water values.

14.4 SUMMARY

Two field experiments were conducted at El-Karada Research Station, Kafr El-sheik Governorate in Egypt, during two summer seasons of 2008 and 2009, to study the effects of irrigation regimes (irrigating soybean plants at 25, 50 and 75% of available soil moisture depletion, ASMD) and biofertilizers (inoculation soybean seeds with phosphoriene, Rizobial and their mixture, comparing with recommended NPK) on yield and its components, as well as, water relations of soybean plants. The experimental design was split plot with three replicates. The main plots were assigned to irrigation regimes, while the subplots were used for fertilizer treatments.

Results showed that both of irrigation treatments and biofertilizers had highly significant effect on yield and its attributes of soybean plants. Irrigation at 50% of ASMD gave the highest values of plant height, number of branches per plant, number of pods per plant, pods weight per plant, seed weight per plant, weight per 1000 seeds and seed yield (kg/fed.), in both growing seasons. Meanwhile, the lowest values of these parameters were recorded under irrigation at 75% ASMD, in both seasons. Data also showed that "phosphorien + rhizobial inoculation" produced tallest plants. However, the remaining variables were highest with application of recommended NPK, followed by rhizobial inoculation in both growing seasons comparing to phosphorien.

Seasonal applied water (as an average of two seasons) was 3451.4, 3006.9 and 2515.5 m^3/feddan for irrigation at 25, 50 and 75% ASMD, respectively. Irrigating soybean plants at 25% ASMD gave highest seasonal water consumption (2272.9 m^3/fed), followed by irrigation at 50% ASMD (2074.2 m^3/fed), while the lowest seasonal water consumption was recorded under irrigation at 75% ASMD (1859.2 m^3/fed). The highest values of crop water use efficiency (0.96 and 0.84 Kg/m^3) and water productivity (0.66 and 0.58 kg of seed yield /m^3 of WA) were achieved under irrigation at 50% ASMD, in both seasons, respectively. Meanwhile, irrigating soybean plants at 75% ASMD recorded the lowest values of CWUE (0.56 and 0.39 kg/m^3) and water productivity (0.38 and 0.29 kg/m^3) in the 1st and 2nd seasons, respectively.

KEYWORDS

- **available soil moisture depletion**
- **biofertilizers**
- **crop water use efficiency**
- **deficit irrigation**
- **furrow irrigation**
- **irrigation scheduling**
- **irrigation regime**
- **nitrogen**
- **phosphorien inoculation**
- **rhizobial inoculation**
- **seed yield**
- **soybean**
- **water consumption**
- **water deficit**
- **water productivity**

REFERENCES

1. Balasubramanian, G., Chari, A. V. (1983). Effect of irrigation scheduling on grain yield and nitrogen use efficiency of irrigated wheat at kadawa and Bakura. *Northern Nigeria Fertilizer Research, 4(3),* 201–210.
2. Cox, W. J., Jolliff, O. D. (1986). Growth and yield of sunflower and soybean under soil water deficits. *Agron. J. 78*, 226–230.

3. Douka, C. F., Nychas, A. F., Xenoulis, A. C. (1986). Nitrogen in soybean as influenced by cultivar and differed rhizobial strains. *Biol. Fertile Soils, 2,* 113–118.
4. Duncan, D. B. (1955). Multiple ranges and Multiple F-test. Biometrics, 11, 1–42.
5. Eck, H. V., Mathers, A. C., Musick, J. T. (1987). Plant water stress at various growth stages of growth and yield of soybean. *Field Crops Research, 17,* 1–16.
6. FAO (http://ftp.fao.org/docrep/fao/010/ah876e00.pdf), (2007). *Food Outlook: Global Market Analysis*. November.
7. Foroud, N., Mundel, H. H., Sa Indon, G., Entz, T. (1993). Effect of level and timing of moisture stress on soybean yield components. *Irrigat. Sci., 13,* 149–155.
8. Ghobrial, W. N., Abady, D., Abdel-Kodoos, R. Y., Hegazy, M. H. (1995). Nitrogen fixation potential as affected by *B. Japonicume* and soybean cultivars for maximizing soybean productivity. *Annals Agric. Science Ain Shams University Cairo, 40(1),* 117–128.
9. Hansen, V. W., Israelson, O. W., Stringharm, Q. E. (1979). *Irrigation Principles and Practices*. 4th ed., John Wiley and Sons, New York.
10. Heathernly, L. G. (1983). Response of soybean cultivars to irrigation of a clay soil. *Agron. J., 75,* 859–864.
11. Jensen, M. E. (1983). *Design and Operation Of Farm Irrigation Systems*. Am. Soc. Agric. Eng. (ASABE), Michigan, USA.
12. Jiamin, L., Shinobu, I., Zhaohu, L., Egrinya Eneji, A. (2005). Optimizing irrigation scheduling of winter wheat in the North China plain. *Agricultural Water Management, 76(1),* 8–23.
13. Karam, F., Masaad, Therese, S., Mounzer, O.. Rouphal, Y. (2005). Evaporation and seed yield of field grown soybean under deficit irrigation conditions. *Agric. Water Management, 75(3),* 226–244.
14. Klute, A. (1986). Methods of Soil Analysis (Part I). *Am. Soc. of Agron.*, Madison, Wisconsin, USA.
15. Moursi, E. A., Azizi, M. A., Mehessen, A. A., Kassab, M. M. (2009). Optimum water requirements of soybean in the clay soil under rhizobial inoculation and urea fertilization. *J. Agricultural Research Kafrel-Sheik University, 33(1),* 328–344.
16. Page, A. L., Millur, R. H., Keeney, D. R. (1982). *Methods of Soil Analysis, Part II: Chemical and Microbiological Properties.* Am. Soc. of Agron, Madison, WI, USA.
17. Salassi, M. E., Musick, J. A., Heatherly, L. G., Hamill, J. G. (1984). An economic analysis of soybean yield response to irrigation of Mississippi river Delta soils. *Mississippi Agric., 60,* 928–935.
18. Snedecor, G. A., Cochram, W. C. (1967). *Statistical Methods*. 6th ed. Oxford and IBH Publishing Co., Calcutta, India.
19. Speck, J. E., EL-More, R. W., Eisenhauer, D. E., Klocke, N. W. (1989). Growth stage scheduling criteria for sprinkler – irrigated soybeans. *Irrigation Science, 10*, 99–111.
20. Stegman, E. C., Schatz, B. G., Gardner, J. C. (1990). Yield sensitivities of short season soybean to irrigation management. *Irrigation Science, 11,* 111–119.
21. Zhen, L., Zi Wei, De Land Wen, Long Li, (2004). Dry period irrigation and fertilizer application affected water use and yield of spring wheat in semiarid regions. *Agricultural*

CHAPTER 15

IRRIGATION SCHEDULING AND WATER USE EFFICIENCY IN SUNFLOWER PRODUCTION: NORTH NILE DELTA

A. S. EL-HENAWY, EMAN M. K. E. SOLTAN, and M. R. KHALIFA

CONTENTS

*In this chapter, area is in units of *feddans*. A *feddan* (Arabic: فدّان, *faddān*) is a unit of area. It is used in Egypt, Sudan and Syria. The feddan is not an SI unit and in Classical Arabic, the word means 'a yoke of oxen': implying the area of ground that could be tilled by them in a certain time. In Egypt the feddan is the only nonmetric unit, which remained in use following the switch to the metric system. One feddan = 24 kirat = 60 meter × 70 meter = 4200 square meters (m^2) = 0.42 hectares = 1.038 acres.

15.1 INTRODUCTION

In Egypt, due to the severe shortage in edible oil and due to the rapid population increase, sunflower has received a special attention. At present, Egypt imports about 80–85% of its annual requirements of edible vegetable oils. The present gap between the domestic production and demand for edible oil can be reduced by initiating several investigations about the effects of fertilization, sowing dates and irrigation treatments on maximizing the productivity of sunflower under local climatic conditions. Because of the water limited water supply in Egypt, we all should strive hard to implement effective rationalization of irrigation at farm level. Gad El-Rab et al. [9] indicated that sunflower plants under drought conditions decreased plant height, head diameter and water consumptive use. According to Casadebaig et al. [4], minimization of water loss is a major aspect of drought tolerance and can be achieved through the lowering of either leaf area expansion rate or transpiration rate per unit leaf area (stomata conductance). Even limited irrigation water, applied at different growth stages of sunflower, can significantly increase the seed yield, especially during three growth stages: heading, flowering, milking stages and at 50% ray flower stage [10]. Soleimanzadeh et al. [17] reported that plant height, head diameter, number of seeds per head, weight per 1000 seeds, biological yield, seed yield, harvest index and oil yield under drought stress were declined. It has been reported that harvest index decreased with increasing water stress [18].

This chapter discusses the research results to evaluate the effects of escaping (withholding or skip) irrigation at different stages of sunflower on growth attributes, photosynthesis pigments, crop water use efficiency and irrigation water productivity, in North Nile Delta area of Egypt.

15.2 MATERIALS AND METHODS

15.2.1 EXPERIMENTAL SITE

This study was conducted at El-Karda Water Management Station Farm, Kafr El-Sheikh Governorate during two successive summer seasons of 2008 and 2009. The characteristics of clay textured soil were: clay 51.7%, silt 26.1%, sand 21.2%, EC 2.59 dSm^{-1} in soil paste extract, pH 8.05, organic matter 1.38%, field capacity 44.7%, wilting point 24.2%. Randomized complete block design with three replications was used in both seasons. The irrigation treatments were:

- T_1 Conventional irrigation every 15 days (control).
- T_2 Escaping irrigation at 30 days after sowing (DAS), (3rd irrigation).
- T_3 Escaping irrigation at 45 DAS, (4th irrigation).
- T_4 Escaping irrigation at 60 DAS, (5th irrigation).
- T_5 Escaping irrigation at 75 DAS, (6th irrigation).

Each plot area was 42 m^2 with 10 ridges. Each ridge was 7 m long and 60 cm apart. Plots were isolated by a ditch of 1.5 m in width to avoid lateral movement of water. Sunflower cv. Sakha 53 seeds were hill seeded on March 15th in 2008 and March 19th in 2009. Sunflower was harvested on July 7 and 17 in both seasons, respectively.

15.2.2 CULTURAL PRACTICES

In both seasons, phosphorous fertilizer in the form of calcium super phosphate (15.5% P_2O_5) was applied at the rate of 30 kg P_2O_5/fed. during land preparation. Nitrogen was added in the form of urea (46% N) at the rate of 40 kg N/fed in two equal doses before the first and second irrigation, respectively. Potassium was added in the form of potassium sulfate (48% K_2O) at the rate of 24 kg K_2O/fed. Cultivation practices for growing sunflower were conducted according to the recommendations by Ministry of Agriculture and Land Reclamation [14].

15.2.3 MEASUREMENTS OF GROWTH PARAMETERS

The growth parameters of sunflower were measured according to the formula mentioned by Hunt [12]. The leaf area (LA) per plant (dm^2) of three samples at 60, 75 and 90 DAS was measured by leaf area index instrument [15]. Relative growth rate (RGR), crop growth rate (CGR) and net assimilation rate (NAR) were calculated, using following formulas:

$$CGR = (w_2 - w_1) / (t_2 - t_1), \text{ g/m}^2\text{/week} \quad (1)$$

$$NAR = (w_2 - w_1) (\log A_2 - \log A_1) / (A_2 - A_1) (t_2 - t_1), \text{ g/m}^2\text{/week} \quad (2)$$

$$RGR = (\log w_2 - \log w_1) / (t_2 - t_1), \text{ g/g/week} \quad (3)$$

where: w_1, A_1 at time t_1 and w_2, A_2 at time t_2 refer to dry weight and leaf area per week.

The amount of chlorophyll pigments (chlorophyll A and chlorophyll B) was determined using spectrophotometer and calculated according to Sadasivam and Manickam [16]. Carotenoid was determined according to Wang et al. [20] at 60, 75 and 90 DAS.

About 10 guarded plants were randomly taken from the fourth inner ridges to determine plant height and head diameter.

15.2.4 IRRIGATION MANAGEMENT IN SUNFLOWER CULTIVATION

Irrigation depth was applied based on the discharge rate through a weir using the following equation:

$$Q = 1.84\ LH^{1.5} \tag{4}$$

where: Q = rate of discharge, m^3/sec; L = length edge of weir, cm; H = Height of water above edge of weir, cm.

Water consumptive use (WCU) was determined by the soil moisture depletion method. Soil samples at 0–60 cm depth were taken before each irrigation using auger and after 48 h from each irrigation. Moisture content in the soil samples was determined gravimetrically and calculated on weight basis to calculate the WCU using the following equation [13]:

$$WCU = [(\theta_2 - \theta_1) \div 100] \times \rho_a \times D \times 4200 \tag{5}$$

where: WCU = Amount of water consumptive use (m^3/fed); θ_2 = soil moisture content in % after irrigation; θ_1 = soil moisture content in % before the next irrigation; ρ_a = soil bulk density (Mg/m^3); D = soil depth in m.

Crop water use efficiency (CWUE, kg/m^3) was calculated as a ratio of yield to WCU according to Doorenbos and Pruitt [5]:

$$CWUE = [(\text{yield in kg/feddan}) \div (\text{WCU in } m^3/\text{fed})] \tag{6}$$

Water productivity (WP), kg/m^3 was calculated according to Ali et al. [1]:

$$WP = [(\text{yield in kg/feddan}) \div (\text{water applied in } m^3/\text{fed})] \tag{7}$$

15.2.5 STATISTICAL ANALYSIS

The data were subjected to analysis of variance according to Gomez and Gomez [11]. Treatment means were compared by Duncan's Multiple Range Test [6]. All statistical analysis was performed using analysis of variance technique by means of "MSTATC" computer software package.

15.3 RESULTS AND DISCUSSIONS

15.3.1 EFFECTS OF ESCAPING IRRIGATION (SKIP OR WITHHOLDING) ON GROWTH OF SUNFLOWER

Table 15.1 indicates the effects of escaping irrigation on plant height and head diameter. Irrigation treatments had significant effect on plant height and head diameter in the 1st season. However, these traits were significantly affected by escaping irrigation in 2nd season. The highest values of plant height (cm) were recorded under T_2 and were 161.8 in 2008 and 164.8 in 2009. The highest values of head diameter (cm) were recorded under T_1 and were 17.6 in 2008 and 15.7 in 2009. The lowest values were obtained under T_5 for the two traits in both seasons.

TABLE 15.1 Effects of Five Irrigation Treatments on Plant Height, Head Diameter and Seed Yield of Sunflower Crop, During 2008 and 2009 Seasons

Performance parameter	F-test	Treatments				
		T_1	T_2	T_3	T_4	T_5
Season 2008						
Plant height, cm	**ns**	153.6^{a}	161.8^{a}	145.8^{a}	143.4^{a}	141.2^{a}
Head diameter, cm	**ns**	17.6^{a}	16.9^{a}	17.6^{a}	17.8^{a}	17.2^{a}
Seed yield, ton/fed.	*	1.316^{a}	1.132^{b}	1.239^{ab}	1.420^{a}	1.227^{ab}
Season 2009						
Plant height, cm	*	164.3^{a}	164.8^{a}	154.3^{ab}	151.1^{ab}	136.8^{b}
Head diameter, cm	**	15.7^{a}	15.5^{a}	13.8^{ab}	13.6^{b}	14.9^{ab}
Seed yield, ton/fed.	*	1.317^{a}	1.122^{b}	1.229^{b}	1.419^{a}	1.327^{a}

*, ** and ns indicate $p < 0.05$, $p < 0.01$ and not significant, respectively. Means of each factor designed by the same letter are not significantly different at 5% level using Duncan's MRT.

These results indicated that escaping irrigation during late vegetative growth (T_5) results in reduced plant height but increased root depth. Adequate water during the late vegetative period is required for proper bud development. The flowering period is most sensitive to water deficits that may cause considerable yield decrease, since fewer flower come to full development [2, 3].

Table 15.2 shows a significant effect due to irrigation treatments on leaf area (dm^2) of sunflower in both seasons. Leaf area was increased at 75 DAS and then was declined slightly at 90 DAS, in both seasons. This is mainly due to the production of new leaves, because more leaves implies better growth of sunflower plant. The highest values of leaf area were obtained under T_3 (escaping irrigation at 45 DAS) followed by T_5 in the first season, while it was under T_3 followed by T_2 in the second season.

TABLE 15.2 Effects of Withholding Irrigation on Growth Parameters and Leaf Area of Sunflower in Both Growing Seasons

Treatments	RGR g/g/week		CGR g/m²/week		NAR g/m²/week		LA dm²/plant		
	Growing period						DAS		
	60–75	75–90	60–75	75–90	60–75	75–90	60	75	90
Season 2008									
T1	0.32	0.27	124.4	235.8	29.5	56.1	48.6	55.0	47.3
T2	0.20	0.16	142.5	221.6	41.2	57.9	47.5	42.8	35.3
T3	0.20	0.17	110.2	253.2	18.4	66.1	53.3	54.1	39.0
T4	0.26	0.23	79.9	261.2	21.0	81.2	44.2	44.6	33.6
T5	0.30	0.25	109.6	190.8	16.8	81.9	50.7	47.5	42.1
F - test	ns	*	ns	*	*	ns	*	*	*
Season 2009									
T1	0.48	0.38	58.6	100.4	42.0	110.0	44.9	54.2	42.6
T2	0.37	0.20	108.1	180.2	47.2	85.3	57.1	65.6	62.4
T3	0.19	0.38	150.1	212.7	123.6	175.4	66.2	70.2	65.8
T4	0.52	0.16	108.6	144.1	37.8	150.3	48.3	50.4	44.0
T5	0.61	0.19	100.2	131.7	85.5	160.6	47.8	51.6	46.5
F- test	*	*	*	*	*	ns	*	*	*

*, ** and ns indicate $p < 0.05$, $p < 0.01$ and not significant, respectively. Means of each factor designed by the same letter are not significantly different at 5% level using Duncan's MRT.

RGR = relative growth rate, CGR= crop growth rate, NAR= net assimilation rate and LA = leaf area.

Results in Table 15.2 also indicate that CGR and NAR values were higher in the second period (75–90 DAS) than the first one (60–75 DAS) in both growing seasons for all irrigation treatments. It was also observed that values of CGR and NAR were significantly affected by five irrigation treatments in both seasons. The highest values of CGR and NAR were obtained under T_4 followed by T_3 in the first season, while it was under T_3 followed by T_5 in the second season. On the other hand, RGR values were also significantly affected by irrigation treatments in both seasons. Similar results were obtained by Rawson and Turner [15] and El-Kady [7].

TABLE 15.3 Effects of Escaping Irrigation on Chlorophyll-A (Ch. A, mg/dm^2 of LA), Chlorophyll-B (Ch. B, mg/dm^2 of LA) and Caroteniods (Crt, mg/dm^2 of LA) of Sunflower Leaves, During Two Growing Seasons (2008 and 2009)

Treatment	DAS	Season 2008			Season 2009		
		Ch. A	Ch. B	Crt.	Ch. A	Ch. B	Crt.
T1		0.796	0.511	0.251	3.587	1.309	0.823
T2		1.216	0.559	0.414	3.148	1.535	0.663
T3	60	1.424	0.674	0.472	4.086	1.435	0.717
T4		2.684	0.923	0.882	3.975	1.289	0.933
T5		2.835	0.975	0.937	4.236	1.356	0.816
F-test		**	**	**	ns	ns	ns
T1		0.950	0.766	0.543	3.892	0.333	1.241
T2		1.947	0.985	0.674	2.905	0.539	0.851
T3	75	1.860	0.837	0.714	2.565	0.881	0.583
T4		2.399	0.903	0.881	3.026	1.008	0.781
T5		2.494	0.921	0.739	2.780	1.190	0.595
F-test		**	ns	**	ns	ns	ns
T1		1.715	0.972	0.707	2.390	3.135	0.141
T2		2.448	1.233	1.210	3.041	2.474	0.137
T3	90	2.268	0.882	0.974	2.061	2.458	0.165
T4		2.047	0.790	0.890	2.536	3.005	0.159
T5		1.882	0.795	0.870	3.721	2.792	0.381
F-test		ns	*	**	ns	ns	ns

*, ** and ns indicate $p < 0.05$, $p < 0.01$ and not significant, respectively. Means of each factor designed by the same letter are not significantly different at 5% level using Duncan's MRT.

Table 15.3 presents effects of escaping irrigation on chlorophyll (A and B) and carotenoids in sunflower leaves at three growth stages (60, 75 and 90 DAS). Data show that values of chlorophyll (A and B) and carotenoids at different growth stages were significantly affected by the irrigation treatments in the first season, while these were not significantly affected in the second season. Also, values of chlorophyll (A and B) and carotenoids concentration were decreased with the age of plant in both growing seasons. The highest values of leaf chlorophyll (A and B) and carotenoids concentration were obtained under T_5 (escaping irrigation at 75 DAS), followed by T_4 (escaping irrigation at 60 DAS). In the first season, chlorophyll A was 2.835 mg/dm^2 LA, chlorophyll B was 0.975 mg/dm^2 LA and Carotenoids was 0.937

mg/dm^2 LA. In the second season, chlorophyll A was 4.236 mg/dm^2 LA, chlorophyll B was 1.356 mg/dm^2 LA and carotenoids were 0.816 mg/dm^2 LA. On the other hand, values of chlorophyll (A and B) and carotenoids concentration under T_4 and T_5 were higher than those under T_1 (irrigation at every 15 days, control) in the first season. In the second season, values of chlorophyll (A and B) at 75 DAS were higher under T_1 compared to T_4 and T_5. These findings are in agreement with those obtained by El-Kady [7] and Gaafar and El-Wakil [8].

15.3.2 EFFECTS OF ESCAPING IRRIGATION ON CROP WATER RELATIONS

Table 15.4 indicates the effects of escaping irrigation (skip or withholding) on water applied, water saving, water consumptive use (WUC), crop water use efficiency (CWUE) and irrigation water productivity (IWP). The treatment T_1 (control) in the first season recorded the highest values of the water application (2823.9 m^3/fed.) and WUC (2232.2 m^3/fed), and water application of 2795 m^3/fed.) and WUC of 2264 m^3/fed) in 2009, respectively. While T_4 and T_5 recorded the lowest values of the amount of water applied and WUC. It may be due to withholding the fifth irrigation for T_4 and the sixth irrigation for T_5, when the plants are in physiological maturity and require high amount of water with increasing plant age. Escaping the sixth irrigation under T_5 induced the highest values of water saving and was 19.2 in 2008 and 17.8% in 2009. While, the lowest water saving was 13.3 in 2008 and 12.9% in 2009, under T2, respectively. The WUC followed the same trend as of water applied.

TABLE 15.4 Effects of Irrigation Escaping on Water Applied (WA), Water Saving, Water Consumptive Use (WCU), Crop Water Use Efficiency (CWUE) and Irrigation Water Productivity (WP), During 2008 and 2009

Parameters	Treatments					
	T1	T2	T3	T4	T5	
Season 2008						
	1st irrigation	690	690	690	690	690
	2nd irrigation	404	404	404	404	404
	3rd irrigation	366	—	348	346	340
WA (m^3/fed.)	4th irrigation	465	461	—	438	432
	5th irrigation	460	458	447	—	416
	6th irrigation	439	434	415	418	—
	Total	2824	2447	2304	2296	2282
Water saving, (m^3/fed.)		—	377	520	528	542
Water saving, %		—	13.3	18.4	18.7	19.2

TABLE 15.4 *(Continued)*

WCU (m^3/fed.)		2232	2050	1956	1893	1879
CWUE (Kg/m^3)		0.59	0.55	0.63	0.75	0.65
WP (Kg/m^3)		0.47	0.46	0.54	0.63	0.54
Season 2009						
WA (m^3/fed.)	1st irrigation	680	680	680	680	680
	2nd irrigation	400	400	400	400	400
	3rd irrigation	370	—	350	357	360
	4th irrigation	460	465	—	439	425
	5th irrigation	455	456	450	—	435
	6th irrigation	430	435	440	425	—
	Total	2795	2435	2320	2301	2299
Water saving, (m^3/fed.)		—	360	475	494	496
Water saving, %		—	12.9	17.0	17.7	17.8
WCU (m^3/fed.)		2264	1925	1799	1791	1778
CWUE (Kg/m^3)		0.58	0.58	0.68	0.79	0.75
WP (Kg/m^3)		0.47	0.46	0.53	0.61	0.58

Data also showed that the highest value of CWUE was 0.75 in 2008 and 0.793 kg/m^3 in 2009, under T_4. The highest value of water productivity was 0.618 in 2008 and 0.617 kg/m^3 in 2009 under T_4. While the lowest values of CWUE were 0.552 in 2008 and 0.588 kg/m^3 in 2009, under T_2; and the lowest values of water productivity was 0.463 in 2008 and 0.465 kg/m^3 in 2009 under T_2. This may be attributed to the higher seed yield and lower amount of WA and WCU under T_4 than that under T_2. Other investigators have found that WUE was greater in stressed treatments than that in the well irrigated plots (control), while Stone et al. [19] and Goksoy et al. [10] found that WUE did not significantly change when irrigation amount was increased.

15.4 CONCLUSIONS

In North Nile Delta at Kafr El-sheik Governorate Area of Egypt, escaping irrigation at different times affected the plant growth attributes of sunflower. These effects were more pronounced in the 2nd season than that in 1st season, because all the growth attributes in this study were significantly affected by the irrigation escaping particularly at 45 or 60 DAS.

From the view point of water, irrigation escaping for sunflower crop at 60 DAS (the 5th irrigation) is the best treatment, since it saved water by about 18.2% during two seasons and had the highest values of CWUE (0.77 kg/m^3) and WP (0.62 kg/m^3)

compared to the control treatment that included irrigating the sunflower crop every 15 days without irrigation escaping.

15.5 SUMMARY

Two field experiments were conducted at El-Karada water management station, Kafr El-Sheikh governorate, Egypt, during two successive seasons of 2008 and 2009. The irrigation treatments included five treatments: conventional irrigation @ every 15 days (T_1), escaping irrigation at the age of 30 days from sowing = 3rd irrigation (T_2), escaping irrigation at the age of 45 days from sowing = 4th irrigation (T_3), escaping irrigation at the age of 60 days from sowing = 5th irrigation (T_4) and escaping irrigation at the age of 75 days from sowing = 6th irrigation (T_5).

The results showed that T_4 had the highest values of CWUE (0.77 kg/m^3) and irrigation water productivity (0.62 kg/m^3) in two seasons. From the view point of irrigation water saving, T_4 & T_5 recorded the highest values, with nearly the same amount, 511 and 519 m^3/feddan, an average of the two seasons, respectively. Data also revealed that irrigation escaping dates had a significant effect on leaf area, crop growth rate, net assimilation rate and relative growth rate in both growing seasons. The highest values of these parameters were obtained under T_3 followed by T_4 in both growing seasons at (75–90) days after sowing. Also, chlorophyll (A and B) and carotenoids concentration were significantly affected by the irrigation escaping dates and the highest values were recorded under T_4 and T_5 at 60 days after sowing, in both seasons. It can be concluded that the 5th irrigation (T_4) is the best water management practice for sunflower production since it saved water by 18.2% compared to the control treatment and had the highest values of crop water use efficiency and water productivity.

KEYWORDS

- **carotenoids**
- **chlorophyll A**
- **chlorophyll B**
- **days after sowing**
- **escaping irrigation**
- **Feddan**
- **furrow irrigation**
- **growing period**
- **growth attributes**
- **growth rate**

- **irrigation**
- **net assimilation rate**
- **Nile Delta**
- **relative growth rate**
- **sunflower water saving**
- **water efficiency**
- **water management**
- **water productivity**
- **water relations**
- **water use efficiency**

REFERENCES

1. Ali, M. H., Hogue, M. R., Hassan, A. A., Khair, A. (2007). Effects of deficit irrigation on yield, water productivity and economic returns of wheat. *Agric. Water Management, 92*, 151–161.
2. Ali, Md. H., Shui L. T. (2009). Potential evapotranspiration model for Muda Irrigation Project, Malaysia. *Water Resour. Manage., 23,* 57–69.
3. Beyazgul, M., Kayam, Y., Engelsman, F. (2000). Estimation methods for crop water requirements in the Gediz Basin of western Turkey. *Journal of Hydrology, 229*, 19–26.
4. Casadebaig, P., Debaeke, P., Lecoeur, J. (2008). Thresholds for leaf expansion and transpiration response to soil water deficit in a range of sunflower genotypes. *European Journal of Agronomy, 28,* 646–654
5. Doorenbos, J., Pruitt, W. O. (1977). *Crop Water Requirements*. FAO Irrigation and Drainage Paper 24, Food and Agriculture Organization of the United Nations, Rome.
6. Duncan, B. D. (1955). Multiple range and multiple F-Test Biometric, 11, 1–42.
7. El-Kady, F. Abd El-Aziz, (1987). *Physiological Studies on Sunflower*. Ph.D. thesis, Faculty of Agriculture Kafr El-Sheikh, Tanta University
8. Gaafar, S. A., El-Wakil, A. M. (1986). Water Requirements For Sunflower. *Agricultural Research Rev., 64(4),* 567–578.
9. Gad El-Rab, G. M., Anter, N. G., Mahmoud, S. A. (1993). Effect of drought conditions at different growth periods on sunflower yield and water use. *J. Agric. Sci, Mansoura University, 18(8),* 2476–(2482).
10. Goksoy, A. T., Demir, A. O., Turan, Z. M., Dagustu, N. (2004). Responses of sunflower (*Helianthus annuus L.*) to full and limited irrigation at different growth stage. *Field Crops Res., 87,* 167–178.
11. Gomez, K. A., Gomez, A. A. (1984). *Statistical Procedures for Agricultural Research.* 2nd Ed. John Wiley and Sons, New York.
12. Hunt, R. (1990). *Basic Growth Analysis Plant Growth Analysis for Beginners*. Academic Press, London, U.K.
13. Israelson, O. W., Hansen, V. E. (1980). *Irrigation Principles and Practices*. 4th ed. John Wiley & Sons Inc. New York.
14. Ministry of Agriculture and Land Reclamation, (2006). *Sunflower*. Technical Bulletin No. 1020, Central Admin. of Agric. Exten., ARC, Giza, Egypt

15. Rawson, H. M., Turner, N. C. (1983). Irrigation timing and relationship between leaf area and yield in sunflower. *Indian J. Agron., 25,* 645–650.
16. Sadasivam, S., Manicham, A. (2005). *Biochemical Methods for Agricultural Sciences.* 2nd Ed., New Age International Pub. Limited. New Delhi, India
17. Soleimanzadeh, H., Habibi, D., Ardakani, M. R., Paknejad, F., Rejali, F. (2010). Response of sunflower (*Helianthus Annuus* L.) todrought stress under different potassium levels. *World Applied Sci J., 8(4),* 443–448.
18. Soriano, M. A., Orgaz, F., Villalobos, F. J., Federes, E. (2004). Efficiency of water use of early plantings of sunflower. *Euro. J. Agron., 21,* 465–476.
19. Stone, L. R., Schlege, A. J., Gwin, R. E., Khan, A. H. (1996). Response of corn, grain sorghum and sunflower to irrigation in the high plains of Kansas. *Agric. Water Manag., 30,* 251–259.
20. Wang, B. J. (2005). Antioxidant activity of Bupleurum fractions fractionated by supercritical CO_2, Lebensm. *Wiss. Technol., 38,* 281.

CHAPTER 16

IRRIGATION WATER MANAGEMENT FOR SUNFLOWER PRODUCTION: NORTH NILE DELTA, EGYPT

A. S. EL-HENAWY and EMAN M. K. E. SOLTAN

CONTENTS

In this chapter, area is in units of *feddans*. A *feddan* (Arabic: فدّان, *faddān*) is a unit of area. It is used in Egypt, Sudan and Syria. The feddan is not an SI unit and in Classical Arabic, the word means 'a yoke of oxen', implying the area of ground that could be tilled by oxen in a certain time. In Egypt the feddan is the only nonmetric unit which remained in use following the switch to the metric system. One fed. = 24 kirat = 60 m × 70 m = 4200 m^2 = 0.42 hectares.

1.00 L.E. = 0.14 US$. Locally, the abbreviation LE or L.E. which stands for *livre égyptienne* (French for Egyptian pound) is frequently used. E£ and £E are rarely used. The ISO 4217 code is EGP. The Egyptian pound (Arabic: جنيه مصري *Genēh Maṣri* Egyptian Arabic pronunciation: [ge'ne: (h)'masmɑsˁɾi] or in Alexandrian accent: *Geni Maṣri* ['geni 'mɑsˁɾi]) (sign: E£ or ج.م; code: EGP) is the currency of Egypt. It is divided into 100 piastres, or ersh (قرش [ʔeɾʃ]; *plural* قروش [ʔʊ'ɾu:ʃ]; German: *Groschen*), or 1000 millimes (Arabic: مليم [mæl'li:m]; French: *Millime*).

16.1 INTRODUCTION

Sunflower is one of the most promising oil crops in Egypt. It is proposed to close up the gap of oil consumption. At present, Egypt imports about 80–85% of annual consumption of edible vegetable oils. A possible solution to the present gap between the domestic production and demand for edible oil can be achieved by conducting numerous investigations on the effects of fertilization, sowing dates and irrigation treatments on maximizing the productivity of sunflower under local climatic conditions. Because of the water limitations in Egypt, we should do our best towards effective rationalization of irrigation at farm level. Several investigators have studied the effects of irrigation levels on yield and its components of sunflower crop. Deficit irrigation (DI) has been considered worldwide as a way of maximizing water use efficiency (WUE) by eliminating irrigations that have little impact on yield [7, 8, 12]. Moreover, Kirnak et al. [13] indicated that yield loss, which may result from deficit irrigation, can be offset by the benefit of reduced water use. Stone et al. [17] reported that when water is limiting, water stress can be scheduled during milking stages, while during flowering water stress should be avoided. In that sense, Tan et al. [18] and Rinaldi [15] found that irrigation at flowering produced the highest net income in sunflower production. Karam et al. [10] indicated that irrigation shortage at early and mid-flowering in sunflower should be avoided while it can be acceptable at seed formation.

The objective of this study was to manage the irrigation water for sunflower production by withholding of some irrigation events (deficit irrigation) during the season. Results are also presented on the yield and its components and yield quality as well as economic returns of sunflower production.

16.2 MATERIALS AND METHODS

This investigation was conducted at El-Karda Water Management Station Farm, Kafr El-Sheikh Governorate in Egypt during two successive summer seasons of 2008 and 2009. Kafr El-Sheikh is located at 31°07′N latitude and 30°52′ E longitude and an elevation of 6 m above mean sea level. The soil texture at the site is clay. The soil characteristics were: clay 51.7%, silt 26.1%, sand 21.2%, EC 2.59 dSm^{-1} in soil paste extract, pH 8.05, organic matter 1.38%, field capacity 44.7%, wilting point 24.2%. Randomized complete block design with three replications was used. The irrigation treatments were:

T_1 Conventional irrigation during the growing season, every 15 days (control).

T_2 Escaping irrigation after 30 days after sowing (DAS), 3rd irrigation.

T_3 Escaping irrigation after 45 DAS, (4th irrigation).

T_4 Escaping irrigation after 60 DAS, (5th irrigation).

T_5 Escaping irrigation after 75 DAS, (6th irrigation).

Each plot area was 42 m^2 including 10 ridges of 7 m long and 0.60 cm apart. Plots were isolated by ditches of 1.5 m in width to avoid lateral movement of water. Sunflower seed cultivar Sakha 53 was sown on March 15th of 2008 and March 19th of 2009 at hills 20 cm apart on one side of ridges and harvested on July 7 and 17 in both seasons, respectively.

In both seasons, phosphorous fertilizer in the form of calcium super phosphate (15.5% P_2O_5) was applied at the rate of 30 kg P_2O_5/fed during land preparation. Nitrogen was added in the form of urea (46% N) at the rate of 40 kgN/fed in two equal doses before the first and second irrigations, respectively. Potassium was added in the form of potassium sulfate (48% K_2O) at the rate of 24 kg K_2O/fed. Thinning practices were conducted after 21 days after sowing to allow one plant per hill. Other practices for growing sunflower were conducted as recommended by Ministry of Agriculture and Land Reclamation [14].

Ten guarded plants were randomly taken from the fourth inner ridge to determine yield components. Sunflower seed was obtained from central area of each treatment to avoid any border effect.

The following parameters were measured: 100-seed weight, seed yield per plant, seed oil percent, seed yield in kg/fed and oil yield in kg/fed. Seed oil percent was determined using soxhlet extraction unit as reported by AOAC [2]. Seed oil yield was calculated by multiplying seed yield in kg/fed by seed oil percent.

Irrigation application was based on the discharge that was measured with a weir. The weir discharge (Table 16.1) was calculated by using the following equation:

$$Q = 1.84\ LH^{1.5} \quad (1)$$

where: Q = rate of discharge, m^3/sec; L = length edge of weir, cm; and H = height of water above edge of weir, cm.

The obtained data were subjected to analysis of variance according to Gomez and Gomez [9]. Means of treatments were compared by Duncan's Multiple Range Test [6]. All statistical analysis was performed using analysis of variance technique by means of "MSTATC" computer software package.

TABLE 16.1 Total Irrigation Water Applied (m^3/fed.) in Sunflower Crop, Under Five Treatments, During Two Growing Seasons 2008 and 2009.

Growing season	Treatments				
	T_1	T_2	T_3	T_4	T_5
	Total irrigation water applied (m^3/fed.)				
2008	2823.9	2447.4	2303.9	2296.4	2282.4
2009	2795.0	2435.0	2320.0	2301.0	2299.0

16.3 RESULTS AND DISCUSSIONS

16.3.1 EFFECT OF IRRIGATION WATER ESCAPING ON YIELD AND YIELD COMPONENTS OF SUNFLOWER CROP

Table 16.2 shows the effects of irrigation water escaping on weight of 100 seeds, seed yield, oil percent and oil yield.

Weight of 100 seed was significantly affected by irrigation water escaping treatments in 1st and 2nd season. The highest values (7.15 and 6.06 g) were found under T_4 in the two growing seasons 2008 and 2009, respectively. While, the lowest weights (5.99 and 5.14 g) were found under T_2 and T_3 in the same growing seasons, respectively. The lowest yield recorded under T_2 and T_3 can be attributed to that irrigation escaping was occurred during the flowering and the seed formation stages. These results follow the same trend of Doorenbos and Kassam [5], who revealed that seed formation is the next most sensitive period to water deficit, causing severe reduction in both yield and oil content.

TABLE 16.2 Effects of Irrigation Escaping on Yield and Yield Components of Sunflower Crop, During 2008 and 2009.

Treatments	Seed weight g/100 seeds	Seed yield, ton/fed.	Oil content in seeds, %	Oil yield, Kg/fed.
Season 2008				
T_1	6.12^b	1.316^a	41.35	544.2^a
T_2	5.99^b	1.132^b	41.45	469.2^b
T_3	7.04^a	1.239^{ab}	41.76	513.7^b
T_4	7.15^a	1.420^a	41.88	594.7^a
T_5	6.46^{ab}	1.227^{ab}	41.06	503.8^{ab}
F-test	**	*	ns	*
Season 2009				
T_1	5.74^a	1.317^a	40.93^a	514.4^{ab}
T_2	5.81^a	1.122^b	41.70^a	529.5^a
T_3	5.14^b	1.229^b	38.73^b	499.4^b
T_4	6.06^a	1.419^a	41.73^a	551.3^a
T_5	5.83^a	1.327^a	39.72^b	479.2^b
F-test	*	*	*	*

* = significant at $p < 0.05$; ** = significant at $= p < 0.01$; ns = not significant. Means are not significantly different at 5% level, if each value designed by the same letter: using Duncan's MRT.

Seed yield (ton/fed.) was significantly affected by irrigation water escaping treatments. The highest values of seed yield were obtained under T_4 and the lowest ones were obtained under T_2 in the two growing seasons. These results were in agreement with those obtained by Browne [4], who showed that yield losses are generally greatest when water stress occurs in the period 20 days prior to flowering. He also reported that seed yield increased by 30% from irrigation at 2 weeks after mid-flowering. The highest yield under T4 can be attributed to the irrigation escaping at 60 DAS (the 5th irrigation), which may trigger the physiological processes that actually increase yield [16]. Severe water deficits during the early vegetative growth may result in reduced plant height but may increase root depth. Adequate water during the late vegetative period is required for proper bud development. The flowering period is the most sensitive to water deficits, which cause considerable yield decrease since fewer flower come to full development [1, 3].

Oil percent in sunflower seeds indicates the yield quality. It was not significantly affected by irrigation water escaping treatments in the 1st season, but was significantly affected in the 2nd season. The highest values of oil percent were obtained under T_4 in two seasons.

Oil yield was significantly affected by irrigation water treatments in the two growing seasons. The highest values of oil yield were obtained under T_4 in two seasons. The lowest values of oil yield were obtained under T_2 and T_5 in 2008 and 2009, respectively. Kazemeini et al. [11] showed that irrigation levels significantly affected seed yield and oil percentage. Their results indicated that deficit irrigation, during the critical growth period, should be avoided.

16.3.2 EFFECT OF IRRIGATION WATER ESCAPING ON NET INCOME AND ECONOMIC FEASIBILITY OF SUNFLOWER CROP

Table 16.3 presents the total cost of sunflower production in two growing seasons 2008 and 2009. The cost of production included fixed and variable cost. Fixed costs are similar in all treatments, while variable cost is higher in T_1 (control) than in the other treatments by an amount of 20 Egyptian Pound (LE), which represents the cost of excess irrigation events (one irrigation). Variable cost in 1st season was higher than the 2nd season by about 140 LE due to the increase in price of chemical fertilizers.

The effect of irrigation water escaping on net income, costs, net return, net return of water unit and economic efficiency are presented in Table 16.4. The economic return was calculated assuming that the price of one kg of sunflower seeds was 6 LE in 1st season and 7 LE in 2nd season (average price in local market).

Data indicated that the highest value of net income was under T_4 (8600 and 9937.4 LE) in the 1st and 2nd seasons, respectively. Net return is a suitable parameter to express the success of sunflower cropping, because it eliminates the yield costs. T_4 give the highest value of net return (4390 and 5867.4 LE) in the first and the second seasons, respectively.

TABLE 16.3 Total Cost of Sunflower Production (LE*/ fed.), During Seasons 2008 and 2009

Input	Treatments				
	T_1	T_2	T_3	T_4	T_5
	Total cost of sunflower production (LE*/ fed.)				
Seasons 2008					
Rent of land	2500	2500	2500	2500	2500
Land preparation	300	300	300	300	300
Seeds and seeding	200	200	200	200	200
Irrigation	120	100	100	100	100
Hand hoeing and weed control	140	140	140	140	140
Fertilization	570	570	570	570	570
Harvesting	400	400	400	400	400
Total	4230	4210	4210	4210	4210
Season 2009					
Rent of land	2500	2500	2500	2500	2500
Land preparation	300	300	300	300	300
Seeds and seeding	200	200	200	200	200
Irrigation	120	100	100	100	100
Hand hoeing and weed control	140	140	140	140	140
Fertilization	430	430	430	430	430
Harvesting	400	400	400	400	400
Total	4090	4070	4070	4070	4070

* 1 fed.= 4200 m^2; 1.00 US$ ≡ 5.43 LE in 2008; 1.00 US$ = 5.54 LE in 2009;
All currency converter is based on annual average.

Data also showed that the highest values of the net return from the water unit were obtained in the 4th treatment (1.91 and 2.55 LE/m^3 water), as well as, the economic efficiency (1.043 and 1.443). This is due to the highest productivity in both seasons in the 4th treatments.

TABLE 16.4 Effect of Irrigation Escaping on Net Income and Economic Feasibility of Production of Sunflower Crop, During 2008 and 2009

Treatments	Net income L.E/fed.	Total costs, L.E/fed	Net return, L.E/fed.	Net return of water unit, LE/m^3	Economic efficiency
Season 2008					
T_1	7976.0	4230	3746	1.33	0.886
T_2	6822.0	4210	2662	1.09	0.632
T_3	7514.0	4210	3304	1.43	0.785
T_4	8600.0	4210	4390	1.91	1.043
T_5	7442.0	4210	3232	1.42	0.768
Season 2009					
T_1	9224.3	4090	5214.3	1.84	1.275
T_2	7923.7	4070	3853.6	1.58	0.947
T_3	8676.9	4070	4686.9	2.02	1.152
T_4	9937.4	4070	5867.4	2.55	1.442
T_5	8591.0	4070	4521.8	1.97	1.111

* 1 fed.= 4200 m^2; 1.00 US$ ≡ 5.43 LE in 2008;

1.00 US$ = 5.54 LE in 2009; All currency converter is based on annual average.

16.4 CONCLUSIONS

It is concluded that escaping irrigation at 60 days after sowing, during the physiological maturity stage (T_4) is the best treatment compared with the other treatments. It increased oil and seed yield and achieved the highest net return and economic efficiency.

16.5 SUMMARY

Two field experiments were conducted at El-Karada water management station, Kafr El- Sheikh Governorate, Egypt, during two successive summer seasons 2008 and 2009, to study the effect of irrigation water management on sunflower production through irrigation escaping of some irrigation events, during the growing season: on yield and yield components and quality of sunflower crop. Randomized complete block design with three replications was used. The irrigation treatments included five treatments: conventional irrigation along the growing season every 15 days (T_1), escaping irrigation at the age of 30 days from sowing = 3rd irrigation (T_2), escaping irrigation at the age of 45 days from sowing = 4th irrigation (T_3), escaping irrigation at the age of 60 days from sowing = 5th irrigation (T_4) and escaping irrigation at the age of 75 days from sowing = 6th irrigation (T_5).

The highest values of seed yield, oil percent, oil yield and weight per 100 seeds were obtained under T_4 as well as the highest net return of water unit and economic efficiency. Therefore, escaping sunflower irrigation at the age of 60 days from sowing (the 5th irrigation) is recommended to maximize sunflower production under the condition of studied area.

KEYWORDS

- **deficit irrigation**
- **economic efficiency**
- **edible oils**
- **Egypt**
- **Egyptian pound**
- **escape irrigation**
- **Evapotranspiration**
- **FAO**
- **feddan**
- **furrow irrigation**
- **net return of water unit**
- **North Nile Delta**
- **oil yield**
- **seed weight**
- **sunflower yield**
- **Turkey**

REFERENCES

1. Ali Md., H., Shui, L. T. (2009). Potential evapotranspiration model for Muda Irrigation Project, Malaysia. *Water Resour. Manage. 23*, 57–69.
2. A. O. A. C. (2005). *Official Methods of Analysis of AOAC International*. 18th Ed., AOAC International, Gaithersburg, MD, USA.

3. Beyazgul, M., Kayam, Y., Engelsman, F. (2000). Estimation methods for crop water requirements in the Gediz Basin of western Turkey. *J. Hydrology, 229*, 19–26.
4. Browne, C. L. (1977). Effect of date of final irrigation on yield and yield components of sunflowers in semiarid environment. *Aust. J. Exp. Agri. Anim. Hush., 17*, 482–488.
5. Doorenbos, J., Kassam, A. H. (1979). Yield response to water. Irrigation and Drainage Paper 33. In: Landon, J. R. (ed.). *Booker tropical soil manual*. Longman Inc., New York, U.S.A.
6. Duncan, B. D. (1955). Multiple Range and Multiple F. Tests. *Biometrics, 11*, 1–42.
7. English, M. J., Raja, S. N. (1996). Perspectives on deficit irrigation. *Agric. Water Manag., 32*, 1–14.
8. English, M. J. (1990). Deficit irrigation. I. Analytical framework. *J. Amer. Soc. Civil Eng., 116(IR3), 399*–412.
9. Gomez, K. A., A. A. Gomez, (1984). *Statistical Procedures for Agricultural Research*. 2nd edition, John Wiley and Sons, New York.
10. Karam, F., Lahoud, R., Masaad, R., Kabalan, R., Breidi, J., Chalita, C., Rouphael, Y. (2007). Evapotranspiration, seed yield and water use efficiency of drip irrigated sunflower under full and deficit irrigation conditions. *Agricultural Water Management, 90(3),* 213–223.
11. Kazemeini, S. A., Edalat, M., Shekoofa, A. (2009). Interaction effects of deficit irrigation and row spacing on sunflower (Helianthus annuus L.) growth, seed yield and oil yield. *African J. Agricultural Research 4 (111),* 165–(1170).
12. Kirda, C., Moutont, P., Hera, C., Nielsen, D. R. (1999). *Crop yield response to deficit irrigation*. Kluwer Academic Publisher, Dordrecht.
13. Kirnak, H., Tas, I., Kaya, C., Higgs, D. (2002). Effects of deficit irrigation on growth, yield and fruit quality of eggplant under semiarid conditions. *Aust. J. Agric. Res., 53,* 1367–1373.
14. Ministry of Agriculture and Land Reclamation, (2006). Sunflower. Technical Bulletin No. 1020, Central Admin. of Agric. Exten., ARC, Giza, Egypt.
15. Rinaldi, M. (2001). Application of EPIC Model for Irrigation Scheduling in Southern Italy. *Agric. Water Manag., 49,* 185–196.
16. Smith, M., Kivumbi D., Heng L. K. (2002). Use of the FAO-CROPWAT model in deficit irrigation studies. *Deficit Irrigation Practices*, FAO Water Report No. 22, Rome.
17. Stone, L. R., Schlege, A. J., Gwin, R. E., Khan, A. H. (1996). Response of corn, grain sorghum and sunflower to irrigation in the high plains of Kansas *Agric. Water Manag., 30*, 251–259.
18. Tan, S., Beyazgul, M., Avcieri, Z., Kayam, Y., Kaya. H. G. (2000). Effect of irrigation at various growth stage on some economic characters of first crop sunflower. *J. Aegean, Agric. Research Institute, 10,* 1–34.

APPENDICES

(Modified and reprinted with permission from Megh R. Goyal (2012). Appendixes. Pages 317–332. In: *Management of Drip/Trickle or Micro Irrigation*, edited by Megh R. Goyal, New Jersey, USA: Apple Academic Press Inc.)

APPENDIX A

CONVERSION SI AND NON-SI UNITS

To convert the Column 1 in the Column 2	Column 1	Column 2	To convert the Column 2 in the Column 1
	Unit	Unit	
Multiply by	SI	Non-SI	Multiply by

LINEAR

0.621 ------	kilometer, km (10^3 m)	miles, mi ------------------	1.609
1.094 ------	meter, m	yard, yd -------------------	0.914
3.28 -------	meter, m	feet, ft ----------------------	0.304
3.94×10^{-2} ----	millimeter, mm (10^{-3})	inch, in -------------------	25.4

SQUARES

2.47 -------	hectare, he	acre ---------------------	0.405
2.47 -------	square kilometer, km^2	acre ---------------------	4.05×10^{-3}
0.386 --------	square kilometer, km^2	square mile, mi^2 ------------	2.590
2.47×10^{-4} ----	square meter, m^2	acre ---------------------	4.05×10^{-3}
10.76 --------	square meter, m^2	square feet, ft^2 --------------	9.29×10^{-2}
1.55×10^{-3} ----	mm^2	square inch, in^2 --------------	645

CUBICS

9.73×10^{-3} ----	cubic meter, m^3	inch-acre -----------------	102.8
35.3 --------	cubic meter, m^3	cubic-feet, ft^3 ----------------	2.83×10^{-2}
6.10×10^4 ----	cubic meter, m^3	cubic inch, in^3 -------------	1.64×10^{-5}
2.84×10^{-2} ----	liter, L (10^{-3} m^3)	bushel, bu ------------------	35.24

1.057 -------- liter, L	liquid quarts, qt ------------	0.946
3.53×10^{-2} ---- liter, L	cubic feet, ft^3 --------------	28.3
0.265 -------- liter, L	gallon --------------------	3.78
33.78 -------- liter, L	fluid ounce, oz -------------	2.96×10^{-2}
2.11 ------- liter, L	fluid dot, dt ---------------	0.473

WEIGHT

2.20×10^{-3} ---- gram, g (10^{-3} kg)	pound, --------------------	454
3.52×10^{-2} ---- gram, g (10^{-3} kg)	ounce, oz ------------------	28.4
2.205 ------ kilogram, kg	pound, lb -----------------	0.454
10^{-2} ------- kilogram, kg	quintal (metric), q ----------	100
1.10×10^{-3} ---- kilogram, kg	ton (2000 lbs), ton ----------	907
1.102 ------ mega gram, mg	ton (US), ton --------------	0.907
1.102 ------ metric ton, t	ton (US), ton --------------	0.907

YIELD AND RATE

0.893 ------- kilogram per hectare	pound per acre ------------	1.12
7.77×10^{-2} --- kilogram per cubic meter	pound per fanega ----------	12.87
1.49×10^{-2} --- kilogram per hectare	pound per acre, 60 lb -----	67.19
1.59×10^{-2} --- kilogram per hectare	pound per acre, 56 lb -----	62.71
1.86×10^{-2} --- kilogram per hectare	pound per acre, 48 lb -----	53.75
0.107 ------- liter per hectare	galloon per acre ---------	9.35
893 ---------- ton per hectare	pound per acre ----------	1.12×10^{-3}
893 ---------- mega gram per hectare	pound per acre ----------	1.12×10^{-3}
0.446------- ton per hectare	ton (2000 lb) per acre -----	2.24
2.24 ---------- meter per second	mile per hour ------------	0.447

SPECIFIC SURFACE

10 ---------square meter per kilogram	square centimeter per gram ------------------	0.1
10^3 ----------square meter per kilogram	square millimeter per gram ------------------	10^{-3}

PRESSURE

9.90 ----------megapascal, MPa	atmosphere --------------	0.101
10 ---------megapascal	bar -------------------	0.1
1.0 ----------megagram per cubic	gram per cubic	

	meter	centimeter --------------	1.00
2.09×10^{-2}	----pascal, Pa	pound per square feet ------	47.9
1.45×10^{-4}	----pascal, Pa	pound per square inch -----	6.90×10^{3}

To convert the Column 1 in the Column 2	Column 1	Column 2	To convert the Column 2 in the Column 1
	Unit	Unit	
Multiply by	SI	Non-SI	Multiply by

TEMPERATURE

1.00 (K-273)	--- Kelvin, K	centigrade, °C --------	1.00 (C+273)
(1.8 C + 32)	--- centigrade, °C	Fahrenheit, °F --------	(F-32)/1.8

ENERGY

9.52×10^{-4}	----Joule J	BTU ------------------	1.05×10^{3}
0.239	--------Joule, J	calories, cal -----------	4.19
0.735	--------Joule, J	feet-pound -----------	1.36
2.387×10^{5}	---Joule per square meter	calories per square centimeter ---	4.19×10^{4}
10^{5}	----------Newton, N	dynes -----------------	10^{-5}

WATER REQUIREMENTS

9.73×10^{-3}	---cubic meter	inch acre ---------------	102.8
9.81×10^{-3}	---cubic meter per hour	cubic feet per second ------	101.9
4.40	----------cubic meter per hour	galloon (US) per minute ----	0.227
8.11	----------hectare-meter	acre-feet ---------------	0.123
97.28	-------hectare-meter	acre-inch ----------------	1.03×10^{-2}
8.1×10^{-2}	----hectare centimeter	acre-feet ---------------	12.33

CONCENTRATION

1	------------centimol per kilogram	milliequivalents per 100 grams --------------	1
0.1	---------gram per kilogram	percent ----------------	10
1	------------milligram per kilogram	parts per million ---------	1

NUTRIENTS FOR PLANTS

2.29	-------- P	P_2O_5 --------------------	0.437
1.20	-------- K	K_2O -------------------	0.830

1.39 -------- Ca	CaO --------------------	0.715
1.66 -------- Mg	MgO ------------------	0.602

NUTRIENT EQUIVALENTS

		Conversion	Equivalent
Column A	Column B	A to B	B to A
N	NH3	1.216	0.822
	NO3	4.429	0.226
	KNO3	7.221	0.1385
	Ca(NO3)2	5.861	0.171
	(NH4)2SO4	4.721	0.212
	NH4NO3	5.718	0.175
	(NH4)2 HPO4	4.718	0.212
P	P2O5	2.292	0.436
	PO4	3.066	0.326
	KH2PO4	4.394	0.228
	(NH4)2 HPO4	4.255	0.235
	H3PO4	3.164	0.316
K	K2O	1.205	0.83
	KNO3	2.586	0.387
	KH2PO4	3.481	0.287
	Kcl	1.907	0.524
	K2SO4	2.229	0.449
Ca	CaO	1.399	0.715
	Ca(NO3)2	4.094	0.244
	CaCl2 × 6H2O	5.467	0.183
	CaSO4 × 2H2O	4.296	0.233
Mg	MgO	1.658	0.603
	MgSO4 × 7H2O	1.014	0.0986
S	H2SO4	3.059	0.327
	(NH4)2 SO4	4.124	0.2425
	K2SO4	5.437	0.184
	MgSO4 × 7H2O	7.689	0.13
	CaSO4 × 2H2O	5.371	0.186

APPENDIX B

PIPE AND CONDUIT FLOW

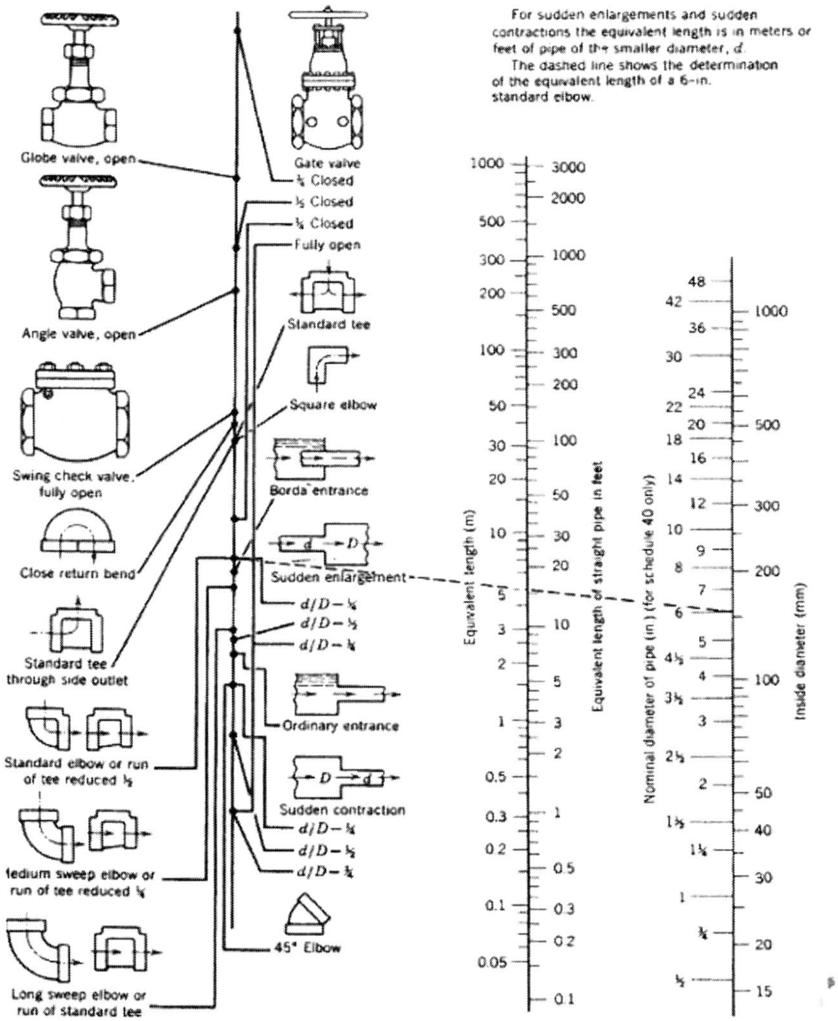

APPENDIX C

PERCENTAGE OF DAILY SUNSHINE HOURS: FOR NORTH AND SOUTH HEMISPHERES

Latitude	Jan	Feb	Mar	Apr	May	Jun	Jul	Aug	Sep	Oct	Nov	Dec
NORTH												
0	8.50	7.66	8.49	8.21	8.50	8.22	8.50	8.49	8.21	8.50	8.22	8.50
5	8.32	7.57	8.47	3.29	8.65	8.41	8.67	8.60	8.23	8.42	8.07	8.30
10	8.13	7.47	8.45	8.37	8.81	8.60	8.86	8.71	8.25	8.34	7.91	8.10
15	7.94	7.36	8.43	8.44	8.98	8.80	9.05	8.83	8.28	8.20	7.75	7.88
20	7.74	7.25	8.41	8.52	9.15	9.00	9.25	8.96	8.30	8.18	7.58	7.66
25	7.53	7.14	8.39	8.61	9.33	9.23	9.45	9.09	8.32	8.09	7.40	7.52
30	7.30	7.03	8.38	8.71	9.53	9.49	9.67	9.22	8.33	7.99	7.19	7.15
32	7.20	6.97	8.37	8.76	9.62	9.59	9.77	9.27	8.34	7.95	7.11	7.05
34	7.10	6.91	8.36	8.80	9.72	9.70	9.88	9.33	8.36	7.90	7.02	6.92
36	6.99	6.85	8.35	8.85	9.82	9.82	9.99	9.40	8.37	7.85	6.92	6.79
38	6.87	6.79	8.34	8.90	9.92	9.95	10.1	9.47	3.38	7.80	6.82	6.66
40	6.76	6.72	8.33	8.95	10.0	10.1	10.2	9.54	8.39	7.75	6.72	7.52
42	6.63	6.65	8.31	9.00	10.1	10.2	10.4	9.62	8.40	7.69	6.62	6.37
44	6.49	6.58	8.30	9.06	10.3	10.4	10.5	9.70	8.41	7.63	6.49	6.21
46	6.34	6.50	8.29	9.12	10.4	10.5	10.6	9.79	8.42	7.57	6.36	6.04
48	6.17	6.41	8.27	9.18	10.5	10.7	10.8	9.89	8.44	7.51	6.23	5.86
50	5.98	6.30	8.24	9.24	10.7	10.9	11.0	10.0	8.35	7.45	6.10	5.64
52	5.77	6.19	8.21	9.29	10.9	11.1	11.2	10.1	8.49	7.39	5.93	5.43
54	5.55	6.08	8.18	9.36	11.0	11.4	11.4	10.3	8.51	7.20	5.74	5.18
56	5.30	5.95	8.15	9.45	11.2	11.7	11.6	10.4	8.53	7.21	5.54	4.89
58	5.01	5.81	8.12	9.55	11.5	12.0	12.0	10.6	8.55	7.10	4.31	4.56
60	4.67	5.65	8.08	9.65	11.7	12.4	12.3	10.7	8.57	6.98	5.04	4.22
SOUTH												
0	8.50	7.66	8.49	8.21	8.50	8.22	8.50	8.49	8.21	8.50	8.22	8.50
5	8.68	7.76	8.51	8.15	8.34	8.05	8.33	8.38	8.19	8.56	8.37	8.68
10	8.86	7.87	8.53	8.09	8.18	7.86	8.14	8.27	8.17	8.62	8.53	8.88
15	9.05	7.98	8.55	8.02	8.02	7.65	7.95	8.15	8.15	8.68	8.70	9.10
20	9.24	8.09	8.57	7.94	7.85	7.43	7.76	8.03	8.13	8.76	8.87	9.33
25	9.46	8.21	8.60	7.74	7.66	7.20	7.54	7.90	8.11	8.86	9.04	9.58

30	9.70	8.33	8.62	7.73	7.45	6.96	7.31	7.76	8.07	8.97	9.24	9.85
32	9.81	8.39	8.63	7.69	7.36	6.85	7.21	7.70	8.06	9.01	9.33	9.96
34	9.92	8.45	8.64	7.64	7.27	6.74	7.10	7.63	8.05	9.06	9.42	10.1
36	10.0	8.51	8.65	7.59	7.18	6.62	6.99	7.56	8.04	9.11	9.35	10.2
38	10.2	8.57	8.66	7.54	7.08	6.50	6.87	7.49	8.03	9.16	9.61	10.3
40	10.3	8.63	8.67	7.49	6.97	6.37	6.76	7.41	8.02	9.21	9.71	10.5
42	10.4	8.70	8.68	7.44	6.85	6.23	6.64	7.33	8.01	9.26	9.8	10.6
44	10.5	8.78	8.69	7.38	6.73	6.08	6.51	7.25	7.99	9.31	9.94	10.8
46	10.7	8.86	8.90	7.32	6.61	5.92	6.37	7.16	7.96	9.37	10.1	11.0

APPENDIX D

PSYCHOMETRIC CONSTANT (Γ) FOR DIFFERENT ALTITUDES (Z)

$$\gamma = 10^{-3}\ [(C_p.P) \div (\varepsilon.\lambda)] = (0.00163) \times [P \div \lambda]$$

γ, psychrometric constant [kPa C^{-1}] c_p, specific heat of moist air = 1.013 [kJ $kg^{-1\,0}C^{-1}$] P, atmospheric pressure [kPa].

ε, ratio molecular weight of water vapor/dry air = 0.622 λ, latent heat of vaporization [MJ kg^{-1}] = 2.45 MJ kg^{-1} at 20°C.

Z (m)	**γ kPa/°C**	**z (m)**	**γ kPa/°C**	**z (m)**	**γ kPa/°C**	**z (m)**	**γ kPa/°C**
0	0.067	1000	0.060	2000	0.053	3000	0.047
100	0.067	1100	0.059	2100	0.052	3100	0.046
200	0.066	1200	0.058	2200	0.052	3200	0.046
300	0.065	1300	0.058	2300	0.051	3300	0.045
400	0.064	1400	0.057	2400	0.051	3400	0.045
500	0.064	1500	0.056	2500	0.050	3500	0.044
600	0.063	1600	0.056	2600	0.049	3600	0.043
700	0.062	1700	0.055	2700	0.049	3700	0.043
800	0.061	1800	0.054	2800	0.048	3800	0.042
900	0.061	1900	0.054	2900	0.047	3900	0.042
1000	0.060	2000	0.053	3000	0.047	4000	0.041

APPENDIX E

SATURATION VAPOR PRESSURE [e_s] FOR DIFFERENT TEMPERATURES (T)

Vapor pressure function = e_s = [0.6108]*exp{[17.27*T]/[T + 237.3]}							
T °C	**e_s kPa**	**T °C**	**e_s kPa**	**T °C**	**e_s kPa**	**T °C**	**e_s kPa**
1.0	0.657	**13.0**	1.498	**25.0**	3.168	**37.0**	6.275
1.5	0.681	**13.5**	1.547	**25.5**	3.263	**37.5**	6.448
2.0	0.706	**14.0**	1.599	**26.0**	3.361	**38.0**	6.625
2.5	0.731	**14.5**	1.651	**26.5**	3.462	**38.5**	6.806
3.0	0.758	**15.0**	1.705	**27.0**	3.565	**39.0**	6.991
3.5	0.785	**15.5**	1.761	**27.5**	3.671	**39.5**	7.181
4.0	0.813	**16.0**	1.818	**28.0**	3.780	**40.0**	7.376
4.5	0.842	**16.5**	1.877	**28.5**	3.891	**40.5**	7.574
5.0	0.872	**17.0**	1.938	**29.0**	4.006	**41.0**	7.778
5.5	0.903	**17.5**	2.000	**29.5**	4.123	**41.5**	7.986
6.0	0.935	**18.0**	2.064	**30.0**	4.243	**42.0**	8.199
6.5	0.968	**18.5**	2.130	**30.5**	4.366	**42.5**	8.417
7.0	1.002	**19.0**	2.197	**31.0**	4.493	**43.0**	8.640
7.5	1.037	**19.5**	2.267	**31.5**	4.622	**43.5**	8.867
8.0	1.073	**20.0**	2.338	**32.0**	4.755	**44.0**	9.101
8.5	1.110	**20.5**	2.412	**32.5**	4.891	**44.5**	9.339
9.0	1.148	**21.0**	2.487	**33.0**	5.030	**45.0**	9.582
9.5	1.187	**21.5**	2.564	**33.5**	5.173	**45.5**	9.832
10.0	1.228	**22.0**	2.644	**34.0**	5.319	**46.0**	10.086
10.5	1.270	**22.5**	2.726	**34.5**	5.469	**46.5**	10.347
11.0	1.313	**23.0**	2.809	**35.0**	5.623	**47.0**	10.613
11.5	1.357	**23.5**	2.896	**35.5**	5.780	**47.5**	10.885
12.0	1.403	**24.0**	2.984	**36.0**	5.941	**48.0**	11.163
12.5	1.449	**24.5**	3.075	**36.5**	6.106	**48.5**	11.447

APPENDIX F

SLOPE OF VAPOR PRESSURE CURVE (Δ) FOR DIFFERENT TEMPERATURES (T)

$\Delta = [4098.\ e^0(T)] \div [T + 237.3]^2$
$= 2504\{\exp[(17.27T) \div (T + 237.2)]\} \div [T + 237.3]^2$

T °C	Δ kPa/°C	T °C	Δ kPa/°C	T °C	Δ kPa/°C	T °C	Δ kPa/°C
1.0	0.047	13.0	0.098	25.0	0.189	37.0	0.342
1.5	0.049	13.5	0.101	25.5	0.194	37.5	0.350
2.0	0.050	14.0	0.104	26.0	0.199	38.0	0.358
2.5	0.052	14.5	0.107	26.5	0.204	38.5	0.367
3.0	0.054	15.0	0.110	27.0	0.209	39.0	0.375
3.5	0.055	15.5	0.113	27.5	0.215	39.5	0.384
4.0	0.057	16.0	0.116	28.0	0.220	40.0	0.393
4.5	0.059	16.5	0.119	28.5	0.226	40.5	0.402
5.0	0.061	17.0	0.123	29.0	0.231	41.0	0.412
5.5	0.063	17.5	0.126	29.5	0.237	41.5	0.421
6.0	0.065	18.0	0.130	30.0	0.243	42.0	0.431
6.5	0.067	18.5	0.133	30.5	0.249	42.5	0.441
7.0	0.069	19.0	0.137	31.0	0.256	43.0	0.451
7.5	0.071	19.5	0.141	31.5	0.262	43.5	0.461
8.0	0.073	20.0	0.145	32.0	0.269	44.0	0.471
8.5	0.075	20.5	0.149	32.5	0.275	44.5	0.482
9.0	0.078	21.0	0.153	33.0	0.282	45.0	0.493
9.5	0.080	21.5	0.157	33.5	0.289	45.5	0.504
10.0	0.082	22.0	0.161	34.0	0.296	46.0	0.515
10.5	0.085	22.5	0.165	34.5	0.303	46.5	0.526
11.0	0.087	23.0	0.170	35.0	0.311	47.0	0.538
11.5	0.090	23.5	0.174	35.5	0.318	47.5	0.550
12.0	0.092	24.0	0.179	36.0	0.326	48.0	0.562
12.5	0.095	24.5	0.184	36.5	0.334	48.5	0.574

APPENDIX G

NUMBER OF THE DAY IN THE YEAR (JULIAN DAY)

Day	Jan	Feb	Mar	Apr	May	Jun	Jul	Aug	Sep	Oct	Nov	Dec
1	1	32	60	91	121	152	182	213	244	274	305	335
2	2	33	61	92	122	153	183	214	245	275	306	336
3	3	34	62	93	123	154	184	215	246	276	307	337
4	4	35	63	94	124	155	185	216	247	277	308	338
5	5	36	64	95	125	156	186	217	248	278	309	339
6	6	37	65	96	126	157	187	218	249	279	310	340
7	7	38	66	97	127	158	188	219	250	280	311	341
8	8	39	67	98	128	159	189	220	251	281	312	342
9	9	40	68	99	129	160	190	221	252	282	313	343
10	10	41	69	100	130	161	191	222	253	283	314	344
11	11	42	70	101	131	162	192	223	254	284	315	345
12	12	43	71	102	132	163	193	224	255	285	316	346
13	13	44	72	103	133	164	194	225	256	286	317	347
14	14	45	73	104	134	165	195	226	257	287	318	348
15	15	46	74	105	135	166	196	227	258	288	319	349
16	16	47	75	106	136	167	197	228	259	289	320	350
17	17	48	76	107	137	168	198	229	260	290	321	351
18	18	49	77	108	138	169	199	230	261	291	322	352
19	19	50	78	109	139	170	200	231	262	292	323	353
20	20	51	79	110	140	171	201	232	263	293	324	354
21	21	52	80	111	141	172	202	233	264	294	325	355
22	22	53	81	112	142	173	203	234	265	295	326	356
23	23	54	82	113	143	174	204	235	266	296	327	357
24	24	55	83	114	144	175	205	236	267	297	328	358
25	25	56	84	115	145	176	206	237	268	298	329	359
26	26	57	85	116	146	177	207	238	269	299	330	360
27	27	58	86	117	147	178	208	239	270	300	331	361
28	28	59	87	118	148	179	209	240	271	301	332	362
29	29	(60)	88	119	149	180	210	241	272	302	333	363
30	30	—	89	120	150	181	211	242	273	303	334	364
31	31	—	90	—	151	—	212	243	—	304	—	365

APPENDIX H

STEFAN-BOLTZMANN LAW AT DIFFERENT TEMPERATURES (T)

$[\sigma^*(T_K)^4] = [4.903 \times 10^{-9}]$, MJ K^{-4} m^{-2} day^{-1}
where, $T_K = \{T[°C] + 273.16\}$

T	$\sigma^*(T_K)^4$	T	$\sigma^*(T_K)^4$	T	$\sigma^*(T_K)^4$
Units					
°C	MJ m^{-2} d^{-1}	°C	MJ m^{-2} d^{-1}	°C	MJ m^{-2} d^{-1}
1.0	27.70	**17.0**	34.75	**33.0**	43.08
1.5	27.90	**17.5**	34.99	**33.5**	43.36
2.0	28.11	**18.0**	35.24	**34.0**	43.64
2.5	28.31	**18.5**	35.48	**34.5**	43.93
3.0	28.52	**19.0**	35.72	**35.0**	44.21
3.5	28.72	**19.5**	35.97	**35.5**	44.50
4.0	28.93	**20.0**	36.21	**36.0**	44.79
4.5	29.14	**20.5**	36.46	**36.5**	45.08
5.0	29.35	**21.0**	36.71	**37.0**	45.37
5.5	29.56	**21.5**	36.96	**37.5**	45.67
6.0	29.78	**22.0**	37.21	**38.0**	45.96
6.5	29.99	**22.5**	37.47	**38.5**	46.26
7.0	30.21	**23.0**	37.72	**39.0**	46.56
7.5	30.42	**23.5**	37.98	**39.5**	46.85
8.0	30.64	**24.0**	38.23	**40.0**	47.15
8.5	30.86	**24.5**	38.49	**40.5**	47.46
9.0	31.08	**25.0**	38.75	**41.0**	47.76
9.5	31.30	**25.5**	39.01	**41.5**	48.06
10.0	31.52	**26.0**	39.27	**42.0**	48.37
10.5	31.74	**26.5**	39.53	**42.5**	48.68
11.0	31.97	**27.0**	39.80	**43.0**	48.99
11.5	32.19	**27.5**	40.06	**43.5**	49.30
12.0	32.42	**28.0**	40.33	**44.0**	49.61
12.5	32.65	**28.5**	40.60	**44.5**	49.92
13.0	32.88	**29.0**	40.87	**45.0**	50.24
13.5	33.11	**29.5**	41.14	**45.5**	50.56
14.0	33.34	**30.0**	41.41	**46.0**	50.87

14.5	33.57	**30.5**	41.69	**46.5**	51.19
15.0	33.81	**31.0**	41.96	**47.0**	51.51
15.5	34.04	**31.5**	42.24	**47.5**	51.84
16.0	34.28	**32.0**	42.52	**48.0**	52.16
16.5	34,52	**32.5**	42.80	**48.5**	52.49

APPENDIX I

THERMODYNAMIC PROPERTIES OF AIR AND WATER

1. Latent Heat of Vaporization (λ)

$$\lambda = [2.501-(2.361 \times 10^{-3})\ T]$$

where, λ = latent heat of vaporization [MJ kg^{-1}]; and T = air temperature [°C].

The value of the latent heat varies only slightly over normal temperature ranges. A single value may be taken (for ambient temperature = 20°C): λ = 2.45 MJ kg^{-1}.

2. Atmospheric Pressure (P)

$$\mathbf{P = P_o\ [\{T_{Ko}-\alpha(Z-Z_o)\ \} \div \{T_{Ko}\}]^{(g/(\alpha.R))}}$$

where, P, atmospheric pressure at elevation *z* [kPa]

P_o, atmospheric pressure at sea level = 101.3 [kPa]

z, elevation [m]

z_o, elevation at reference level [m]

g, gravitational acceleration = 9.807 [m s^{-2}]

R, specific gas constant == 287 [J kg^{-1} K^{-1}]

α, constant lapse rate for moist air = 0.0065 [K m^{-1}]

T_{Ko}, reference temperature [K] at elevation z_o = 273.16 + T

T, means air temperature for the time period of calculation [°C]

When assuming P_o = 101.3 [kPa] at z_o = 0, and T_{Ko} = 293 [K] for T = 20 [°C], above equation reduces to:

$$\mathbf{P = 101.3[(293-0.0065Z)\ (293)]^{5.26}}$$

3. Atmospheric Density (ρ)

$\rho = [1000P] \div [T_{Kv}\ R] = [3.486P] \div [T_{Kv}]$, and $T_{Kv} = T_K[1-0.378(e_a)/P]^{-1}$

where, ρ, atmospheric density [kg m^{-3}]

R, specific gas constant = 287 [J $kg^{-1\ K-1}$]

$T_{Kv,}$ virtual temperature [K]

$T_{K,}$ absolute temperature [K]: $T_K = 273.16 + T$ [°C]
$e_{a,}$ actual vapor pressure [kPa]
T, mean daily temperature for 24-h calculation time steps.

For average conditions (e_a in the range 1–5 kPa and P between 80–100 kPa), T_{Kv} can be substituted by: $T_{Kv} \approx 1.01\ (T + 273)$

4. Saturation Vapor Pressure function (e_s)

$$e_s = [0.6108]*\exp\{[17.27*T]/[T + 237.3]\}$$

where, e_s, saturation vapor pressure function [kPa]
T, air temperature [°C]

5. Slope Vapor Pressure Curve (Δ)

$$\Delta = [4098.\ e°(T)] \div [T + 237.3]^2$$
$$= 2504\{\exp[(17.27T) \div (T + 237.2)]\} \div [T + 237.3]^2$$

where, Δ, slope vapor pressure curve [kPa C^{-1}]
T, air temperature [°C]
$e^0(T)$, saturation vapor pressure at temperature T [kPa]

In 24-h calculations, Δ is calculated using mean daily air temperature. In hourly calculations T refers to the hourly mean, T_{hr}.

6. Psychrometric Constant (γ)

$$\gamma = 10^{-3}\ \mathbf{[(C_p.P) \div (\varepsilon.\lambda)] = (0.00163) \times [P \div \lambda]}$$

where, γ, psychrometric constant [kPa C^{-1}]
c_p, specific heat of moist air = 1.013 [kJ $kg^{-1 0}C^{-1}$]
P, atmospheric pressure [kPa]: Eqs. (2) or (4)
ε, ratio molecular weight of water vapor/dry air = 0.622
λ, latent heat of vaporization [MJ kg^{-1}]

7. Dew Point Temperature (T_{dew})

When data is not available, T_{dew} can be computed from e_a by:

$$T_{dew} = [\{116.91 + 237.3\text{Log}_e(e_a)\} \div \{16.78–\text{Log}_e(e_a)\}]$$

Where, T_{dew}, dew point temperature [°C]
e_a, actual vapor pressure [kPa]

For the case of measurements with the Assmann psychrometer, T_{dew} can be calculated from:

$$\mathbf{T_{dew} = (112 + 0.9T_{wet})[e_a \div (e^0\ T_{wet})]^{0.125}–[112–0.1T_{wet}]}$$

8. Short Wave Radiation on a Clear-Sky Day (R_{so})

The calculation of R_{so} is required for computing net long wave radiation and for checking calibration of pyranometers and integrity of R_{so} data. A good approximation for R_{so} for daily and hourly periods is:

$$R_{so} = (0.75 + 2 \times 10^{-5} z)R_a$$

where, z, station elevation [m]

R_a, extraterrestrial radiation [MJ m^{-2} d^{-1}]

Equation is valid for station elevations less than 6000 m having low air turbidity. The equation was developed by linearizing Beer's radiation extinction law as a function of station elevation and assuming that the average angle of the sun above the horizon is about 50°.

For areas of high turbidity caused by pollution or airborne dust or for regions where the sun angle is significantly less than 50° so that the path length of radiation through the atmosphere is increased, an adoption of Beer's law can be employed where P is used to represent atmospheric mass:

$$\mathbf{R_{so} = (R_a) \exp[(-0.0018P) \div (K_t \sin(\Phi))]}$$

where, K_t, turbidity coefficient, $0 < K_t \leq 1.0$, where $K_t = 1.0$ for clean air and $K_t = 1.0$ for extremely turbid, dusty or polluted air

P, atmospheric pressure [kPa]

Φ, angle of the sun above the horizon [rad]

R_a, extraterrestrial radiation [MJ m^{-2} d^{-1}]

For hourly or shorter periods, Φ is calculated as:

$\sin \Phi = \sin \varphi \sin \delta + \cos \varphi \cos \delta \cos \omega$

where, φ, latitude [rad]

δ, solar declination [rad] (Eq. (24) in Chapter 3)

ω, solar time angle at midpoint of hourly or shorter period [rad]

For 24-hour periods, the mean daily sun angle, weighted according to R_a, can be approximated as:

$$\mathbf{\sin(\Phi_{24}) = \sin[0.85 + 0.3\, \varphi \sin\{(2\pi J/365) - 1.39\} - 0.42\, \varphi^2]}$$

where, Φ_{24}, average Φ during the daylight period, weighted according to R_a [rad]

φ, latitude [rad]

J, day in the year

The Φ_{24} variable is used to represent the average sun angle during daylight hours and has been weighted to represent integrated 24-h transmission effects on 24-h R_{so} by the atmosphere. Φ_{24} should be limited to ≥ 0. In some situations, the estimation for R_{so} can be improved by modifying to consider the effects of water vapor on short wave absorption, so that:

$R_{so} = (K_B + K_D) R_a$, where,

$$\mathbf{K_B = 0.98exp[\{(-0.00146P) \div (K_t \sin \Phi)\} - 0.091\{w/\sin \Phi\}^{0.25}]}$$

where, K_B, the clearness index for direct beam radiation

K_D, the corresponding index for diffuse beam radiation

$K_D = 0.35–0.33\ K_B$ for $K_B \geq 0.15$

$K_D = 0.18 + 0.82\ K_B$ for $K_B < 0.15$

R_a, extraterrestrial radiation [MJ m^{-2} d^{-1}]

K_t, turbidity coefficient, $0 < K_t \leq 1.0$ where $K_t = 1.0$ for clean air and $K_t = 1.0$ for extremely turbid, dusty or polluted air.

P, atmospheric pressure [kPa]

Φ, angle of the sun above the horizon [rad]

W, perceptible water in the atmosphere [mm] = $0.14\ e_a\ P + 2.1$

e_a, actual vapor pressure [kPa]

P, atmospheric pressure [kPa]

APPENDIX J

PSYCHROMETRIC CHART AT SEA LEVEL

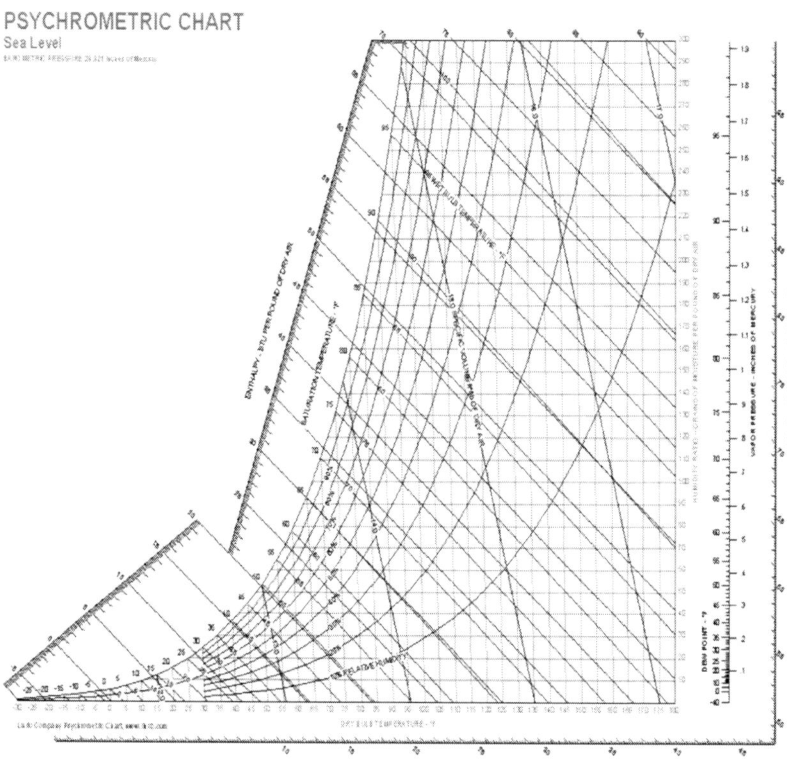

INDEX

D

I

J

K

L

M

N

O

P

R

S

Y

Z